Undergraduate Texts in Mathematics

Undergraduate Texts in Mathematics

Series Editors:

Sheldon Axler
San Francisco State University, San Francisco, CA, USA

Kenneth Ribet
University of California, Berkeley, CA, USA

Advisory Board:

Colin C. Adams, *Williams College, Williamstown, MA, USA*
Alejandro Adem, *University of British Columbia, Vancouver, BC, Canada*
Ruth Charney, *Brandeis University, Waltham, MA, USA*
Irene M. Gamba, *The University of Texas at Austin, Austin, TX, USA*
Roger E. Howe, *Yale University, New Haven, CT, USA*
David Jerison, *Massachusetts Institute of Technology, Cambridge, MA, USA*
Jeffrey C. Lagarias, *University of Michigan, Ann Arbor, MI, USA*
Jill Pipher, *Brown University, Providence, RI, USA*
Fadil Santosa, *University of Minnesota, Minneapolis, MN, USA*
Amie Wilkinson, *University of Chicago, Chicago, IL, USA*

Undergraduate Texts in Mathematics are generally aimed at third- and fourth-year undergraduate mathematics students at North American universities. These texts strive to provide students and teachers with new perspectives and novel approaches. The books include motivation that guides the reader to an appreciation of interrelations among different aspects of the subject. They feature examples that illustrate key concepts as well as exercises that strengthen understanding.

For further volumes:
http://www.springer.com/series/666

Kenneth A. Ross

Elementary Analysis

The Theory of Calculus

Second Edition

In collaboration with Jorge M. López, University of
Puerto Rico, Río Piedras

 Springer

Kenneth A. Ross
Department of Mathematics
University of Oregon
Eugene, OR, USA

ISSN 0172-6056
ISBN 978-1-4939-0128 -9   ISBN 978-1-4614-6271-2 (eBook)
DOI 10.1007/978-1-4614-6271-2
Springer New York Heidelberg Dordrecht London

Mathematics Subject Classification: 26-01, 00-01, 26A06, 26A24, 26A27, 26A42

© Springer Science+Business Media New York 2013
Softcover reprint of the hardcover 2nd edition 2013
This work is subject to copyright. All rights are reserved by the Publisher, whether the whole or part
of the material is concerned, specifically the rights of translation, reprinting, reuse of illustrations,
recitation, broadcasting, reproduction on microfilms or in any other physical way, and transmission or
information storage and retrieval, electronic adaptation, computer software, or by similar or dissim-
ilar methodology now known or hereafter developed. Exempted from this legal reservation are brief
excerpts in connection with reviews or scholarly analysis or material supplied specifically for the purpose of being entered and executed on a computer system, for exclusive use by the purchaser of the
work. Duplication of this publication or parts thereof is permitted only under the provisions of the
Copyright Law of the Publisher's location, in its current version, and permission for use must always
be obtained from Springer. Permissions for use may be obtained through RightsLink at the Copyright
Clearance Center. Violations are liable to prosecution under the respective Copyright Law.
The use of general descriptive names, registered names, trademarks, service marks, etc. in this publication does not imply, even in the absence of a specific statement, that such names are exempt from
the relevant protective laws and regulations and therefore free for general use.
While the advice and information in this book are believed to be true and accurate at the date of
publication, neither the authors nor the editors nor the publisher can accept any legal responsibility for
any errors or omissions that may be made. The publisher makes no warranty, express or implied, with
respect to the material contained herein.

Printed on acid-free paper

Springer is part of Springer Science+Business Media (www.springer.com)

Preface

Preface to the First Edition A study of this book, and especially the exercises, should give the reader a thorough understanding of a few basic concepts in analysis such as continuity, convergence of sequences and series of numbers, and convergence of sequences and series of functions. An ability to read and write proofs will be stressed. A precise knowledge of definitions is essential. The beginner should memorize them; such memorization will help lead to understanding.

Chapter 1 sets the scene and, except for the completeness axiom, should be more or less familiar. Accordingly, readers and instructors are urged to move quickly through this chapter and refer back to it when necessary. The most critical sections in the book are §§7–12 in Chap. 2. If these sections are thoroughly digested and understood, the remainder of the book should be smooth sailing.

The first four chapters form a unit for a short course on analysis. I cover these four chapters (except for the enrichment sections and §20) in about 38 class periods; this includes time for quizzes and examinations. For such a short course, my philosophy is that the students are relatively comfortable with derivatives and integrals but do not really understand sequences and series, much less sequences and series of functions, so Chaps. 1–4 focus on these topics. On two or three occasions, I draw on the Fundamental Theorem of Calculus or the Mean Value Theorem, which appears later in the book, but of course these important theorems are at least discussed in a standard calculus class.

In the early sections, especially in Chap. 2, the proofs are very detailed with careful references for even the most elementary facts. Most sophisticated readers find excessive details and references a hindrance (they break the flow of the proof and tend to obscure the main ideas) and would prefer to check the items mentally as they proceed. Accordingly, in later chapters, the proofs will be somewhat less detailed, and references for the simplest facts will often be omitted. This should help prepare the reader for more advanced books which frequently give very brief arguments.

Mastery of the basic concepts in this book should make the analysis in such areas as complex variables, differential equations, numerical analysis, and statistics more meaningful. The book can also serve as a foundation for an in-depth study of real analysis given in books such as [4, 33, 34, 53, 62, 65] listed in the bibliography.

Readers planning to teach calculus will also benefit from a careful study of analysis. Even after studying this book (or writing it), it will not be easy to handle questions such as "What is a number?" but at least this book should help give a clearer picture of the subtleties to which such questions lead.

The enrichment sections contain discussions of some topics that I think are important or interesting. Sometimes the topic is dealt with lightly, and suggestions for further reading are given. Though these sections are not particularly designed for classroom use, I hope that some readers will use them to broaden their horizons and see how this material fits in the general scheme of things.

I have benefitted from numerous helpful suggestions from my colleagues Robert Freeman, William Kantor, Richard Koch, and John Leahy and from Timothy Hall, Gimli Khazad, and Jorge López. I have also had helpful conversations with my wife Lynn concerning grammar and taste. Of course, remaining errors in grammar and mathematics are the responsibility of the author.

Several users have supplied me with corrections and suggestions that I've incorporated in subsequent printings. I thank them all, including Robert Messer of Albion College, who caught a subtle error in the proof of Theorem 12.1.

Preface to the Second Edition After 32 years, it seemed time to revise this book. Since the first edition was so successful, I have retained the format and material from the first edition. The numbering of theorems, examples, and exercises in each section will be the same, and new material will be added to some of the sections. Every rule has an exception, and this rule is no exception. In §11, a theorem (Theorem 11.2) has been added, which allows the simplification of four almost-identical proofs in the section: Examples 3 and 4, Theorem 11.7 (formerly Corollary 11.4), and Theorem 11.8 (formerly Theorem 11.7).

Where appropriate, the presentation has been improved. See especially the proof of the Chain Rule 28.4, the shorter proof of Abel's Theorem 26.6, and the shorter treatment of decimal expansions in §16. Also, a few examples have been added, a few exercises have been modified or added, and a couple of exercises have been deleted.

Here are the main additions to this revision. The proof of the irrationality of e in §16 is now accompanied by an elegant proof that π is also irrational. Even though this is an "enrichment" section, it is especially recommended for those who teach or will teach pre-college mathematics. The Baire Category Theorem and interesting consequences have been added to the enrichment §21. Section 31, on Taylor's Theorem, has been overhauled. It now includes a discussion of Newton's method for approximating zeros of functions, as well as its cousin, the secant method. Proofs are provided for theorems that guarantee when these approximation methods work. Section 35 on Riemann-Stieltjes integrals has been improved and expanded. A new section, §38, contains an example of a continuous nowhere-differentiable function and a theorem that shows "most" continuous functions are nowhere differentiable. Also, each of §§22, 32, and 33 has been modestly enhanced.

It is a pleasure to thank many people who have helped over the years since the first edition appeared in 1980. This includes David M. Bloom, Robert B. Burckel, Kai Lai Chung, Mark Dalthorp (grandson), M. K. Das (India), Richard Dowds, Ray Hoobler,

Richard M. Koch, Lisa J. Madsen, Pablo V. Negrón Marrero (Puerto Rico), Rajiv Monsurate (India), Theodore W. Palmer, Jürg Rätz (Switzerland), Peter Renz, Karl Stromberg, and Jesús Sueiras (Puerto Rico).

Special thanks go to my collaborator, Jorge M. López, who provided a huge amount of help and support with the revision. Working with him was also a lot of fun. My plan to revise the book was supported from the beginning by my wife, Ruth Madsen Ross. Finally, I thank my editor at Springer, Kaitlin Leach, who was attentive to my needs whenever they arose.

Especially for the Student: Don't be dismayed if you run into material that doesn't make sense, for whatever reason. It happens to all of us. Just tentatively accept the result as true, set it aside as something to return to, and forge ahead. Also, don't forget to use the Index or Symbols Index if some terminology or notation is puzzling.

Contents

1

CHAPTER

Introduction

The underlying space for all the analysis in this book is the set of
real numbers. In this chapter we set down some basic properties of
this set. These properties will serve as our axioms in the sense that
it is possible to derive all the properties of the real numbers using
only these axioms. However, we will avoid getting bogged down in
this endeavor. Some readers may wish to refer to the appendix on
set notation.

§1 The Set ℕ of Natural Numbers

We denote the set $\{1, 2, 3, \ldots\}$ of all *positive integers* by ℕ. Each
positive integer n has a successor, namely $n + 1$. Thus the successor
of 2 is 3, and 37 is the successor of 36. You will probably agree that
the following properties of ℕ are obvious; at least the first four are.

N1. 1 belongs to ℕ.
N2. If n belongs to ℕ, then its successor $n + 1$ belongs to ℕ.
N3. 1 is not the successor of any element in ℕ.

K.A. Ross, *Elementary Analysis: The Theory of Calculus*,
Undergraduate Texts in Mathematics, DOI 10.1007/978-1-4614-6271-2_1,
© Springer Science+Business Media New York 2013

N4. If n and m in $\mathbb{N}$ have the same successor, then $n = m$.

N5. A subset of $\mathbb{N}$ which contains 1, and which contains $n + 1$ whenever it contains n, must equal $\mathbb{N}$.

Properties N1 through N5 are known as the *Peano Axioms* or *Peano Postulates*. It turns out most familiar properties of $\mathbb{N}$ can be proved based on these five axioms; see [8] or [39].

Let's focus our attention on axiom N5, the one axiom that may not be obvious. Here is what the axiom is saying. Consider a subset S of $\mathbb{N}$ as described in N5. Then 1 belongs to S. Since S contains $n + 1$ whenever it contains n, it follows that S contains $2 = 1 + 1$. Again, since S contains $n + 1$ whenever it contains n, it follows that S contains $3 = 2 + 1$. Once again, since S contains $n + 1$ whenever it contains n, it follows that S contains $4 = 3+1$. We could continue this monotonous line of reasoning to conclude S contains any number in $\mathbb{N}$. Thus it seems reasonable to conclude $S = \mathbb{N}$. It is this reasonable conclusion that is asserted by axiom N5.

Here is another way to view axiom N5. Assume axiom N5 is false. Then $\mathbb{N}$ contains a set S such that

(i) $1 \in S$,
(ii) If $n \in S$, then $n + 1 \in S$,

and yet $S \neq \mathbb{N}$. Consider the smallest member of the set $\{n \in \mathbb{N} : n \notin S\}$, call it n_0. Since (i) holds, it is clear $n_0 \neq 1$. So n_0 is a successor to some number in $\mathbb{N}$, namely $n_0 - 1$. We have $n_0 - 1 \in S$ since n_0 is the smallest member of $\{n \in \mathbb{N} : n \notin S\}$. By (ii), the successor of $n_0 - 1$, namely n_0, is also in S, which is a contradiction. This discussion may be plausible, but we emphasize that we have not proved axiom N5 using the successor notion and axioms N1 through N4, because we implicitly used two unproven facts. We assumed every nonempty subset of $\mathbb{N}$ contains a least element and we assumed that if $n_0 \neq 1$ then n_0 is the successor to some number in $\mathbb{N}$.

Axiom N5 is the basis of mathematical induction. Let $P_1, P_2, P_3, \ldots$ be a list of statements or propositions that may or may not be true. The principle of mathematical induction asserts all the statements $P_1, P_2, P_3, \ldots$ are true provided

(I_1) P_1 is true,
(I_2) P_{n+1} is true whenever P_n is true.

We will refer to (I_1), i.e., the fact that P_1 is true, as the basis for induction and we will refer to (I_2) as the induction step. For a sound proof based on mathematical induction, properties (I_1) and (I_2) must both be verified. In practice, (I_1) will be easy to check.

Example 1
Prove $1 + 2 + \cdots + n = \frac{1}{2}n(n+1)$ for positive integers n. $\qquad\square$

Solution
Our nth proposition is

$$P_n: \text{``} 1 + 2 + \cdots + n = \frac{1}{2}n(n+1). \text{''}$$

Thus P_1 asserts $1 = \frac{1}{2} \cdot 1(1+1)$, P_2 asserts $1 + 2 = \frac{1}{2} \cdot 2(2+1)$, P_{37} asserts $1 + 2 + \cdots + 37 = \frac{1}{2} \cdot 37(37+1) = 703$, etc. In particular, P_1 is a true assertion which serves as our basis for induction.

For the induction step, suppose P_n is true. That is, we suppose

$$1 + 2 + \cdots + n = \frac{1}{2}n(n+1)$$

is true. Since we wish to prove P_{n+1} from this, we add $n+1$ to both sides to obtain

$$1 + 2 + \cdots + n + (n+1) = \frac{1}{2}n(n+1) + (n+1)$$
$$= \frac{1}{2}[n(n+1) + 2(n+1)] = \frac{1}{2}(n+1)(n+2)$$
$$= \frac{1}{2}(n+1)((n+1)+1).$$

Thus P_{n+1} holds if P_n holds. By the principle of mathematical induction, we conclude P_n is true for all n. $\qquad\square$

We emphasize that prior to the last sentence of our solution we did *not* prove "P_{n+1} is true." We merely proved an implication: "if P_n is true, then P_{n+1} is true." In a sense we proved an infinite number of assertions, namely: P_1 is true; if P_1 is true then P_2 is true; if P_2 is true then P_3 is true; if P_3 is true then P_4 is true; etc. Then we applied mathematical induction to conclude P_1 is true, P_2 is true, P_3 is true, P_4 is true, etc. We also confess that formulas like the one just proved are easier to prove than to discover. It can be a tricky matter to guess such a result. Sometimes results such as this are discovered by trial and error.

Example 2

All numbers of the form $5^n - 4n - 1$ are divisible by 16.   □

Solution

More precisely, we show $5^n - 4n - 1$ is divisible by 16 for each n in $\mathbb{N}$. Our nth proposition is

$$P_n: \text{``}5^n - 4n - 1 \quad \text{is divisible by} \quad 16.\text{''}$$

The basis for induction P_1 is clearly true, since $5^1 - 4 \cdot 1 - 1 = 0$. Proposition P_2 is also true because $5^2 - 4 \cdot 2 - 1 = 16$, but note we didn't need to check this case before proceeding to the induction step. For the induction step, suppose P_n is true. To verify P_{n+1}, the trick is to write

$$5^{n+1} - 4(n+1) - 1 = 5(5^n - 4n - 1) + 16n.$$

Since $5^n - 4n - 1$ is a multiple of 16 by the induction hypothesis, it follows that $5^{n+1} - 4(n+1) - 1$ is also a multiple of 16. In fact, if $5^n - 4n - 1 = 16m$, then $5^{n+1} - 4(n+1) - 1 = 16 \cdot (5m + n)$. We have shown P_n implies P_{n+1}, so the induction step holds. An application of mathematical induction completes the proof.   □

Example 3

Show $|\sin nx| \leq n|\sin x|$ for all positive integers n and all real numbers x.   □

Solution

Our nth proposition is

$$P_n: \text{``}|\sin nx| \leq n|\sin x| \quad \text{for all real numbers} \quad x.\text{''}$$

The basis for induction is again clear. Suppose P_n is true. We apply the addition formula for sine to obtain

$$|\sin(n+1)x| = |\sin(nx + x)| = |\sin nx \cos x + \cos nx \sin x|.$$

Now we apply the Triangle Inequality and properties of the absolute value [see Theorems 3.7 and 3.5] to obtain

$$|\sin(n+1)x| \leq |\sin nx| \cdot |\cos x| + |\cos nx| \cdot |\sin x|.$$

Since $|\cos y| \leq 1$ for all y we see that

$$|\sin(n+1)x| \leq |\sin nx| + |\sin x|.$$

Now we apply the induction hypothesis P_n to obtain

$$|\sin(n+1)x| \le n|\sin x| + |\sin x| = (n+1)|\sin x|.$$

Thus P_{n+1} holds. Finally, the result holds for all n by mathematical induction. □

Exercises

1.1 Prove $1^2 + 2^2 + \cdots + n^2 = \frac{1}{6}n(n+1)(2n+1)$ for all positive integers n.

1.2 Prove $3 + 11 + \cdots + (8n - 5) = 4n^2 - n$ for all positive integers n.

1.3 Prove $1^3 + 2^3 + \cdots + n^3 = (1 + 2 + \cdots + n)^2$ for all positive integers n.

1.4 (a) Guess a formula for $1 + 3 + \cdots + (2n - 1)$ by evaluating the sum for $n = 1, 2, 3,$ and 4. [For $n = 1$, the sum is simply 1.]

 (b) Prove your formula using mathematical induction.

1.5 Prove $1 + \frac{1}{2} + \frac{1}{4} + \cdots + \frac{1}{2^n} = 2 - \frac{1}{2^n}$ for all positive integers n.

1.6 Prove $(11)^n - 4^n$ is divisible by 7 when n is a positive integer.

1.7 Prove $7^n - 6n - 1$ is divisible by 36 for all positive integers n.

1.8 The principle of mathematical induction can be extended as follows. A list $P_m, P_{m+1}, \ldots$ of propositions is true provided (i) P_m is true, (ii) P_{n+1} is true whenever P_n is true and $n \ge m$.

 (a) Prove $n^2 > n + 1$ for all integers $n \ge 2$.

 (b) Prove $n! > n^2$ for all integers $n \ge 4$. [Recall $n! = n(n-1)\cdots 2 \cdot 1$; for example, $5! = 5 \cdot 4 \cdot 3 \cdot 2 \cdot 1 = 120$.]

1.9 (a) Decide for which integers the inequality $2^n > n^2$ is true.

 (b) Prove your claim in (a) by mathematical induction.

1.10 Prove $(2n + 1) + (2n + 3) + (2n + 5) + \cdots + (4n - 1) = 3n^2$ for all positive integers n.

1.11 For each $n \in \mathbb{N}$, let P_n denote the assertion "$n^2 + 5n + 1$ is an even integer."

 (a) Prove P_{n+1} is true whenever P_n is true.

 (b) For which n is P_n actually true? What is the moral of this exercise?

1.12 For $n \in \mathbb{N}$, let $n!$ [read "n factorial"] denote the product $1 \cdot 2 \cdot 3 \cdots n$. Also let $0! = 1$ and define

$$\binom{n}{k} = \frac{n!}{k!(n-k)!} \quad \text{for} \quad k = 0, 1, \ldots, n. \tag{1.1}$$

The *binomial theorem* asserts that

$$(a+b)^n = \binom{n}{0}a^n + \binom{n}{1}a^{n-1}b + \binom{n}{2}a^{n-2}b^2 + \cdots + \binom{n}{n-1}ab^{n-1} + \binom{n}{n}b^n$$

$$= a^n + na^{n-1}b + \frac{1}{2}n(n-1)a^{n-2}b^2 + \cdots + nab^{n-1} + b^n.$$

(a) Verify the binomial theorem for $n = 1, 2$, and 3.

(b) Show $\binom{n}{k} + \binom{n}{k-1} = \binom{n+1}{k}$ for $k = 1, 2, \ldots, n$.

(c) Prove the binomial theorem using mathematical induction and part (b).

§2   The Set $\mathbb{Q}$ of Rational Numbers

Small children first learn to add and to multiply positive integers. After subtraction is introduced, the need to expand the number system to include 0 and negative integers becomes apparent. At this point the world of numbers is enlarged to include the set $\mathbb{Z}$ of all *integers*. Thus we have $\mathbb{Z} = \{0, 1, -1, 2, -2, \ldots\}$.

Soon the space $\mathbb{Z}$ also becomes inadequate when division is introduced. The solution is to enlarge the world of numbers to include all fractions. Accordingly, we study the space $\mathbb{Q}$ of all *rational numbers*, i.e., numbers of the form $\frac{m}{n}$ where $m, n \in \mathbb{Z}$ and $n \neq 0$. Note that $\mathbb{Q}$ contains all terminating decimals such as $1.492 = \frac{1,492}{1,000}$. The connection between decimals and real numbers is discussed in 10.3 on page 58 and in §16. The space $\mathbb{Q}$ is a highly satisfactory algebraic system in which the basic operations addition, multiplication, subtraction and division can be fully studied. No system is perfect, however, and $\mathbb{Q}$ is inadequate in some ways. In this section we will consider the defects of $\mathbb{Q}$. In the next section we will stress the good features of $\mathbb{Q}$ and then move on to the system of real numbers.

The set $\mathbb{Q}$ of rational numbers is a very nice algebraic system until one tries to solve equations like $x^2 = 2$. It turns out that no rational

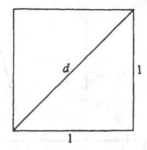

FIGURE 2.1

number satisfies this equation, and yet there are good reasons to believe some kind of number satisfies this equation. Consider, for example, a square with sides having length one; see Fig. 2.1. If d is the length of the diagonal, then from geometry we know $1^2 + 1^2 = d^2$, i.e., $d^2 = 2$. Apparently there is a positive length whose square is 2, which we write as $\sqrt{2}$. But $\sqrt{2}$ cannot be a rational number, as we will show in Example 2. Of course, $\sqrt{2}$ can be approximated by rational numbers. There are rational numbers whose squares are close to 2; for example, $(1.4142)^2 = 1.99996164$ and $(1.4143)^2 = 2.00024449$.

It is evident that there are lots of rational numbers and yet there are "gaps" in ℚ. Here is another way to view this situation. Consider the graph of the polynomial $x^2 - 2$ in Fig. 2.2. Does the graph of $x^2 - 2$ cross the x-axis? We are inclined to say it does, because when we draw the x-axis we include "all" the points. We allow no "gaps." But notice that the graph of $x^2 - 2$ slips by all the rational numbers on the x-axis. The x-axis is our picture of the number line, and the set of rational numbers again appears to have significant "gaps."

There are even more exotic numbers such as π and e that are not rational numbers, but which come up naturally in mathematics. The number π is basic to the study of circles and spheres, and e arises in problems of exponential growth.

We return to $\sqrt{2}$. This is an example of what is called an algebraic number because it satisfies the equation $x^2 - 2 = 0$.

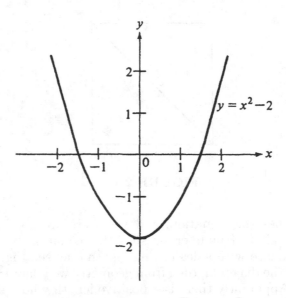

FIGURE 2.2

2.1 Definition.
A number is called an *algebraic number* if it satisfies a polynomial equation

$$c_n x^n + c_{n-1} x^{n-1} + \cdots + c_1 x + c_0 = 0$$

where the coefficients $c_0, c_1, \ldots, c_n$ are integers, $c_n \neq 0$ and $n \geq 1$.

Rational numbers are always algebraic numbers. In fact, if $r = \frac{m}{n}$ is a rational number $[m, n \in \mathbb{Z}$ and $n \neq 0]$, then it satisfies the equation $nx - m = 0$. Numbers defined in terms of $\sqrt{}$, $\sqrt[3]{}$, etc. [or fractional exponents, if you prefer] and ordinary algebraic operations on the rational numbers are invariably algebraic numbers.

Example 1
$\frac{4}{17}$, $\sqrt{3}$, $\sqrt[3]{17}$, $\sqrt{2 + \sqrt[3]{5}}$ and $\sqrt{\frac{4 - 2\sqrt{3}}{7}}$ are algebraic numbers. In fact, $\frac{4}{17}$ is a solution of $17x - 4 = 0$, $\sqrt{3}$ is a solution of $x^2 - 3 = 0$, and $\sqrt[3]{17}$ is a solution of $x^3 - 17 = 0$. The expression $a = \sqrt{2 + \sqrt[3]{5}}$ means $a^2 = 2 + \sqrt[3]{5}$ or $a^2 - 2 = \sqrt[3]{5}$ so that $(a^2 - 2)^3 = 5$. Therefore we

have $a^6 - 6a^4 + 12a^2 - 13 = 0$, which shows $a = \sqrt{2 + \sqrt[3]{5}}$ satisfies the polynomial equation $x^6 - 6x^4 + 12x^2 - 13 = 0$.

Similarly, the expression $b = \sqrt{\frac{4-2\sqrt{3}}{7}}$ leads to $7b^2 = 4 - 2\sqrt{3}$, hence $2\sqrt{3} = 4 - 7b^2$, hence $12 = (4 - 7b^2)^2$, hence $49b^4 - 56b^2 + 4 = 0$. Thus b satisfies the polynomial equation $49x^4 - 56x^2 + 4 = 0$. $\qquad\square$

The next theorem may be familiar from elementary algebra. It is the theorem that justifies the following remarks: the only possible rational solutions of $x^3 - 7x^2 + 2x - 12 = 0$ are $\pm 1, \pm 2, \pm 3, \pm 4, \pm 6, \pm 12$, so the only possible (rational) monomial factors of $x^3 - 7x^2 + 2x - 12$ are $x - 1$, $x + 1$, $x - 2$, $x + 2$, $x - 3$, $x + 3$, $x - 4$, $x + 4$, $x - 6$, $x + 6$, $x - 12$, $x + 12$. We won't pursue these algebraic problems; we merely make these observations in the hope they will be familiar.

The next theorem also allows one to prove algebraic numbers that do not look like rational numbers are usually not rational numbers. Thus $\sqrt{4}$ is obviously a rational number, while $\sqrt{2}$, $\sqrt{3}$, $\sqrt{5}$, etc. turn out to be nonrational. See the examples following the theorem. Also, compare Exercise 2.7. Recall that an integer k is a *factor* of an integer m or *divides* m if $\frac{m}{k}$ is also an integer.

If the next theorem seems complicated, first read the special case in Corollary 2.3 and Examples 2–5.

2.2 Rational Zeros Theorem.
Suppose $c_0, c_1, \ldots, c_n$ are integers and r is a rational number satisfying the polynomial equation

$$c_n x^n + c_{n-1} x^{n-1} + \cdots + c_1 x + c_0 = 0 \qquad (1)$$

where $n \geq 1$, $c_n \neq 0$ and $c_0 \neq 0$. Let $r = \frac{c}{d}$ where c, d are integers having no common factors and $d \neq 0$. Then c divides c_0 and d divides c_n.

In other words, the only rational *candidates* for solutions of (1) have the form $\frac{c}{d}$ where c divides c_0 and d divides c_n.

Proof
We are given

$$c_n \left(\frac{c}{d}\right)^n + c_{n-1} \left(\frac{c}{d}\right)^{n-1} + \cdots + c_1 \left(\frac{c}{d}\right) + c_0 = 0.$$

We multiply through by d^n and obtain

$$c_n c^n + c_{n-1} c^{n-1} d + c_{n-2} c^{n-2} d^2 + \cdots + c_2 c^2 d^{n-2} + c_1 c d^{n-1} + c_0 d^n = 0. \tag{2}$$

If we solve for $c_0 d^n$, we obtain

$$c_0 d^n = -c[c_n c^{n-1} + c_{n-1} c^{n-2} d + c_{n-2} c^{n-3} d^2 + \cdots + c_2 c d^{n-2} + c_1 d^{n-1}].$$

It follows that c divides $c_0 d^n$. But c and d^n have no common factors, so c divides c_0. This follows from the basic fact that if an integer c divides a product ab of integers, and if c and b have no common factors, then c divides a. See, for example, Theorem 1.10 in [50].

Now we solve (2) for $c_n c^n$ and obtain

$$c_n c^n = -d[c_{n-1} c^{n-1} + c_{n-2} c^{n-2} d + \cdots + c_2 c^2 d^{n-3} + c_1 c d^{n-2} + c_0 d^{n-1}].$$

Thus d divides $c_n c^n$. Since c^n and d have no common factors, d divides c_n. ∎

2.3 Corollary.
Consider the polynomial equation

$$x^n + c_{n-1} x^{n-1} + \cdots + c_1 x + c_0 = 0,$$

where the coefficients $c_0, c_1, \ldots, c_{n-1}$ are integers and $c_0 \neq 0$.[1] *Any rational solution of this equation must be an integer that divides c_0.*

Proof
In the Rational Zeros Theorem 2.2, the denominator of r must divide the coefficient of x^n, which is 1 in this case. Thus r is an integer and it divides c_0. ∎

Example 2
$\sqrt{2}$ is not a rational number. □

Proof
By Corollary 2.3, the only rational numbers that could possibly be solutions of $x^2 - 2 = 0$ are $\pm 1, \pm 2$. [Here $n = 2$, $c_2 = 1$, $c_1 = 0$, $c_0 = -2$. So the rational solutions have the form $\frac{c}{d}$ where c divides

---

[1] Polynomials like this, where the highest power has coefficient 1, are called *monic* polynomials.

$c_0 = -2$ and d divides $c_2 = 1$.] One can substitute each of the four numbers $\pm 1, \pm 2$ into the equation $x^2 - 2 = 0$ to quickly eliminate them as possible solutions of the equation. Since $\sqrt{2}$ is a solution of $x^2 - 2 = 0$, it cannot be a rational number. ∎

Example 3
$\sqrt{17}$ is not a rational number. □

Proof
The only possible rational solutions of $x^2 - 17 = 0$ are $\pm 1, \pm 17$, and none of these numbers are solutions. ∎

Example 4
$\sqrt[3]{6}$ is not a rational number. □

Proof
The only possible rational solutions of $x^3 - 6 = 0$ are $\pm 1, \pm 2, \pm 3, \pm 6$. It is easy to verify that none of these eight numbers satisfies the equation $x^3 - 6 = 0$. ∎

Example 5
$a = \sqrt{2 + \sqrt[3]{5}}$ is not a rational number. □

Proof
In Example 1 we showed a is a solution of $x^6 - 6x^4 + 12x^2 - 13 = 0$. By Corollary 2.3, the only possible rational solutions are $\pm 1, \pm 13$. When $x = 1$ or -1, the left hand side of the equation is -6 and when $x = 13$ or -13, the left hand side of the equation turns out to equal $4{,}657{,}458$. This last computation could be avoided by using a little common sense. Either observe a is "obviously" bigger than 1 and less than 13, or observe

$$13^6 - 6 \cdot 13^4 + 12 \cdot 13^2 - 13 = 13(13^5 - 6 \cdot 13^3 + 12 \cdot 13 - 1) \neq 0$$

since the term in parentheses cannot be zero: it is one less than some multiple of 13. ∎

Example 6
$b = \sqrt{\frac{4 - 2\sqrt{3}}{7}}$ is not a rational number. □

Proof

In Example 1 we showed b is a solution of $49x^4 - 56x^2 + 4 = 0$. By Theorem 2.2, the only possible rational solutions are

$$\pm 1, \pm 1/7, \pm 1/49, \pm 2, \pm 2/7, \pm 2/49, \pm 4, \pm 4/7, \pm 4/49.$$

To complete our proof, all we need to do is substitute these 18 candidates into the equation $49x^4 - 56x^2 + 4 = 0$. This prospect is so discouraging, however, that we choose to find a more clever approach. In Example 1, we also showed $12 = (4 - 7b^2)^2$. Now if b were rational, then $4 - 7b^2$ would also be rational [Exercise 2.6], so the equation $12 = x^2$ would have a rational solution. But the only possible rational solutions to $x^2 - 12 = 0$ are $\pm 1, \pm 2, \pm 3, \pm 4, \pm 6, \pm 12$, and these all can be eliminated by mentally substituting them into the equation. We conclude $4 - 7b^2$ cannot be rational, so b cannot be rational. ∎

As a practical matter, many or all of the rational candidates given by the Rational Zeros Theorem can be eliminated by approximating the quantity in question. It is nearly obvious that the values in Examples 2 through 5 are not integers, while all the rational candidates are. The number b in Example 6 is approximately 0.2767; the nearest rational candidate is $+2/7$ which is approximately 0.2857.

It should be noted that not all irrational-looking expressions are actually irrational. See Exercise 2.7.

2.4 Remark.

While admiring the efficient Rational Zeros Theorem for finding rational zeros of polynomials with integer coefficients, you might wonder how one would find other zeros of these polynomials, or zeros of other functions. In §31, we will discuss the most well-known method, called Newton's method, and its cousin, the secant method. That discussion can be read now; only the proof of the theorem uses material from §31.

Exercises

2.1 Show $\sqrt{3}$, $\sqrt{5}$, $\sqrt{7}$, $\sqrt{24}$, and $\sqrt{31}$ are not rational numbers.

2.2 Show $\sqrt[3]{2}$, $\sqrt[7]{5}$ and $\sqrt[4]{13}$ are not rational numbers.

2.3 Show $\sqrt{2 + \sqrt{2}}$ is not a rational number.

2.4 Show $\sqrt[3]{5 - \sqrt{3}}$ is not a rational number.

2.5 Show $[3 + \sqrt{2}]^{2/3}$ is not a rational number.

2.6 In connection with Example 6, discuss why $4 - 7b^2$ is rational if b is rational.

2.7 Show the following irrational-looking expressions are actually rational numbers: (a) $\sqrt{4 + 2\sqrt{3}} - \sqrt{3}$, and (b) $\sqrt{6 + 4\sqrt{2}} - \sqrt{2}$.

2.8 Find all rational solutions of the equation $x^8 - 4x^5 + 13x^3 - 7x + 1 = 0$.

§3 The Set ℝ of Real Numbers

The set $\mathbb{Q}$ is probably the largest system of numbers with which you really feel comfortable. There are some subtleties but you have learned to cope with them. For example, $\mathbb{Q}$ is not simply the set of symbols m/n, where $m, n \in \mathbb{Z}, n \neq 0$, since we regard some pairs of different looking fractions as equal. For example, $\frac{2}{4}$ and $\frac{3}{6}$ represent the same element of $\mathbb{Q}$. A rigorous development of $\mathbb{Q}$ based on $\mathbb{Z}$, which in turn is based on $\mathbb{N}$, would require us to introduce the notion of equivalence classes. In this book we assume a familiarity with and understanding of $\mathbb{Q}$ as an algebraic system. However, in order to clarify exactly what we need to know about $\mathbb{Q}$, we set down some of its basic axioms and properties.

The basic algebraic operations in $\mathbb{Q}$ are addition and multiplication. Given a pair a, b of rational numbers, the sum $a + b$ and the product ab also represent rational numbers. Moreover, the following properties hold.

A1. $a + (b + c) = (a + b) + c$ for all a, b, c.
A2. $a + b = b + a$ for all a, b.
A3. $a + 0 = a$ for all a.
A4. For each a, there is an element $-a$ such that $a + (-a) = 0$.
M1. $a(bc) = (ab)c$ for all a, b, c.
M2. $ab = ba$ for all a, b.

M3. $a \cdot 1 = a$ for all a.

M4. For each $a \neq 0$, there is an element a^{-1} such that $aa^{-1} = 1$.

DL $a(b + c) = ab + ac$ for all a, b, c.

Properties A1 and M1 are called the *associative laws*, and properties A2 and M2 are the *commutative laws*. Property DL is the *distributive law*; this is the least obvious law and is the one that justifies "factorization" and "multiplying out" in algebra. A system that has more than one element and satisfies these nine properties is called a *field*. The basic algebraic properties of $\mathbb{Q}$ can proved solely on the basis of these field properties. We will not pursue this topic in any depth, but we illustrate our claim by proving some familiar properties in Theorem 3.1 below.

The set $\mathbb{Q}$ also has an order structure $\leq$ satisfying

O1. Given a and b, either $a \leq b$ or $b \leq a$.

O2. If $a \leq b$ and $b \leq a$, then $a = b$.

O3. If $a \leq b$ and $b \leq c$, then $a \leq c$.

O4. If $a \leq b$, then $a + c \leq b + c$.

O5. If $a \leq b$ and $0 \leq c$, then $ac \leq bc$.

Property O3 is called the *transitive law*. This is the characteristic property of an ordering. A field with an ordering satisfying properties O1 through O5 is called an *ordered field*. Most of the algebraic and order properties of $\mathbb{Q}$ can be established for an ordered field. We will prove a few of them in Theorem 3.2 below.

The mathematical system on which we will do our analysis will be the set $\mathbb{R}$ of all *real numbers*. The set $\mathbb{R}$ will include all rational numbers, all algebraic numbers, π, e, and more. It will be a set that can be drawn as the real number line; see Fig. 3.1. That is, every real number will correspond to a point on the number line, and every point on the number line will correspond to a real number. In particular, unlike $\mathbb{Q}$, $\mathbb{R}$ will not have any "gaps." We will also see that real numbers have decimal expansions; see 10.3 on page 58 and §16. These remarks help describe $\mathbb{R}$, but we certainly have not defined $\mathbb{R}$ as a precise mathematical object. It turns out that $\mathbb{R}$ can be defined entirely in terms of the set $\mathbb{Q}$ of rational numbers; we indicate in the enrichment §6 one way this can be done. But then it is a long and tedious task to show how to add and multiply the

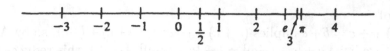

FIGURE 3.1

objects defined in this way and to show that the set ℝ, with these operations, satisfies all the familiar algebraic and order properties we expect to hold for ℝ. To develop ℝ properly from ℚ in this way and to develop ℚ properly from ℕ would take us several chapters. This would defeat the purpose of this book, which is to accept ℝ as a mathematical system and to study some important properties of ℝ and functions on ℝ. Nevertheless, it is desirable to specify exactly what properties of ℝ we are assuming.

Real numbers, i.e., elements of ℝ, can be added together and multiplied together. That is, given real numbers a and b, the sum $a+b$ and the product ab also represent real numbers. Moreover, these operations satisfy the field properties A1 through A4, M1 through M4, and DL. The set ℝ also has an order structure $\leq$ that satisfies properties O1 through O5. Thus, like ℚ, ℝ is an ordered field.

In the remainder of this section, we will obtain some results for ℝ that are valid in any ordered field. In particular, these results would be equally valid if we restricted our attention to ℚ. These remarks emphasize the similarities between ℝ and ℚ. We have not yet indicated how ℝ can be distinguished from ℚ as a mathematical object, although we have asserted that ℝ has no "gaps." We will make this observation much more precise in the next section, and then we will give a "gap filling" axiom that finally will distinguish ℝ from ℚ.

3.1 Theorem.
The following are consequences of the field properties:
  (i) $a + c = b + c$ *implies* $a = b$;
 (ii) $a \cdot 0 = 0$ *for all* a;
(iii) $(-a)b = -ab$ *for all* a, b;
(iv) $(-a)(-b) = ab$ *for all* a, b;
 (v) $ac = bc$ *and* $c \neq 0$ *imply* $a = b$;
(vi) $ab = 0$ *implies either* $a = 0$ *or* $b = 0$;
    for $a, b, c \in ℝ$.

Proof

(i) $a + c = b + c$ implies $(a + c) + (-c) = (b + c) + (-c)$, so by A1, we have $a + [c + (-c)] = b + [c + (-c)]$. By A4, this reduces to $a + 0 = b + 0$, so $a = b$ by A3.

(ii) We use A3 and DL to obtain $a \cdot 0 = a \cdot (0 + 0) = a \cdot 0 + a \cdot 0$, so $0 + a \cdot 0 = a \cdot 0 + a \cdot 0$. By (i) we conclude $0 = a \cdot 0$.

(iii) Since $a + (-a) = 0$, we have $ab + (-a)b = [a + (-a)] \cdot b = 0 \cdot b = 0 = ab + (-(ab))$. From (i) we obtain $(-a)b = -(ab)$.

(iv) **and** (v) are left to Exercise 3.3.

(vi) If $ab = 0$ and $b \neq 0$, then $0 = b^{-1} \cdot 0 = 0 \cdot b^{-1} = (ab) \cdot b^{-1} = a(bb^{-1}) = a \cdot 1 = a$. ∎

3.2 Theorem.

The following are consequences of the properties of an ordered field:

(i) *If $a \leq b$, then $-b \leq -a$;*

(ii) *If $a \leq b$ and $c \leq 0$, then $bc \leq ac$;*

(iii) *If $0 \leq a$ and $0 \leq b$, then $0 \leq ab$;*

(iv) *$0 \leq a^2$ for all a;*

(v) *$0 < 1$;*

(vi) *If $0 < a$, then $0 < a^{-1}$;*

(vii) *If $0 < a < b$, then $0 < b^{-1} < a^{-1}$;*

for $a, b, c \in \mathbb{R}$.

Note $a < b$ means $a \leq b$ and $a \neq b$.

Proof

(i) Suppose $a \leq b$. By O4 applied to $c = (-a) + (-b)$, we have $a + [(-a) + (-b)] \leq b + [(-a) + (-b)]$. It follows that $-b \leq -a$.

(ii) If $a \leq b$ and $c \leq 0$, then $0 \leq -c$ by (i). Now by O5 we have $a(-c) \leq b(-c)$, i.e., $-ac \leq -bc$. From (i) again, we see $bc \leq ac$.

(iii) If we put $a = 0$ in property O5, we obtain: $0 \leq b$ and $0 \leq c$ imply $0 \leq bc$. Except for notation, this is exactly assertion (iii).

(iv) For any a, either $a \geq 0$ or $a \leq 0$ by O1. If $a \geq 0$, then $a^2 \geq 0$ by (iii). If $a \leq 0$, then we have $-a \geq 0$ by (i), so $(-a)^2 \geq 0$, i.e., $a^2 \geq 0$.

(v) Is left to Exercise 3.4.

(vi) Suppose $0 < a$ but $0 < a^{-1}$ fails. Then we must have $a^{-1} \leq 0$ and $0 \leq -a^{-1}$. Now by (iii) $0 \leq a(-a^{-1}) = -1$, so that $1 \leq 0$, contrary to (v).

(vii) Is left to Exercise 3.4. ∎

Another important notion that should be familiar is that of absolute value.

3.3 Definition.
We define

$$|a| = a \quad \text{if} \quad a \geq 0 \quad \text{and} \quad |a| = -a \quad \text{if} \quad a \leq 0.$$

$|a|$ is called the *absolute value of a.*

Intuitively, the absolute value of a represents the distance between 0 and a, but in fact we will *define* the idea of "distance" in terms of the "absolute value," which in turn was defined in terms of the ordering.

3.4 Definition.
For numbers a and b we define $\text{dist}(a, b) = |a - b|$; $\text{dist}(a, b)$ represents the *distance between a and b.*

The basic properties of the absolute value are given in the next theorem.

3.5 Theorem.
(i) $|a| \geq 0$ *for all* $a \in \mathbb{R}$.
(ii) $|ab| = |a| \cdot |b|$ *for all* $a, b \in \mathbb{R}$.
(iii) $|a + b| \leq |a| + |b|$ *for all* $a, b \in \mathbb{R}$.

Proof
(i) is obvious from the definition. [The word "obvious" as used here signifies the reader should be able to quickly see why the result is true. Certainly if $a \geq 0$, then $|a| = a \geq 0$, while $a < 0$ implies $|a| = -a > 0$. We will use expressions like "obviously" and "clearly" in place of very simple arguments, but we will not use these terms to obscure subtle points.]

(ii) There are four easy cases here. If $a \geq 0$ and $b \geq 0$, then $ab \geq 0$, so $|a| \cdot |b| = ab = |ab|$. If $a \leq 0$ and $b \leq 0$, then $-a \geq 0$, $-b \geq 0$ and $(-a)(-b) \geq 0$ so that $|a| \cdot |b| = (-a)(-b) = ab = |ab|$. If $a \geq 0$ and $b \leq 0$, then $-b \geq 0$ and $a(-b) \geq 0$ so that $|a| \cdot |b| = a(-b) = -(ab) = |ab|$. If $a \leq 0$ and $b \geq 0$, then $-a \geq 0$ and $(-a)b \geq 0$ so that $|a| \cdot |b| = (-a)b = -ab = |ab|$.

(iii) The inequalities $-|a| \leq a \leq |a|$ are obvious, since either $a = |a|$ or else $a = -|a|$. Similarly $-|b| \leq b \leq |b|$. Now four applications of O4 yield

$$-|a| + (-|b|) \leq a + b \leq |a| + b \leq |a| + |b|$$

so that

$$-(|a| + |b|) \leq a + b \leq |a| + |b|.$$

This tells us $a + b \leq |a| + |b|$ and also $-(a + b) \leq |a| + |b|$. Since $|a + b|$ is equal to either $a + b$ or $-(a + b)$, we conclude $|a + b| \leq |a| + |b|$. ∎

3.6 Corollary.
$\mathrm{dist}(a, c) \leq \mathrm{dist}(a, b) + \mathrm{dist}(b, c)$ *for all* $a, b, c \in \mathbb{R}$.

Proof
We can apply inequality (iii) of Theorem 3.5 to $a - b$ and $b - c$ to obtain $|(a - b) + (b - c)| \leq |a - b| + |b - c|$ or $\mathrm{dist}(a, c) = |a - c| \leq |a - b| + |b - c| \leq \mathrm{dist}(a, b) + \mathrm{dist}(b, c)$. ∎

The inequality in Corollary 3.6 is very closely related to an inequality concerning points **a**, **b**, **c** in the plane, and the latter inequality can be interpreted as a statement about triangles: the length of a side of a triangle is less than or equal to the sum of the lengths of the other two sides. See Fig. 3.2. For this reason, the inequality in Corollary 3.6 and its close relative (iii) in Theorem 3.5 are often called the *Triangle Inequality*.

3.7 Triangle Inequality.
$|a + b| \leq |a| + |b|$ *for all* a, b.

A useful variant of the triangle inequality is given in Exercise 3.5(b).

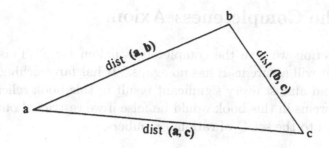

FIGURE 3.2

Exercises

3.1 (a) Which of the properties A1–A4, M1–M4, DL, O1–O5 fail for $\mathbb{N}$?

  (b) Which of these properties fail for $\mathbb{Z}$?

3.2 (a) The commutative law A2 was used in the proof of (ii) in Theorem 3.1. Where?

  (b) The commutative law A2 was also used in the proof of (iii) in Theorem 3.1. Where?

3.3 Prove (iv) and (v) of Theorem 3.1.

3.4 Prove (v) and (vii) of Theorem 3.2.

3.5 (a) Show $|b| \leq a$ if and only if $-a \leq b \leq a$.

  (b) Prove $||a| - |b|| \leq |a - b|$ for all $a, b \in \mathbb{R}$.

3.6 (a) Prove $|a + b + c| \leq |a| + |b| + |c|$ for all $a, b, c \in \mathbb{R}$. *Hint:* Apply the triangle inequality twice. Do *not* consider eight cases.

  (b) Use induction to prove

$$|a_1 + a_2 + \cdots + a_n| \leq |a_1| + |a_2| + \cdots + |a_n|$$

  for n numbers $a_1, a_2, \ldots, a_n$.

3.7 (a) Show $|b| < a$ if and only if $-a < b < a$.

  (b) Show $|a - b| < c$ if and only if $b - c < a < b + c$.

  (c) Show $|a - b| \leq c$ if and only if $b - c \leq a \leq b + c$.

3.8 Let $a, b \in \mathbb{R}$. Show if $a \leq b_1$ for every $b_1 > b$, then $a \leq b$.

§4 The Completeness Axiom

In this section we give the completeness axiom for $\mathbb{R}$. This is the axiom that will assure us $\mathbb{R}$ has no "gaps." It has far-reaching consequences and almost every significant result in this book relies on it. Most theorems in this book would be false if we restricted our world of numbers to the set $\mathbb{Q}$ of rational numbers.

4.1 Definition.
Let S be a nonempty subset of $\mathbb{R}$.
 (a) If S contains a largest element s_0 [that is, s_0 belongs to S and $s \leq s_0$ for all $s \in S$], then we call s_0 the *maximum of S* and write $s_0 = \max S$.
 (b) If S contains a smallest element, then we call the smallest element the *minimum of S* and write it as $\min S$.

Example 1
 (a) Every finite nonempty subset of $\mathbb{R}$ has a maximum and a minimum. Thus

$$\max\{1,2,3,4,5\} = 5 \quad \text{and} \quad \min\{1,2,3,4,5\} = 1,$$
$$\max\{0,\pi,-7,e,3,4/3\} = \pi \quad \text{and} \quad \min\{0,\pi,-7,e,3,4/3\} = -7,$$
$$\max\{n \in \mathbb{Z} : -4 < n \leq 100\} = 100 \quad \text{and}$$
$$\min\{n \in \mathbb{Z} : -4 < n \leq 100\} = -3.$$

 (b) Consider real numbers a and b where $a < b$. The following notation will be used throughout

$$[a,b] = \{x \in \mathbb{R} : a \leq x \leq b\}, \qquad (a,b) = \{x \in \mathbb{R} : a < x < b\},$$
$$[a,b) = \{x \in \mathbb{R} : a \leq x < b\}, \qquad (a,b] = \{x \in \mathbb{R} : a < x \leq b\}.$$

 $[a,b]$ is called a *closed interval*, (a,b) is called an *open interval*, while $[a,b)$ and $(a,b]$ are called *half-open* or *semi-open intervals*. Observe $\max[a,b] = b$ and $\min[a,b] = a$. The set (a,b) has no maximum and no minimum, since the endpoints a and b do not belong to the set. The set $[a,b)$ has no maximum, but a is its minimum.
 (c) The sets $\mathbb{Z}$ and $\mathbb{Q}$ have no maximum or minimum. The set $\mathbb{N}$ has no maximum, but $\min \mathbb{N} = 1$.

(d) The set $\{r \in \mathbb{Q} : 0 \le r \le \sqrt{2}\}$ has a minimum, namely 0, but no maximum. This is because $\sqrt{2}$ does not belong to the set, but there are rationals in the set arbitrarily close to $\sqrt{2}$.

(e) Consider the set $\{n^{(-1)^n} : n \in \mathbb{N}\}$. This is shorthand for the set

$$\{1^{-1}, 2, 3^{-1}, 4, 5^{-1}, 6, 7^{-1}, \ldots\} = \{1, 2, \tfrac{1}{3}, 4, \tfrac{1}{5}, 6, \tfrac{1}{7}, \ldots\}.$$

The set has no maximum and no minimum. ☐

4.2 Definition.

Let S be a nonempty subset of $\mathbb{R}$.

(a) If a real number M satisfies $s \le M$ for all $s \in S$, then M is called an *upper bound of S* and the set S is said to be *bounded above*.

(b) If a real number m satisfies $m \le s$ for all $s \in S$, then m is called a *lower bound of S* and the set S is said to be *bounded below*.

(c) The set S is said to be *bounded* if it is bounded above and bounded below. Thus S is bounded if there exist real numbers m and M such that $S \subseteq [m, M]$.

Example 2

(a) The maximum of a set is always an upper bound for the set. Likewise, the minimum of a set is always a lower bound for the set.

(b) Consider a, b in $\mathbb{R}$, $a < b$. The number b is an upper bound for each of the sets $[a, b]$, (a, b), $[a, b)$, $(a, b]$. Every number larger than b is also an upper bound for each of these sets, but b is the smallest or least upper bound.

(c) None of the sets $\mathbb{Z}$, $\mathbb{Q}$ and $\mathbb{N}$ is bounded above. The set $\mathbb{N}$ is bounded below; 1 is a lower bound for $\mathbb{N}$ and so is any number less than 1. In fact, 1 is the largest or greatest lower bound.

(d) Any nonpositive real number is a lower bound for $\{r \in \mathbb{Q} : 0 \le r \le \sqrt{2}\}$ and 0 is the set's greatest lower bound. The least upper bound is $\sqrt{2}$.

(e) The set $\{n^{(-1)^n} : n \in \mathbb{N}\}$ is not bounded above. Among its many lower bounds, 0 is the greatest lower bound. ☐

We now formalize two notions that have already appeared in Example 2.

4.3 Definition.

Let S be a nonempty subset of $\mathbb{R}$.
  (a) If S is bounded above and S has a least upper bound, then we will call it the *supremum of* S and denote it by $\sup S$.
  (b) If S is bounded below and S has a greatest lower bound, then we will call it the *infimum of* S and denote it by $\inf S$.

Note that, unlike $\max S$ and $\min S$, $\sup S$ and $\inf S$ need not belong to S. Note also that a set can have at most one maximum, minimum, supremum and infimum. Sometimes the expressions "least upper bound" and "greatest lower bound" are used instead of the Latin "supremum" and "infimum" and sometimes $\sup S$ is written lub S and $\inf S$ is written glb S. We have chosen the Latin terminology for a good reason: We will be studying the notions "lim sup" and "lim inf" and this notation is completely standard; no one writes "lim lub" for instance.

Observe that if S is bounded above, then $M = \sup S$ if and only if (i) $s \leq M$ for all $s \in S$, and (ii) whenever $M_1 < M$, there exists $s_1 \in S$ such that $s_1 > M_1$.

Example 3
  (a) If a set S has a maximum, then $\max S = \sup S$. A similar remark applies to sets that have infimums.
  (b) If $a, b \in \mathbb{R}$ and $a < b$, then

$$\sup[a, b] = \sup(a, b) = \sup[a, b) = \sup(a, b] = b.$$

  (c) As noted in Example 2, we have $\inf \mathbb{N} = 1$.
  (d) If $A = \{r \in \mathbb{Q} : 0 \leq r \leq \sqrt{2}\}$, then $\sup A = \sqrt{2}$ and $\inf A = 0$.
  (e) We have $\inf\{n^{(-1)^n} : n \in \mathbb{N}\} = 0$.   □

Example 4
  (a) The set $A = \{\frac{1}{n^2} : n \in \mathbb{N}$ and $n \geq 3\}$ is bounded above and bounded below. It has a maximum, namely $\frac{1}{9}$, but it has no minimum. In fact, $\sup(A) = \frac{1}{9}$ and $\inf(A) = 0$.

(b) The set $B = \{r \in \mathbb{Q} : r^3 \leq 7\}$ is bounded above, by 2 for example. It does not have a maximum, because $r^3 \neq 7$ for all $r \in \mathbb{Q}$, by the Rational Zeros Theorem 2.2. However, $\sup(B) = \sqrt[3]{7}$. The set B is not bounded below; if this isn't obvious, think about the graph of $y = x^3$. Clearly B has no minimum. Starting with the next section, we would write $\inf(B) = -\infty$.

(c) The set $C = \{m + n\sqrt{2} : m, n \in \mathbb{Z}\}$ isn't bounded above or below, so it has no maximum or minimum. We could write $\sup(C) = +\infty$ and $\inf(C) = -\infty$.

(d) The set $D = \{x \in \mathbb{R} : x^2 < 10\}$ is the open interval $(-\sqrt{10}, \sqrt{10})$. Thus it is bounded above and below, but it has no maximum or minimum. However, $\inf(D) = -\sqrt{10}$ and $\sup(D) = \sqrt{10}$. $\qquad\square$

Note that, in Examples 2–4, every set S that is bounded above possesses a least upper bound, i.e., $\sup S$ exists. This is not an accident. Otherwise there would be a "gap" between the set S and the set of its upper bounds.

4.4 Completeness Axiom.
Every nonempty subset S of $\mathbb{R}$ that is bounded above has a least upper bound. In other words, $\sup S$ exists and is a real number.

The completeness axiom for $\mathbb{Q}$ would assert that every nonempty subset of $\mathbb{Q}$, that is bounded above by some rational number, has a least upper bound that is a rational number. The set $A = \{r \in \mathbb{Q} : 0 \leq r \leq \sqrt{2}\}$ is a set of rational numbers and it is bounded above by some rational numbers [3/2 for example], but A has no least upper bound that is a rational number. Thus the completeness axiom does not hold for $\mathbb{Q}$! Incidentally, the set A can be described entirely in terms of rationals: $A = \{r \in \mathbb{Q} : 0 \leq r \text{ and } r^2 \leq 2\}$.

The completeness axiom for sets bounded below comes free.

4.5 Corollary.
Every nonempty subset S of $\mathbb{R}$ that is bounded below has a greatest lower bound $\inf S$.

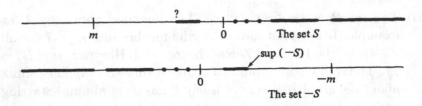

FIGURE 4.1

Proof

Let $-S$ be the set $\{-s : s \in S\}$; $-S$ consists of the negatives of the numbers in S. Since S is bounded below there is an m in $\mathbb{R}$ such that $m \leq s$ for all $s \in S$. This implies $-m \geq -s$ for all $s \in S$, so $-m \geq u$ for all u in the set $-S$. Thus $-S$ is bounded above by $-m$. The Completeness Axiom 4.4 applies to $-S$, so $\sup(-S)$ exists. Figure 4.1 suggests we prove $\inf S = -\sup(-S)$.

Let $s_0 = \sup(-S)$; we need to prove

$$-s_0 \leq s \quad \text{for all} \quad s \in S, \tag{1}$$

and

$$\text{if } t \leq s \quad \text{for all} \quad s \in S, \quad \text{then} \quad t \leq -s_0. \tag{2}$$

The inequality (1) will show $-s_0$ is a lower bound for S, while (2) will show $-s_0$ is the *greatest* lower bound, that is, $-s_0 = \inf S$. We leave the proofs of (1) and (2) to Exercise 4.9. ∎

It is useful to know:

$$\text{if } a > 0, \quad \text{then} \quad \frac{1}{n} < a \quad \text{for some positive integer} \quad n, \tag{*}$$

and

$$\text{if } b > 0, \quad \text{then} \quad b < n \quad \text{for some positive integer} \quad n. \tag{**}$$

These assertions are not as obvious as they may appear. If fact, there exist ordered fields that do not have these properties. In other words, there exists a mathematical system satisfying all the properties A1–A4, M1–M4, DL and O1–O5 in §3 and yet possessing elements $a > 0$ and $b > 0$ such that $a < 1/n$ and $n < b$ *for all n*. On the other hand, such strange elements cannot exist in $\mathbb{R}$ or $\mathbb{Q}$. We next prove this; in view of the previous remarks we *must* expect to use the Completeness Axiom.

4.6 Archimedean Property.
If $a > 0$ and $b > 0$, then for some positive integer n, we have $na > b$.

This tells us that, even if a is quite small and b is quite large, some integer multiple of a will exceed b. Or, to quote [4], given enough time, one can empty a large bathtub with a small spoon. [Note that if we set $b = 1$, we obtain assertion (*), and if we set $a = 1$, we obtain assertion (**).]

Proof
Assume the Archimedean property fails. Then there exist $a > 0$ and $b > 0$ such that $na \leq b$ for all $n \in \mathbb{N}$. In particular, b is an upper bound for the set $S = \{na : n \in \mathbb{N}\}$. Let $s_0 = \sup S$; this is where we are using the completeness axiom. Since $a > 0$, we have $s_0 < s_0 + a$, so $s_0 - a < s_0$. [To be precise, we obtain $s_0 \leq s_0 + a$ and $s_0 - a \leq s_0$ by property O4 and the fact that $a + (-a) = 0$. Then we conclude $s_0 - a < s_0$ since $s_0 - a = s_0$ implies $a = 0$ by Theorem 3.1(i).] Since s_0 is the least upper bound for S, $s_0 - a$ cannot be an upper bound for S. It follows that $s_0 - a < n_0 a$ for some $n_0 \in \mathbb{N}$. This implies $s_0 < (n_0 + 1)a$. Since $(n_0 + 1)a$ is in S, s_0 is not an upper bound for S and we have reached a contradiction. Our assumption that the Archimedean property fails was wrong. ∎

We give one more result that seems obvious from our experience with the real number line, but which cannot be proved for an arbitrary ordered field.

4.7 Denseness of $\mathbb{Q}$.
If $a, b \in \mathbb{R}$ and $a < b$, then there is a rational $r \in \mathbb{Q}$ such that $a < r < b$.

Proof
We need to show $a < \frac{m}{n} < b$ for some integers m and n where $n > 0$, and thus we need

$$an < m < bn. \tag{1}$$

Since $b - a > 0$, the Archimedean property shows there exists an $n \in \mathbb{N}$ such that

$$n(b-a) > 1, \quad \text{and hence} \quad bn - an > 1. \tag{2}$$

From this, it is fairly evident that there is an integer m between an and bn, so that (1) holds. However, the proof that such an m exists is a little delicate. We argue as follows. By the Archimedean property again, there exists an integer $k > \max\{|an|, |bn|\}$, so that

$$-k < an < bn < k.$$

Then the sets $K = \{j \in \mathbb{Z} : -k \leq j \leq k\}$ and $\{j \in K : an < j\}$ are finite, and they are nonempty, since they both contain k. Let $m = \min\{j \in K : an < j\}$. Then $-k < an < m$. Since $m > -k$, we have $m - 1$ in K, so the inequality $an < m - 1$ is false by our choice of m. Thus $m - 1 \leq an$ and, using (2), we have $m \leq an + 1 < bn$. Since $an < m < bn$, (1) holds. ∎

Exercises

4.1 For each set below that is bounded above, list three upper bounds for the set.[2] Otherwise write "NOT BOUNDED ABOVE" or "NBA."

  (a) $[0, 1]$ (b) $(0, 1)$
  (c) $\{2, 7\}$ (d) $\{\pi, e\}$
  (e) $\{\frac{1}{n} : n \in \mathbb{N}\}$ (f) $\{0\}$
  (g) $[0, 1] \cup [2, 3]$ (h) $\cup_{n=1}^{\infty} [2n, 2n + 1]$
  (i) $\cap_{n=1}^{\infty} [-\frac{1}{n}, 1 + \frac{1}{n}]$ (j) $\{1 - \frac{1}{3^n} : n \in \mathbb{N}\}$
  (k) $\{n + \frac{(-1)^n}{n} : n \in \mathbb{N}\}$ (l) $\{r \in \mathbb{Q} : r < 2\}$
  (m) $\{r \in \mathbb{Q} : r^2 < 4\}$ (n) $\{r \in \mathbb{Q} : r^2 < 2\}$
  (o) $\{x \in \mathbb{R} : x < 0\}$ (p) $\{1, \frac{\pi}{3}, \pi^2, 10\}$
  (q) $\{0, 1, 2, 4, 8, 16\}$ (r) $\cap_{n=1}^{\infty} (1 - \frac{1}{n}, 1 + \frac{1}{n})$
  (s) $\{\frac{1}{n} : n \in \mathbb{N} \text{ and } n \text{ is prime}\}$ (t) $\{x \in \mathbb{R} : x^3 < 8\}$
  (u) $\{x^2 : x \in \mathbb{R}\}$ (v) $\{\cos(\frac{n\pi}{3}) : n \in \mathbb{N}\}$
  (w) $\{\sin(\frac{n\pi}{3}) : n \in \mathbb{N}\}$

4.2 Repeat Exercise 4.1 for lower bounds.

4.3 For each set in Exercise 4.1, give its supremum if it has one. Otherwise write "NO sup."

---

[2] An integer $p \geq 2$ is a *prime* provided the only positive factors of p are 1 and p.

4.4 Repeat Exercise 4.3 for infima [plural of infimum].

4.5 Let S be a nonempty subset of $\mathbb{R}$ that is bounded above. Prove if $\sup S$ belongs to S, then $\sup S = \max S$. *Hint*: Your proof should be very short.

4.6 Let S be a nonempty bounded subset of $\mathbb{R}$.

   (a) Prove $\inf S \leq \sup S$. *Hint*: This is almost obvious; your proof should be short.

   (b) What can you say about S if $\inf S = \sup S$?

4.7 Let S and T be nonempty bounded subsets of $\mathbb{R}$.

   (a) Prove if $S \subseteq T$, then $\inf T \leq \inf S \leq \sup S \leq \sup T$.

   (b) Prove $\sup(S \cup T) = \max\{\sup S, \sup T\}$. *Note*: In part (b), do *not* assume $S \subseteq T$.

4.8 Let S and T be nonempty subsets of $\mathbb{R}$ with the following property: $s \leq t$ for all $s \in S$ and $t \in T$.

   (a) Observe S is bounded above and T is bounded below.

   (b) Prove $\sup S \leq \inf T$.

   (c) Give an example of such sets S and T where $S \cap T$ is nonempty.

   (d) Give an example of sets S and T where $\sup S = \inf T$ and $S \cap T$ is the empty set.

4.9 Complete the proof that $\inf S = -\sup(-S)$ in Corollary 4.5 by proving (1) and (2).

4.10 Prove that if $a > 0$, then there exists $n \in \mathbb{N}$ such that $\frac{1}{n} < a < n$.

4.11 Consider $a, b \in \mathbb{R}$ where $a < b$. Use Denseness of $\mathbb{Q}$ 4.7 to show there are infinitely many rationals between a and b.

4.12 Let $\mathbb{I}$ be the set of real numbers that are not rational; elements of $\mathbb{I}$ are called *irrational numbers*. Prove if $a < b$, then there exists $x \in \mathbb{I}$ such that $a < x < b$. *Hint*: First show $\{r + \sqrt{2} : r \in \mathbb{Q}\} \subseteq \mathbb{I}$.

4.13 Prove the following are equivalent for real numbers a, b, c. [*Equivalent* means that either all the properties hold or none of the properties hold.]

   (i) $|a - b| < c$,

   (ii) $b - c < a < b + c$,

(iii) $a \in (b - c, b + c)$.

Hint: Use Exercise 3.7(b).

4.14 Let A and B be nonempty bounded subsets of $\mathbb{R}$, and let $A + B$ be the set of all sums $a + b$ where $a \in A$ and $b \in B$.

(a) Prove $\sup(A+B) = \sup A + \sup B$. *Hint*: To show $\sup A + \sup B \le \sup(A + B)$, show that for each $b \in B$, $\sup(A + B) - b$ is an upper bound for A, hence $\sup A \le \sup(A + B) - b$. Then show $\sup(A + B) - \sup A$ is an upper bound for B.

(b) Prove $\inf(A + B) = \inf A + \inf B$.

4.15 Let $a, b \in \mathbb{R}$. Show if $a \le b + \frac{1}{n}$ for all $n \in \mathbb{N}$, then $a \le b$. Compare Exercise 3.8.

4.16 Show $\sup\{r \in \mathbb{Q} : r < a\} = a$ for each $a \in \mathbb{R}$.

§5   The Symbols $+\infty$ and $-\infty$

The symbols $+\infty$ and $-\infty$ are extremely useful even though they are *not* real numbers. We will often write $+\infty$ as simply ∞. We will adjoin $+\infty$ and $-\infty$ to the set $\mathbb{R}$ and extend our ordering to the set $\mathbb{R} \cup \{-\infty, +\infty\}$. Explicitly, we will agree that $-\infty \le a \le +\infty$ for all a in $\mathbb{R} \cup \{-\infty, \infty\}$. This provides the set $\mathbb{R} \cup \{-\infty, +\infty\}$ with an ordering that satisfies properties O1, O2 and O3 of §3. We emphasize we will *not* provide the set $\mathbb{R} \cup \{-\infty, +\infty\}$ with any algebraic structure. We may use the symbols $+\infty$ and $-\infty$, but we must continue to remember they do not represent real numbers. Do *not* apply a theorem or exercise that is stated for real numbers to the symbols $+\infty$ or $-\infty$.

It is convenient to use the symbols $+\infty$ and $-\infty$ to extend the notation established in Example 1(b) of §4 to unbounded intervals. For real numbers $a, b \in \mathbb{R}$, we adopt the following

$$[a, \infty) = \{x \in \mathbb{R} : a \le x\}, \quad (a, \infty) = \{x \in \mathbb{R} : a < x\},$$
$$(-\infty, b] = \{x \in \mathbb{R} : x \le b\}, \quad (-\infty, b) = \{x \in \mathbb{R} : x < b\}.$$

We occasionally also write $(-\infty, \infty)$ for $\mathbb{R}$. $[a, \infty)$ and $(-\infty, b]$ are called *closed intervals* or *unbounded closed intervals*, while (a, ∞) and

$(-\infty, b)$ are called *open intervals* or *unbounded open intervals*. Consider a nonempty subset S of $\mathbb{R}$. Recall that if S is bounded above, then $\sup S$ exists and represents a real number by the completeness axiom 4.4. We define

$$\sup S = +\infty \quad \text{if } S \text{ is not bounded above.}$$

Likewise, if S is bounded below, then $\inf S$ exists and represents a real number [Corollary 4.5]. And we define

$$\inf S = -\infty \quad \text{if } S \text{ is not bounded below.}$$

For emphasis, we recapitulate:

Let S be any nonempty subset of $\mathbb{R}$. The symbols $\sup S$ and $\inf S$ always make sense. If S is bounded above, then $\sup S$ is a real number; otherwise $\sup S = +\infty$. If S is bounded below, then $\inf S$ is a real number; otherwise $\inf S = -\infty$. Moreover, we have $\inf S \leq \sup S$.

Example 1

For nonempty bounded subsets A and B of $\mathbb{R}$, Exercise 4.14 asserts

$$\sup(A+B) = \sup A + \sup B \quad \text{and} \quad \inf(A+B) = \inf A + \inf B. \quad (1)$$

We verify the first equality is true even if A or B is unbounded, and Exercise 5.7 asks you to do the same for the second equality.

Consider $x \in A + B$, so that $x = a + b$ for some $a \in A$ and $b \in B$. Then $x = a + b \leq \sup A + \sup B$. Since x is any element in $A + B$, $\sup A + \sup B$ is an upper bound for $A + B$; hence $\sup(A + B) \leq \sup A + \sup B$. It remains to show $\sup(A + B) \geq \sup A + \sup B$.

Since the sets here are nonempty, the suprema here are not equal to $-\infty$, so we're not in danger of encountering the undefined sum $-\infty + \infty$. If $\sup A + \sup B = +\infty$, then at least one of the suprema, say $\sup B$, equals $+\infty$. Select some a_0 in A. Then $\sup(A + B) \geq \sup(a_0 + B) = a_0 + \sup B = +\infty$, so (1) holds in this case. Otherwise $\sup A + \sup B$ is finite. Consider $\epsilon > 0$. Then there exists $a \in A$ and $b \in B$, so that $a > \sup A - \frac{\epsilon}{2}$ and $b > \sup B - \frac{\epsilon}{2}$. Then $a + b \in A + B$ and so $\sup(A + B) \geq a + b > \sup A + \sup B - \epsilon$. Since $\epsilon > 0$ is arbitrary, we conclude $\sup(A + B) \geq \sup A + \sup B$. $\square$

The exercises for this section clear up some loose ends. Most of them extend results in §4 to sets that are not necessarily bounded.

Exercises

5.1 Write the following sets in interval notation:

(a) $\{x \in \mathbb{R} : x < 0\}$ (b) $\{x \in \mathbb{R} : x^3 \leq 8\}$

(c) $\{x^2 : x \in \mathbb{R}\}$ (d) $\{x \in \mathbb{R} : x^2 < 8\}$

5.2 Give the infimum and supremum of each set listed in Exercise 5.1.

5.3 Give the infimum and supremum of each unbounded set listed in Exercise 4.1.

5.4 Let S be a nonempty subset of $\mathbb{R}$, and let $-S = \{-s : s \in S\}$. Prove $\inf S = -\sup(-S)$. *Hint*: For the case $-\infty < \inf S$, simply state that this was proved in Exercise 4.9.

5.5 Prove $\inf S \leq \sup S$ for every nonempty subset of $\mathbb{R}$. Compare Exercise 4.6(a).

5.6 Let S and T be nonempty subsets of $\mathbb{R}$ such that $S \subseteq T$. Prove $\inf T \leq \inf S \leq \sup S \leq \sup T$. Compare Exercise 4.7(a).

5.7 Finish Example 1 by verifying the equality involving infimums.

§6 * A Development of $\mathbb{R}$

There are several ways to give a careful development of $\mathbb{R}$ based on $\mathbb{Q}$. We will briefly discuss one of them and give suggestions for further reading on this topic. [See the remarks about enrichment sections in the preface.]

To motivate our development we begin by observing

$$a = \sup\{r \in \mathbb{Q} : r < a\} \quad \text{for each} \quad a \in \mathbb{R};$$

see Exercise 4.16. Note the intimate relationship: $a \leq b$ if and only if $\{r \in \mathbb{Q} : r < a\} \subseteq \{r \in \mathbb{Q} : r < b\}$ and, moreover, $a = b$ if and only if $\{r \in \mathbb{Q} : r < a\} = \{r \in \mathbb{Q} : r < b\}$. Subsets α of $\mathbb{Q}$ having the form $\{r \in \mathbb{Q} : r < a\}$ satisfy these properties:

(i) $\alpha \neq \mathbb{Q}$ and α is not empty,

(ii) If $r \in \alpha$, $s \in \mathbb{Q}$ and $s < r$, then $s \in \alpha$,

(iii) α contains no largest rational.

Moreover, every subset α of $\mathbb{Q}$ that satisfies (i)–(iii) has the form $\{r \in \mathbb{Q} : r < a\}$ for some $a \in \mathbb{R}$; in fact, $a = \sup \alpha$. Subsets α of $\mathbb{Q}$ satisfying (i)–(iii) are called *Dedekind cuts*.

The remarks in the last paragraph relating real numbers and Dedekind cuts are based on our knowledge of $\mathbb{R}$, including the completeness axiom. But they can also motivate a development of $\mathbb{R}$ based solely on $\mathbb{Q}$. In such a development we make no a priori assumptions about $\mathbb{R}$. We assume only that we have the ordered field $\mathbb{Q}$ and that $\mathbb{Q}$ satisfies the Archimedean property 4.6. A Dedekind cut is a subset α of $\mathbb{Q}$ satisfying (i)–(iii). The set $\mathbb{R}$ of real numbers is *defined* as the space of all Dedekind cuts. Thus elements of $\mathbb{R}$ are *defined* as certain subsets of $\mathbb{Q}$. The rational numbers are identified with certain Dedekind cuts in the natural way: each rational s corresponds to the Dedekind cut $s^* = \{r \in \mathbb{Q} : r < s\}$. In this way $\mathbb{Q}$ is regarded as a subset of $\mathbb{R}$, that is, $\mathbb{Q}$ is identified with the set $\mathbb{Q}^* = \{s^* : s \in \mathbb{Q}\}$.

The set $\mathbb{R}$ defined in the last paragraph is given an order structure as follows: if α and β are Dedekind cuts, then we define $\alpha \leq \beta$ to signify $\alpha \subseteq \beta$. Properties O1, O2 and O3 in §3 hold for this ordering. Addition is defined in $\mathbb{R}$ as follows: if α and β are Dedekind cuts, then

$$\alpha + \beta = \{r_1 + r_2 : r_1 \in \alpha \text{ and } r_2 \in \beta\}.$$

It turns out that $\alpha + \beta$ is a Dedekind cut [hence in $\mathbb{R}$] and this definition of addition satisfies properties A1–A4 in §3. Multiplication of Dedekind cuts is a tedious business and has to be defined first for Dedekind cuts $\geq 0^*$. For a naive attempt, see Exercise 6.4. After the product of Dedekind cuts has been defined, the remaining properties of an ordered field can be verified for $\mathbb{R}$. The ordered field $\mathbb{R}$ constructed in this manner from $\mathbb{Q}$ is complete: the completeness property in 4.4 can be *proved* rather than taken as an axiom.

The real numbers are developed from Cauchy sequences in $\mathbb{Q}$ in [31, §5]. A thorough development of $\mathbb{R}$ based on Peano's axioms is given in [39].

Exercises

6.1 Consider $s, t \in \mathbb{Q}$. Show

(a) $s \leq t$ if and only if $s^* \subseteq t^*$;

(b) $s = t$ if and only if $s^* = t^*$;

(c) $(s + t)^* = s^* + t^*$. Note that $s^* + t^*$ is a sum of Dedekind cuts.

6.2 Show that if α and β are Dedekind cuts, then so is $\alpha + \beta = \{r_1 + r_2 : r_1 \in \alpha \text{ and } r_2 \in \beta\}$.

6.3 (a) Show $\alpha + 0^* = \alpha$ for all Dedekind cuts α.

(b) We claimed, without proof, that addition of Dedekind cuts satisfies property A4. Thus if α is a Dedekind cut, there is a Dedekind cut $-\alpha$ such that $\alpha + (-\alpha) = 0^*$. How would you define $-\alpha$?

6.4 Let α and β be Dedekind cuts and define the "product": $\alpha \cdot \beta = \{r_1 r_2 : r_1 \in \alpha \text{ and } r_2 \in \beta\}$.

(a) Calculate some "products" of Dedekind cuts using the Dedekind cuts 0^*, 1^* and $(-1)^*$.

(b) Discuss why this definition of "product" is totally unsatisfactory for defining multiplication in $\mathbb{R}$.

6.5 (a) Show $\{r \in \mathbb{Q} : r^3 < 2\}$ is a Dedekind cut, but $\{r \in \mathbb{Q} : r^2 < 2\}$ is not a Dedekind cut.

(b) Does the Dedekind cut $\{r \in \mathbb{Q} : r^3 < 2\}$ correspond to a rational number in $\mathbb{R}$?

(c) Show $0^* \cup \{r \in \mathbb{Q} : r \geq 0 \text{ and } r^2 < 2\}$ is a Dedekind cut. Does it correspond to a rational number in $\mathbb{R}$?

2

CHAPTER

Sequences

§7   Limits of Sequences

A *sequence* is a function whose domain is a set of the form $\{n \in \mathbb{Z} : n \geq m\}$; m is usually 1 or 0. Thus a sequence is a function that has a specified value for each integer $n \geq m$. It is customary to denote a sequence by a letter such as s and to denote its value at n as s_n rather than $s(n)$. It is often convenient to write the sequence as $(s_n)_{n=m}^{\infty}$ or $(s_m, s_{m+1}, s_{m+2}, \ldots)$. If $m = 1$ we may write $(s_n)_{n \in \mathbb{N}}$ or of course $(s_1, s_2, s_3, \ldots)$. Sometimes we will write (s_n) when the domain is understood or when the results under discussion do not depend on the specific value of m. In this chapter, we will be interested in sequences whose range values are real numbers, i.e., each s_n represents a real number.

Example 1
(a) Consider the sequence $(s_n)_{n \in \mathbb{N}}$ where $s_n = \frac{1}{n^2}$. This is the sequence $(1, \frac{1}{4}, \frac{1}{9}, \frac{1}{16}, \frac{1}{25}, \ldots)$. Formally, of course, this is the function with domain $\mathbb{N}$ whose value at each n is $\frac{1}{n^2}$. The *set* of values is $\{1, \frac{1}{4}, \frac{1}{9}, \frac{1}{16}, \frac{1}{25}, \ldots\}$.

K.A. Ross, *Elementary Analysis: The Theory of Calculus*,
Undergraduate Texts in Mathematics, DOI 10.1007/978-1-4614-6271-2_2,
© Springer Science+Business Media New York 2013

(b) Consider the sequence given by $a_n = (-1)^n$ for $n \geq 0$, i.e., $(a_n)_{n=0}^{\infty}$ where $a_n = (-1)^n$. Note that the first term of the sequence is $a_0 = 1$ and the sequence is $(1, -1, 1, -1, 1, -1, 1, \ldots)$. Formally, this is a function whose domain is $\{0, 1, 2, \ldots\}$ and whose *set* of values is $\{-1, 1\}$.

It is important to distinguish between a sequence and its set of values, since the validity of many results in this book depends on whether we are working with a sequence or a set. We will always use parentheses ( ) to signify a sequence and braces { } to signify a set. The sequence given by $a_n = (-1)^n$ has an infinite number of terms even though their values are repeated over and over. On the other hand, the *set* $\{(-1)^n : n = 0, 1, 2, \ldots\}$ is exactly the set $\{-1, 1\}$ consisting of two numbers.

(c) Consider the sequence $\cos(\frac{n\pi}{3})$, $n \in \mathbb{N}$. The first term of this sequence is $\cos(\frac{\pi}{3}) = \cos 60° = \frac{1}{2}$ and the sequence looks like

$$(\tfrac{1}{2}, -\tfrac{1}{2}, -1, -\tfrac{1}{2}, \tfrac{1}{2}, 1, \tfrac{1}{2}, -\tfrac{1}{2}, -1, -\tfrac{1}{2}, \tfrac{1}{2}, 1, \tfrac{1}{2}, -\tfrac{1}{2}, -1, \ldots).$$

The set of values is $\{\cos(\frac{n\pi}{3}) : n \in \mathbb{N}\} = \{\tfrac{1}{2}, -\tfrac{1}{2}, -1, 1\}$.

(d) If $a_n = \sqrt[n]{n}$, $n \in \mathbb{N}$, the sequence is $(1, \sqrt{2}, \sqrt[3]{3}, \sqrt[4]{4}, \ldots)$. If we approximate values to four decimal places, the sequence looks like

$$(1, 1.4142, 1.4422, 1.4142, 1.3797, 1.3480, 1.3205, 1.2968, \ldots).$$

It turns out that a_{100} is approximately 1.0471 and $a_{1,000}$ is approximately 1.0069.

(e) Consider the sequence $b_n = (1 + \frac{1}{n})^n$, $n \in \mathbb{N}$. This is the sequence $(2, (\frac{3}{2})^2, (\frac{4}{3})^3, (\frac{5}{4})^4, \ldots)$. If we approximate the values to four decimal places, we obtain

$$(2, 2.25, 2.3704, 2.4414, 2.4883, 2.5216, 2.5465, 2.5658, \ldots).$$

Also b_{100} is approximately 2.7048 and $b_{1,000}$ is approximately 2.7169. $\qquad\square$

The "limit" of a sequence (s_n) is a real number that the values s_n are close to for large values of n. For instance, the values of the sequence in Example 1(a) are close to 0 for large n and the values of the sequence in Example 1(d) appear to be close to 1 for large n.

The sequence (a_n) given by $a_n = (-1)^n$ requires some thought. We might say 1 is a limit because in fact $a_n = 1$ for the large values of n that are even. On the other hand, $a_n = -1$ [which is quite a distance from 1] for other large values of n. We need a precise definition in order to decide whether 1 is a limit of $a_n = (-1)^n$. It turns out that our definition will require the values to be close to the limit value for *all* large n, so 1 will *not* be a limit of the sequence $a_n = (-1)^n$.

7.1 Definition.

A sequence (s_n) of real numbers is said to *converge* to the real number s provided that

for each $\epsilon > 0$ there exists a number N such that
$$n > N \quad \text{implies} \quad |s_n - s| < \epsilon. \tag{1}$$

If (s_n) converges to s, we will write $\lim_{n \to \infty} s_n = s$, or $s_n \to s$. The number s is called the *limit* of the sequence (s_n). A sequence that does not converge to some real number is said to *diverge*.

Several comments are in order. First, in view of the Archimedean property, the number N in Definition 7.1 can be taken to be a positive integer if we wish. Second, the symbol ϵ [lower case Greek epsilon] in this definition represents a positive number, not some new exotic number. However, it is traditional in mathematics to use ϵ and δ [lower case Greek delta] in situations where the interesting or challenging values are the small positive values. Third, condition (1) is an infinite number of statements, one for each positive value of ϵ. The condition states that to each $\epsilon > 0$ there corresponds a number N with a certain property, namely $n > N$ implies $|s_n - s| < \epsilon$. The value N depends on the value ϵ, and normally N will have to be large if ϵ is small. We illustrate these remarks in the next example.

Example 2

Consider the sequence $s_n = \frac{3n+1}{7n-4}$. If we write s_n as $\frac{3+\frac{1}{n}}{7-\frac{4}{n}}$ and note $\frac{1}{n}$ and $\frac{4}{n}$ are very small for large n, it seems reasonable to conclude $\lim s_n = \frac{3}{7}$. In fact, this reasoning will be completely valid after we have the limit theorems in §9:

$$\lim s_n = \lim \left[\frac{3 + \frac{1}{n}}{7 - \frac{4}{n}} \right] = \frac{\lim 3 + \lim(\frac{1}{n})}{\lim 7 - 4 \lim(\frac{1}{n})} = \frac{3 + 0}{7 - 4 \cdot 0} = \frac{3}{7}.$$

However, for now we are interested in analyzing exactly what we mean by $\lim s_n = \frac{3}{7}$. By Definition 7.1, $\lim s_n = \frac{3}{7}$ means

for each $\epsilon > 0$ there exists a number N such that
$$n > N \quad \text{implies} \quad \left| \frac{3n+1}{7n-4} - \frac{3}{7} \right| < \epsilon. \tag{1}$$

As ϵ varies, N varies. In Example 2 of the next section we will show that, for this particular sequence, N can be taken to be $\frac{19}{49\epsilon} + \frac{4}{7}$. Using this observation, we find that for ϵ equal to 1, 0.1, 0.01, 0.001, and 0.000001, respectively, N can be taken to be approximately 0.96, 4.45, 39.35, 388.33, and 387,755.67, respectively. Since we are interested only in integer values of n, we may as well drop the fractional part of N. Then we see five of the infinitely many statements given by (1) are:

$$n > 0 \quad \text{implies} \quad \left| \frac{3n+1}{7n-4} - \frac{3}{7} \right| < 1; \tag{2}$$

$$n > 4 \quad \text{implies} \quad \left| \frac{3n+1}{7n-4} - \frac{3}{7} \right| < 0.1; \tag{3}$$

$$n > 39 \quad \text{implies} \quad \left| \frac{3n+1}{7n-4} - \frac{3}{7} \right| < 0.01; \tag{4}$$

$$n > 388 \quad \text{implies} \quad \left| \frac{3n+1}{7n-4} - \frac{3}{7} \right| < 0.001; \tag{5}$$

$$n > 387,755 \quad \text{implies} \quad \left| \frac{3n+1}{7n-4} - \frac{3}{7} \right| < 0.000001. \tag{6}$$

Table 7.1 partially confirms assertions (2) through (6). We could go on and on with these numerical illustrations, but it should be clear we need a more theoretical approach if we are going to *prove* results about limits.   □

Example 3
We return to the examples in Example 1.

TABLE 7.1

n	$s_n = \frac{3n+1}{7n-4}$ is approximately	$\|s_n - \frac{3}{7}\|$ is approximately
2	0.7000	0.2714
3	0.5882	0.1597
4	0.5417	0.1131
5	0.5161	0.0876
6	0.5000	0.0714
40	0.4384	0.0098
400	0.4295	0.0010

(a) $\lim \frac{1}{n^2} = 0$. This will be proved in Example 1 of the next section.

(b) The sequence (a_n) where $a_n = (-1)^n$ does not converge. Thus the expression "$\lim a_n$" is meaningless in this case. We will discuss this example again in Example 4 of the next section.

(c) The sequence $\cos(\frac{n\pi}{3})$ does not converge. See Exercise 8.7.

(d) The sequence $n^{1/n}$ appears to converge to 1. We will prove $\lim n^{1/n} = 1$ in Theorem 9.7(c) on page 48.

(e) The sequence (b_n) where $b_n = (1+\frac{1}{n})^n$ converges to the number e that should be familiar from calculus. The limit $\lim b_n$ and the number e will be discussed further in Example 6 in §16 and in §37. Recall e is approximately 2.7182818. $\qquad\square$

We conclude this section by showing that limits are unique. That is, if $\lim s_n = s$ and $\lim s_n = t$, then we must have $s = t$. In short, the values s_n cannot be getting arbitrarily close to different values for large n. To prove this, consider $\epsilon > 0$. By the definition of limit there exists N_1 so that

$$n > N_1 \quad \text{implies} \quad |s_n - s| < \frac{\epsilon}{2}$$

and there exists N_2 so that

$$n > N_2 \quad \text{implies} \quad |s_n - t| < \frac{\epsilon}{2}.$$

For $n > \max\{N_1, N_2\}$, the Triangle Inequality 3.7 shows

$$|s - t| = |(s - s_n) + (s_n - t)| \leq |s - s_n| + |s_n - t| \leq \frac{\epsilon}{2} + \frac{\epsilon}{2} = \epsilon.$$

This shows $|s - t| < \epsilon$ for all $\epsilon > 0$. It follows that $|s - t| = 0$; hence $s = t$.

Exercises

7.1 Write out the first five terms of the following sequences.
  (a) $s_n = \frac{1}{3n+1}$ (b) $b_n = \frac{3n+1}{4n-1}$
  (c) $c_n = \frac{n}{3^n}$ (d) $\sin(\frac{n\pi}{4})$

7.2 For each sequence in Exercise 7.1, determine whether it converges. If it converges, give its limit. No proofs are required.

7.3 For each sequence below, determine whether it converges and, if it converges, give its limit. No proofs are required.
  (a) $a_n = \frac{n}{n+1}$ (b) $b_n = \frac{n^2+3}{n^2-3}$
  (c) $c_n = 2^{-n}$ (d) $t_n = 1 + \frac{2}{n}$
  (e) $x_n = 73 + (-1)^n$ (f) $s_n = (2)^{1/n}$
  (g) $y_n = n!$ (h) $d_n = (-1)^n n$
  (i) $\frac{(-1)^n}{n}$ (j) $\frac{7n^3+8n}{2n^3-3}$
  (k) $\frac{9n^2-18}{6n+18}$ (l) $\sin(\frac{n\pi}{2})$
  (m) $\sin(n\pi)$ (n) $\sin(\frac{2n\pi}{3})$
  (o) $\frac{1}{n}\sin n$ (p) $\frac{2^{n+1}+5}{2^n-7}$
  (q) $\frac{3^n}{n!}$ (r) $(1 + \frac{1}{n})^2$
  (s) $\frac{4n^2+3}{3n^2-2}$ (t) $\frac{6n+4}{9n^2+7}$

7.4 Give examples of

  (a) A sequence (x_n) of irrational numbers having a limit $\lim x_n$ that is a rational number.

  (b) A sequence (r_n) of rational numbers having a limit $\lim r_n$ that is an irrational number.

7.5 Determine the following limits. No proofs are required, but show any relevant algebra.

  (a) $\lim s_n$ where $s_n = \sqrt{n^2+1} - n$,

  (b) $\lim(\sqrt{n^2+n} - n)$,

  (c) $\lim(\sqrt{4n^2+n} - 2n)$.
      Hint for (a): First show $s_n = \frac{1}{\sqrt{n^2+1}+n}$.

§8 A Discussion about Proofs

In this section we give several examples of proofs using the definition of the limit of a sequence. With a little study and practice, students should be able to do proofs of this sort themselves. We will sometimes refer to a proof as a *formal proof* to emphasize it is a rigorous mathematical proof.

Example 1

Prove $\lim \frac{1}{n^2} = 0$.

Discussion. Our task is to consider an arbitrary $\epsilon > 0$ and show there exists a number N [which will depend on ϵ] such that $n > N$ implies $|\frac{1}{n^2} - 0| < \epsilon$. So we expect our formal proof to begin with "Let $\epsilon > 0$" and to end with something like "Hence $n > N$ implies $|\frac{1}{n^2} - 0| < \epsilon$." In between the proof should specify an N and then verify N has the desired property, namely $n > N$ does indeed imply $|\frac{1}{n^2} - 0| < \epsilon$.

As is often the case with trigonometric identities, we will initially work backward from our desired conclusion, but in the formal proof we will have to be sure our steps are reversible. In the present example, we want $|\frac{1}{n^2} - 0| < \epsilon$ and we want to know how big n must be. So we will operate on this inequality algebraically and try to "solve" for n. Thus we want $\frac{1}{n^2} < \epsilon$. By multiplying both sides by n^2 and dividing both sides by ϵ, we find we want $\frac{1}{\epsilon} < n^2$ or $\frac{1}{\sqrt{\epsilon}} < n$. If our steps are reversible, we see $n > \frac{1}{\sqrt{\epsilon}}$ implies $|\frac{1}{n^2} - 0| < \epsilon$. This suggests we put $N = \frac{1}{\sqrt{\epsilon}}$.

Formal Proof

Let $\epsilon > 0$. Let $N = \frac{1}{\sqrt{\epsilon}}$. Then $n > N$ implies $n > \frac{1}{\sqrt{\epsilon}}$ which implies $n^2 > \frac{1}{\epsilon}$ and hence $\epsilon > \frac{1}{n^2}$. Thus $n > N$ implies $|\frac{1}{n^2} - 0| < \epsilon$. This proves $\lim \frac{1}{n^2} = 0$. ∎

Example 2

Prove $\lim \frac{3n+1}{7n-4} = \frac{3}{7}$.

Discussion. For each $\epsilon > 0$, we need to to decide how big n must be to guarantee $|\frac{3n+1}{7n-4} - \frac{3}{7}| < \epsilon$. Thus we want

$$\left| \frac{21n + 7 - 21n + 12}{7(4n - 4)} \right| < \epsilon \quad \text{or} \quad \left| \frac{19}{7(7n - 4)} \right| < \epsilon.$$

Since $7n - 4 > 0$, we can drop the absolute value and manipulate the inequality further to "solve" for n:

$$\frac{19}{7\epsilon} < 7n - 4 \quad \text{or} \quad \frac{19}{7\epsilon} + 4 < 7n \quad \text{or} \quad \frac{19}{49\epsilon} + \frac{4}{7} < n.$$

Our steps are reversible, so we will put $N = \frac{19}{49\epsilon} + \frac{4}{7}$. Incidentally, we could have chosen N to be any number larger than $\frac{19}{49\epsilon} + \frac{4}{7}$.

Formal Proof

Let $\epsilon > 0$ and let $N = \frac{19}{49\epsilon} + \frac{4}{7}$. Then $n > N$ implies $n > \frac{19}{49\epsilon} + \frac{4}{7}$, hence $7n > \frac{19}{7\epsilon} + 4$, hence $7n - 4 > \frac{19}{7\epsilon}$, hence $\frac{19}{7(7n-4)} < \epsilon$, and hence $|\frac{3n+1}{7n-4} - \frac{3}{7}| < \epsilon$. This proves $\lim \frac{3n+1}{7n-4} = \frac{3}{7}$. ∎

Example 3

Prove $\lim \frac{4n^3+3n}{n^3-6} = 4$.

Discussion. For each $\epsilon > 0$, we need to determine how large n has to be to imply

$$\left| \frac{4n^3 + 3n}{n^3 - 6} - 4 \right| < \epsilon \quad \text{or} \quad \left| \frac{3n + 24}{n^3 - 6} \right| < \epsilon.$$

By considering $n > 1$, we may drop the absolute values; thus we need to find how big n must be to give $\frac{3n+24}{n^3-6} < \epsilon$. This time it would be very difficult to "solve" for or isolate n. Recall we need to find some N such that $n > N$ implies $\frac{3n+24}{n^3-6} < \epsilon$, but we do not need to find the least such N. So we will simplify matters by making estimates. The idea is that $\frac{3n+24}{n^3-6}$ is bounded by some constant times $\frac{n}{n^3} = \frac{1}{n^2}$ for sufficiently large n. To find such a bound we will find an upper bound for the numerator and a lower bound for the denominator. For example, since $3n + 24 \leq 27n$, it suffices for us to get $\frac{27n}{n^3-6} < \epsilon$. To make the denominator smaller and yet a constant multiple of n^3, we note $n^3 - 6 \geq \frac{n^3}{2}$ provided n is sufficiently large; in fact, all we need is $\frac{n^3}{2} \geq 6$ or $n^3 \geq 12$ or $n > 2$. So it suffices to get $\frac{27n}{n^3/2} < \epsilon$ or $\frac{54}{n^2} < \epsilon$ or $n > \sqrt{\frac{54}{\epsilon}}$, provided $n > 2$.

Formal Proof

Let $\epsilon > 0$ and let $N = \max\{2, \sqrt{\frac{54}{\epsilon}}\}$. Then $n > N$ implies $n > \sqrt{\frac{54}{\epsilon}}$, hence $\frac{54}{n^2} < \epsilon$, hence $\frac{27n}{n^3/2} < \epsilon$. Since $n > 2$, we have $\frac{n^3}{2} \leq n^3 - 6$ and also $27n \geq 3n + 24$. Thus $n > N$ implies

$$\frac{3n + 24}{n^3 - 6} \leq \frac{27n}{\frac{1}{2}n^3} = \frac{54}{n^2} < \epsilon,$$

and hence

$$\left| \frac{4n^3 + 3n}{n^3 - 6} - 4 \right| < \epsilon,$$

as desired. ∎

Example 3 illustrates direct proofs of even rather simple limits can get complicated. With the limit theorems of §9 we would just write

$$\lim \left[\frac{4n^3 + 3n}{n^3 - 6} \right] = \lim \left[\frac{4 + \frac{3}{n^2}}{1 - \frac{6}{n^3}} \right] = \frac{\lim 4 + 3 \cdot \lim(\frac{1}{n^2})}{\lim 1 - 6 \cdot \lim(\frac{1}{n^3})} = 4.$$

Example 4

Show that the sequence $a_n = (-1)^n$ does not converge.

Discussion. We will assume $\lim(-1)^n = a$ and obtain a contradiction. No matter what a is, either 1 or -1 will have distance at least 1 from a. Thus the inequality $|(-1)^n - a| < 1$ will not hold for all large n.

Formal Proof

Assume $\lim(-1)^n = a$ for some $a \in \mathbb{R}$. Letting $\epsilon = 1$ in the definition of the limit, we see that there exists N such that

$$n > N \quad \text{implies} \quad |(-1)^n - a| < 1.$$

By considering both an even and an odd $n > N$, we see that

$$|1 - a| < 1 \quad \text{and} \quad |-1 - a| < 1.$$

Now by the Triangle Inequality 3.7

$$2 = |1 - (-1)| = |1 - a + a - (-1)| \leq |1 - a| + |a - (-1)| < 1 + 1 = 2.$$

This absurdity shows our assumption $\lim(-1)^n = a$ must be wrong, so the sequence $(-1)^n$ does not converge. ∎

Example 5

Let (s_n) be a sequence of nonnegative real numbers and suppose $s = \lim s_n$. Note $s \geq 0$; see Exercise 8.9(a). Prove $\lim \sqrt{s_n} = \sqrt{s}$.

   Discussion. We need to consider $\epsilon > 0$ and show there exists N such that

$$n > N \quad \text{implies} \quad |\sqrt{s_n} - \sqrt{s}| < \epsilon.$$

This time we cannot expect to obtain N explicitly in terms of ϵ because of the general nature of the problem. But we can hope to show such N exists. The trick here is to violate our training in algebra and "irrationalize the denominator":

$$\sqrt{s_n} - \sqrt{s} = \frac{(\sqrt{s_n} - \sqrt{s})(\sqrt{s_n} + \sqrt{s})}{\sqrt{s_n} + \sqrt{s}} = \frac{s_n - s}{\sqrt{s_n} + \sqrt{s}}.$$

Since $s_n \to s$ we will be able to make the numerator small [for large n]. Unfortunately, if $s = 0$ the denominator will also be small. So we consider two cases. If $s > 0$, the denominator is bounded below by $\sqrt{s}$ and our trick will work:

$$|\sqrt{s_n} - \sqrt{s}| \leq \frac{|s_n - s|}{\sqrt{s}},$$

so we will select N so that $|s_n - s| < \sqrt{s}\epsilon$ for $n > N$. Note that N exists, since we can apply the definition of limit to $\sqrt{s}\epsilon$ just as well as to ϵ. For $s = 0$, it can be shown directly that $\lim s_n = 0$ implies $\lim \sqrt{s_n} = 0$; the trick of "irrationalizing the denominator" is not needed in this case.

Formal Proof

   Case I: $s > 0$. Let $\epsilon > 0$. Since $\lim s_n = s$, there exists N such that

$$n > N \quad \text{implies} \quad |s_n - s| < \sqrt{s}\epsilon.$$

Now $n > N$ implies

$$|\sqrt{s_n} - \sqrt{s}| = \frac{|s_n - s|}{\sqrt{s_n} + \sqrt{s}} \le \frac{|s_n - s|}{\sqrt{s}} < \frac{\sqrt{s}\epsilon}{\sqrt{s}} = \epsilon.$$

Case II: $s = 0$. This case is left to Exercise 8.3. ∎

Example 6

Let (s_n) be a convergent sequence of real numbers such that $s_n \ne 0$ for all $n \in \mathbb{N}$ and $\lim s_n = s \ne 0$. Prove $\inf\{|s_n| : n \in \mathbb{N}\} > 0$.

Discussion. The idea is that "most" of the terms s_n are close to s and hence not close to 0. More explicitly, "most" of the terms s_n are within $\frac{1}{2}|s|$ of s, hence most s_n satisfy $|s_n| \ge \frac{1}{2}|s|$. This seems clear from Fig. 8.1, but a formal proof will use the triangle inequality.

Formal Proof

Let $\epsilon = \frac{1}{2}|s| > 0$. Since $\lim s_n = s$, there exists N in $\mathbb{N}$ so that

$$n > N \quad \text{implies} \quad |s_n - s| < \frac{|s|}{2}.$$

Now

$$n > N \quad \text{implies} \quad |s_n| \ge \frac{|s|}{2}, \tag{1}$$

since otherwise the triangle inequality would imply

$$|s| = |s - s_n + s_n| \le |s - s_n| + |s_n| < \frac{|s|}{2} + \frac{|s|}{2} = |s|$$

which is absurd. If we set

$$m = \min\left\{\frac{|s|}{2}, |s_1|, |s_2|, \ldots, |s_N|\right\},$$

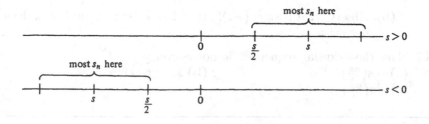

FIGURE 8.1

then we clearly have $m > 0$ and $|s_n| \geq m$ for *all* $n \in \mathbb{N}$ in view of (1). Thus $\inf\{|s_n| : n \in \mathbb{N}\} \geq m > 0$, as desired.   ∎

Formal proofs are required in the following exercises.

Exercises

8.1 Prove the following:
   (a) $\lim \frac{(-1)^n}{n} = 0$ (b) $\lim \frac{1}{n^{1/3}} = 0$
   (c) $\lim \frac{2n-1}{3n+2} = \frac{2}{3}$ (d) $\lim \frac{n+6}{n^2-6} = 0$

8.2 Determine the limits of the following sequences, and then prove your claims.
   (a) $a_n = \frac{n}{n^2+1}$ (b) $b_n = \frac{7n-19}{3n+7}$
   (c) $c_n = \frac{4n+3}{7n-5}$ (d) $d_n = \frac{2n+4}{5n+2}$
   (e) $s_n = \frac{1}{n}\sin n$

8.3 Let (s_n) be a sequence of nonnegative real numbers, and suppose $\lim s_n = 0$. Prove $\lim \sqrt{s_n} = 0$. This will complete the proof for Example 5.

8.4 Let (t_n) be a bounded sequence, i.e., there exists M such that $|t_n| \leq M$ for all n, and let (s_n) be a sequence such that $\lim s_n = 0$. Prove $\lim(s_n t_n) = 0$.

8.5 ★[1]

   (a) Consider three sequences (a_n), (b_n) and (s_n) such that $a_n \leq s_n \leq b_n$ for all n and $\lim a_n = \lim b_n = s$. Prove $\lim s_n = s$. This is called the "squeeze lemma."

   (b) Suppose (s_n) and (t_n) are sequences such that $|s_n| \leq t_n$ for all n and $\lim t_n = 0$. Prove $\lim s_n = 0$.

8.6 Let (s_n) be a sequence in $\mathbb{R}$.

   (a) Prove $\lim s_n = 0$ if and only if $\lim |s_n| = 0$.

   (b) Observe that if $s_n = (-1)^n$, then $\lim |s_n|$ exists, but $\lim s_n$ does not exist.

8.7 Show the following sequences do not converge.
   (a) $\cos(\frac{n\pi}{3})$ (b) $s_n = (-1)^n n$
   (c) $\sin(\frac{n\pi}{3})$

---

[1]This exercise is referred to in several places.

8.8 Prove the following [see Exercise 7.5]:
  (a) $\lim[\sqrt{n^2+1}-n]=0$ (b) $\lim[\sqrt{n^2+n}-n]=\frac{1}{2}$
  (c) $\lim[\sqrt{4n^2+n}-2n]=\frac{1}{4}$

8.9 ★[2] Let (s_n) be a sequence that converges.

  (a) Show that if $s_n \geq a$ for all but finitely many n, then $\lim s_n \geq a$.

  (b) Show that if $s_n \leq b$ for all but finitely many n, then $\lim s_n \leq b$.

  (c) Conclude that if all but finitely many s_n belong to $[a,b]$, then $\lim s_n$ belongs to $[a,b]$.

8.10 Let (s_n) be a convergent sequence, and suppose $\lim s_n > a$. Prove there exists a number N such that $n > N$ implies $s_n > a$.

§9 Limit Theorems for Sequences

In this section we prove some basic results that are probably already familiar to the reader. First we prove convergent sequences are bounded. A sequence (s_n) of real numbers is said to be *bounded* if the set $\{s_n : n \in \mathbb{N}\}$ is a bounded set, i.e., if there exists a constant M such that $|s_n| \leq M$ for all n.

9.1 Theorem.
Convergent sequences are bounded.

Proof
Let (s_n) be a convergent sequence, and let $s = \lim s_n$. Applying Definition 7.1 with $\epsilon = 1$ we obtain N in $\mathbb{N}$ so that

$$n > N \quad \text{implies} \quad |s_n - s| < 1.$$

From the triangle inequality we see $n > N$ implies $|s_n| < |s| + 1$. Define $M = \max\{|s|+1, |s_1|, |s_2|, \ldots, |s_N|\}$. Then we have $|s_n| \leq M$ for all $n \in \mathbb{N}$, so (s_n) is a bounded sequence. ∎

In the proof of Theorem 9.1 we only needed to use property 7.1(1) for a single value of ϵ. Our choice of $\epsilon = 1$ was quite arbitrary.

---

[2]This exercise is referred to in several places.

9.2 Theorem.
If the sequence (s_n) converges to s and k is in $\mathbb{R}$, then the sequence (ks_n) converges to ks. That is, $\lim(ks_n) = k \cdot \lim s_n$.

Proof
We assume $k \neq 0$, since this result is trivial for $k = 0$. Let $\epsilon > 0$ and note we need to show $|ks_n - ks| < \epsilon$ for large n. Since $\lim s_n = s$, there exists N such that

$$n > N \quad \text{implies} \quad |s_n - s| < \frac{\epsilon}{|k|}.$$

Then

$$n > N \quad \text{implies} \quad |ks_n - ks| < \epsilon. \qquad \blacksquare$$

9.3 Theorem.
If (s_n) converges to s and (t_n) converges to t, then (s_n+t_n) converges to $s + t$. That is,

$$\lim(s_n + t_n) = \lim s_n + \lim t_n.$$

Proof
Let $\epsilon > 0$; we need to show

$$|s_n + t_n - (s + t)| < \epsilon \quad \text{for large} \quad n.$$

We note $|s_n + t_n - (s+t)| \leq |s_n - s| + |t_n - t|$. Since $\lim s_n = s$, there exists N_1 such that

$$n > N_1 \quad \text{implies} \quad |s_n - s| < \frac{\epsilon}{2}.$$

Likewise, there exists N_2 such that

$$n > N_2 \quad \text{implies} \quad |t_n - t| < \frac{\epsilon}{2}.$$

Let $N = \max\{N_1, N_2\}$. Then clearly

$$n > N \quad \text{implies} \quad |s_n+t_n-(s+t)| \leq |s_n-s|+|t_n-t| < \frac{\epsilon}{2}+\frac{\epsilon}{2} = \epsilon. \qquad \blacksquare$$

9.4 Theorem.
If (s_n) converges to s and (t_n) converges to t, then $(s_n t_n)$ converges to st. That is,

$$\lim(s_n t_n) = (\lim s_n)(\lim t_n).$$

Discussion. The trick here is to look at the inequality

$$|s_n t_n - st| = |s_n t_n - s_n t + s_n t - st|$$
$$\leq |s_n t_n - s_n t| + |s_n t - st| = |s_n| \cdot |t_n - t| + |t| \cdot |s_n - s|.$$

For large n, $|t_n - t|$ and $|s_n - s|$ are small and $|t|$ is, of course, constant. Fortunately, Theorem 9.1 shows $|s_n|$ is bounded, so we will be able to show $|s_n t_n - s_n t|$ is small.

Proof

Let $\epsilon > 0$. By Theorem 9.1 there is a constant $M > 0$ such that $|s_n| \leq M$ for all n. Since $\lim t_n = t$ there exists N_1 such that

$$n > N_1 \quad \text{implies} \quad |t_n - t| < \frac{\epsilon}{2M}.$$

Also, since $\lim s_n = s$ there exists N_2 such that

$$n > N_2 \quad \text{implies} \quad |s_n - s| < \frac{\epsilon}{2(|t| + 1)}.$$

[We used $\frac{\epsilon}{2(|t|+1)}$ instead of $\frac{\epsilon}{2|t|}$, because t could be 0.] Now if $N = \max\{N_1, N_2\}$, then $n > N$ implies

$$|s_n t_n - st| \leq |s_n| \cdot |t_n - t| + |t| \cdot |s_n - s|$$
$$\leq M \cdot \frac{\epsilon}{2M} + |t| \cdot \frac{\epsilon}{2(|t| + 1)} < \frac{\epsilon}{2} + \frac{\epsilon}{2} = \epsilon. \qquad \blacksquare$$

To handle quotients of sequences, we first deal with reciprocals.

9.5 Lemma.
If (s_n) converges to s, if $s_n \neq 0$ for all n, and if $s \neq 0$, then $(1/s_n)$ converges to $1/s$.

Discussion. We begin by considering the equality

$$\left| \frac{1}{s_n} - \frac{1}{s} \right| = \left| \frac{s - s_n}{s_n s} \right|.$$

For large n, the numerator is small. The only possible difficulty would be if the denominator were also small for large n. This difficulty was solved in Example 6 of §8 where we proved $m = \inf\{|s_n| : n \in \mathbb{N}\} > 0$. Thus

$$\left| \frac{1}{s_n} - \frac{1}{s} \right| \leq \frac{|s - s_n|}{m|s|},$$

and it is clear how our proof should proceed.

Proof
Let $\epsilon > 0$. By Example 6 of §8, there exists $m > 0$ such that $|s_n| \geq m$ for all n. Since $\lim s_n = s$ there exists N such that

$$n > N \quad \text{implies} \quad |s - s_n| < \epsilon \cdot m|s|.$$

Then $n > N$ implies

$$\left| \frac{1}{s_n} - \frac{1}{s} \right| = \frac{|s - s_n|}{|s_n s|} \leq \frac{|s - s_n|}{m|s|} < \epsilon.$$

$\blacksquare$

9.6 Theorem.
Suppose (s_n) converges to s and (t_n) converges to t. If $s \neq 0$ and $s_n \neq 0$ for all n, then (t_n/s_n) converges to t/s.

Proof
By Lemma 9.5, the sequence $(1/s_n)$ converges to $1/s$, so

$$\lim \frac{t_n}{s_n} = \lim \frac{1}{s_n} \cdot t_n = \frac{1}{s} \cdot t = \frac{t}{s}$$

by Theorem 9.4.

$\blacksquare$

The preceding limit theorems and a few standard examples allow one to easily calculate many limits.

9.7 Theorem (Basic Examples).
  (a) $\lim_{n \to \infty} \left(\frac{1}{n^p} \right) = 0$ for $p > 0$.
  (b) $\lim_{n \to \infty} a^n = 0$ if $|a| < 1$.
  (c) $\lim(n^{1/n}) = 1$.
  (d) $\lim_{n \to \infty} (a^{1/n}) = 1$ for $a > 0$.

$\square$

Proof
  (a) Let $\epsilon > 0$ and let $N = \left(\frac{1}{\epsilon} \right)^{1/p}$. Then $n > N$ implies $n^p > \frac{1}{\epsilon}$ and hence $\epsilon > \frac{1}{n^p}$. Since $\frac{1}{n^p} > 0$, this shows $n > N$ implies $\left| \frac{1}{n^p} - 0 \right| < \epsilon$. [The meaning of n^p when p is not an integer will be discussed in §37.]
  (b) We may suppose $a \neq 0$, because $\lim_{n \to \infty} a^n = 0$ is obvious for $a = 0$. Since $|a| < 1$, we can write $|a| = \frac{1}{1+b}$ where $b > 0$. By

the binomial theorem [Exercise 1.12], $(1 + b)^n \geq 1 + nb > nb$, so

$$|a^n - 0| = |a^n| = \frac{1}{(1 + b)^n} < \frac{1}{nb}.$$

Now consider $\epsilon > 0$ and let $N = \frac{1}{\epsilon b}$. Then $n > N$ implies $n > \frac{1}{\epsilon b}$ and hence $|a^n - 0| < \frac{1}{nb} < \epsilon$.

(c) Let $s_n = (n^{1/n}) - 1$ and note $s_n \geq 0$ for all n. By Theorem 9.3 it suffices to show $\lim s_n = 0$. Since $1 + s_n = (n^{1/n})$, we have $n = (1 + s_n)^n$. For $n \geq 2$ we use the binomial expansion of $(1 + s_n)^n$ to conclude

$$n = (1 + s_n)^n \geq 1 + ns_n + \frac{1}{2}n(n-1)s_n^2 > \frac{1}{2}n(n-1)s_n^2.$$

Thus $n > \frac{1}{2}n(n-1)s_n^2$, so $s_n^2 < \frac{2}{n-1}$. Consequently, we have $s_n < \sqrt{\frac{2}{n-1}}$ for $n \geq 2$. A standard argument now shows $\lim s_n = 0$; see Exercise 9.7.

(d) First suppose $a \geq 1$. Then for $n \geq a$ we have $1 \leq a^{1/n} \leq n^{1/n}$. Since $\lim n^{1/n} = 1$, it follows easily that $\lim(a^{1/n}) = 1$; compare Exercise 8.5(a). Suppose $0 < a < 1$. Then $\frac{1}{a} > 1$, so $\lim(\frac{1}{a})^{1/n} = 1$ from above. Lemma 9.5 now shows $\lim(a^{1/n}) = 1$. ∎

Example 1
Prove $\lim s_n = \frac{1}{4}$, where

$$s_n = \frac{n^3 + 6n^2 + 7}{4n^3 + 3n - 4}.$$ □

Solution
We have

$$s_n = \frac{1 + \frac{6}{n} + \frac{7}{n^3}}{4 + \frac{3}{n^2} - \frac{4}{n^3}}.$$

By Theorem 9.7(a) we have $\lim \frac{1}{n} = 0$ and $\lim \frac{1}{n^3} = 0$. Hence by Theorems 9.3 and 9.2 we have

$$\lim \left(1 + \frac{6}{n} + \frac{7}{n^3}\right) = \lim(1) + 6 \cdot \lim \left(\frac{1}{n}\right) + 7 \cdot \lim \left(\frac{1}{n^3}\right) = 1.$$

Similarly, we have

$$\lim \left(4 + \frac{3}{n^2} - \frac{4}{n^3} \right) = 4.$$

Hence Theorem 9.6 implies $\lim s_n = \frac{1}{4}$. □

Example 2
Find $\lim \frac{n-5}{n^2+7}$. □

Solution
Let $s_n = \frac{n-5}{n^2+7}$. We can write s_n as $\frac{1-\frac{5}{n}}{n+\frac{7}{n}}$, but then the denominator does not converge. So we write

$$s_n = \frac{\frac{1}{n} - \frac{5}{n^2}}{1 + \frac{7}{n^2}}.$$

Now $\lim(\frac{1}{n} - \frac{5}{n^2}) = 0$ by Theorems 9.7(a), 9.3 and 9.2. Likewise $\lim(1 + \frac{7}{n^2}) = 1$, so Theorem 9.6 implies $\lim s_n = \frac{0}{1} = 0$. □

Example 3
Find $\lim \frac{n^2+3}{n+1}$. □

Solution
We can write $\frac{n^2+3}{n+1}$ as

$$\frac{n + \frac{3}{n}}{1 + \frac{1}{n}} \quad \text{or} \quad \frac{1 + \frac{3}{n^2}}{\frac{1}{n} + \frac{1}{n^2}}.$$

Both fractions lead to problems: either the numerator does not converge or else the denominator converges to 0. It turns out $\frac{n^2+3}{n+1}$ does not converge and the symbol $\lim \frac{n^2+3}{n+1}$ is undefined, at least for the present; see Example 6. The reader may have the urge to use the symbol $+\infty$ here. Our next task is to make such use of the symbol $+\infty$ legitimate. For a sequence (s_n), $\lim s_n = +\infty$ will signify that the terms s_n are eventually all large. Here is the precise definition. □

9.8 Definition.
For a sequence (s_n), we write $\lim s_n = +\infty$ provided
for each $M > 0$ there is a number N such that
$n > N$ implies $s_n > M$.

In this case we say the sequence *diverges to* $+\infty$.

Similarly, we write $\lim s_n = -\infty$ provided

for each $M < 0$ there is a number N such that

$n > N$ implies $s_n < M$.

Henceforth we will say (s_n) has a *limit* or the *limit exists* provided (s_n) converges or diverges to $+\infty$ or diverges to $-\infty$. In the definition of $\lim s_n = +\infty$ the challenging values of M are large positive numbers: the larger M is the larger N will need to be. In the definition of $\lim s_n = -\infty$ the challenging values of M are "large" negative numbers like $-10,000,000,000$.

Example 4

We have $\lim n^2 = +\infty$, $\lim(-n) = -\infty$, $\lim 2^n = +\infty$ and $\lim(\sqrt{n} + 7) = +\infty$. Of course, many sequences do not have limits $+\infty$ or $-\infty$ even if they are unbounded. For example, the sequences defined by $s_n = (-1)^n n$ and $t_n = n\cos^2(\frac{n\pi}{2})$ are unbounded, but they do not diverge to $+\infty$ or $-\infty$, so the expressions $\lim[(-1)^n n]$ and $\lim[n\cos^2(\frac{n\pi}{2})]$ are meaningless. Note $t_n = n$ when n is even and $t_n = 0$ when n is odd. □

The strategy for proofs involving infinite limits is very much the same as for finite limits. We give some examples.

Example 5

Give a formal proof that $\lim(\sqrt{n} + 7) = +\infty$. □

Discussion. We need to consider an arbitrary $M > 0$ and show there exists N [which will depend on M] such that

$$n > N \quad \text{implies} \quad \sqrt{n} + 7 > M.$$

To see how big N must be we "solve" for n in the inequality $\sqrt{n}+7 > M$. This inequality holds provided $\sqrt{n} > M - 7$ or $n > (M - 7)^2$. Thus we will take $N = (M - 7)^2$.

Formal Proof

Let $M > 0$ and let $N = (M-7)^2$. Then $n > N$ implies $n > (M-7)^2$, hence $\sqrt{n} > M - 7$, hence $\sqrt{n} + 7 > M$. This shows $\lim(\sqrt{n} + 7) = +\infty$. ■

Example 6
Give a formal proof that $\lim \frac{n^2+3}{n+1} = +\infty$; see Example 3. □

Discussion. Consider $M > 0$. We need to determine how large n must be to guarantee $\frac{n^2+3}{n+1} > M$. The idea is to bound the fraction $\frac{n^2+3}{n+1}$ below by some multiple of $\frac{n^2}{n} = n$; compare Example 3 of §8. Since $n^2 + 3 > n^2$ and $n + 1 \le 2n$, we have $\frac{n^2+3}{n+1} > \frac{n^2}{2n} = \frac{1}{2}n$, and it suffices to arrange for $\frac{1}{2}n > M$.

Formal Proof
Let $M > 0$ and let $N = 2M$. Then $n > N$ implies $\frac{1}{2}n > M$, which implies

$$\frac{n^2 + 3}{n + 1} > \frac{n^2}{2n} = \frac{1}{2}n > M.$$

Hence $\lim \frac{n^2+3}{n+1} = +\infty$. ■

The limit in Example 6 would be easier to handle if we could apply a limit theorem. But the limit Theorems 9.2–9.6 do not apply.

WARNING. Do not attempt to apply the limit Theorems 9.2–9.6 to infinite limits. Use Theorem 9.9 or 9.10 below or Exercises 9.9–9.12.

9.9 Theorem.
Let (s_n) and (t_n) be sequences such that $\lim s_n = +\infty$ and $\lim t_n > 0$ [$\lim t_n$ can be finite or $+\infty$]. Then $\lim s_n t_n = +\infty$.

Discussion. Let $M > 0$. We need to show $s_n t_n > M$ for large n. We have $\lim s_n = +\infty$, and we need to be sure the t_n's are bounded away from 0 for large n. We will choose a real number m so that $0 < m < \lim t_n$ and observe $t_n > m$ for large n. Then all we need is $s_n > \frac{M}{m}$ for large n.

Proof
Let $M > 0$. Select a real number m so that $0 < m < \lim t_n$. Whether $\lim t_n = +\infty$ or not, it is clear there exists N_1 such that

$$n > N_1 \quad \text{implies} \quad t_n > m;$$

see Exercise 8.10. Since $\lim s_n = +\infty$, there exists N_2 so that

$$n > N_2 \quad \text{implies} \quad s_n > \frac{M}{m}.$$

Put $N = \max\{N_1, N_2\}$. Then $n > N$ implies $s_n t_n > \frac{M}{m} \cdot m = M$. ∎

Example 7
Use Theorem 9.9 to prove $\lim \frac{n^2+3}{n+1} = +\infty$; see Example 6. □

Solution
We observe $\frac{n^2+3}{n+1} = \frac{n+\frac{3}{n}}{1+\frac{1}{n}} = s_n t_n$ where $s_n = n + \frac{3}{n}$ and $t_n = \frac{1}{1+\frac{1}{n}}$. It
is easy to show $\lim s_n = +\infty$ and $\lim t_n = 1$. So by Theorem 9.9, we
have $\lim s_n t_n = +\infty$. □

Here is another useful theorem.

9.10 Theorem.
*For a sequence (s_n) of positive real numbers, we have $\lim s_n = +\infty$
if and only if $\lim(\frac{1}{s_n}) = 0$.*

Proof
Let (s_n) be a sequence of positive real numbers. We have to show

$$\lim s_n = +\infty \quad \text{implies} \quad \lim \left(\frac{1}{s_n}\right) = 0 \qquad (1)$$

and

$$\lim \left(\frac{1}{s_n}\right) = 0 \quad \text{implies} \quad \lim s_n = +\infty. \qquad (2)$$

In this case the proofs will appear very similar, but the thought
processes will be quite different.

To prove (1), suppose $\lim s_n = +\infty$. Let $\epsilon > 0$ and let $M = \frac{1}{\epsilon}$.
Since $\lim s_n = +\infty$, there exists N such that $n > N$ implies $s_n > M = \frac{1}{\epsilon}$. Therefore $n > N$ implies $\epsilon > \frac{1}{s_n} > 0$, so

$$n > N \quad \text{implies} \quad \left| \frac{1}{s_n} - 0 \right| < \epsilon.$$

That is, $\lim(\frac{1}{s_n}) = 0$. This proves (1).

To prove (2), we abandon the notation of the last paragraph and
begin anew. Suppose $\lim(\frac{1}{s_n}) = 0$. Let $M > 0$ and let $\epsilon = \frac{1}{M}$. Then

$\epsilon > 0$, so there exists N such that $n > N$ implies $|\frac{1}{s_n} - 0| < \epsilon = \frac{1}{M}$. Since $s_n > 0$, we can write

$$n > N \quad \text{implies} \quad 0 < \frac{1}{s_n} < \frac{1}{M}$$

and hence

$$n > N \quad \text{implies} \quad M < s_n.$$

That is, $\lim s_n = +\infty$ and (2) holds.   ∎

Exercises

9.1 Using the limit Theorems 9.2–9.7, prove the following. Justify all steps.
   (a) $\lim \frac{n+1}{n} = 1$ (b) $\lim \frac{3n+7}{6n-5} = \frac{1}{2}$
   (c) $\lim \frac{17n^5+73n^4-18n^2+3}{23n^5+13n^3} = \frac{17}{23}$

9.2 Suppose $\lim x_n = 3$, $\lim y_n = 7$ and all y_n are nonzero. Determine the following limits:
   (a) $\lim(x_n + y_n)$ (b) $\lim \frac{3y_n - x_n}{y_n^2}$

9.3 Suppose $\lim a_n = a$, $\lim b_n = b$, and $s_n = \frac{a_n^3 + 4a_n}{b_n^2 + 1}$. Prove $\lim s_n = \frac{a^3 + 4a}{b^2 + 1}$ carefully, using the limit theorems.

9.4 Let $s_1 = 1$ and for $n \geq 1$ let $s_{n+1} = \sqrt{s_n + 1}$.

   (a) List the first four terms of (s_n).

   (b) It turns out that (s_n) converges. Assume this fact and prove the limit is $\frac{1}{2}(1 + \sqrt{5})$.

9.5 Let $t_1 = 1$ and $t_{n+1} = \frac{t_n^2 + 2}{2t_n}$ for $n \geq 1$. Assume (t_n) converges and find the limit.

9.6 Let $x_1 = 1$ and $x_{n+1} = 3x_n^2$ for $n \geq 1$.

   (a) Show if $a = \lim x_n$, then $a = \frac{1}{3}$ or $a = 0$.

   (b) Does $\lim x_n$ exist? Explain.

   (c) Discuss the apparent contradiction between parts (a) and (b).

9.7 Complete the proof of Theorem 9.7(c), i.e., give the standard argument needed to show $\lim s_n = 0$.

9.8 Give the following when they exist. Otherwise assert "NOT EXIST."
(a) $\lim n^3$ (b) $\lim(-n^3)$
(c) $\lim(-n)^n$ (d) $\lim(1.01)^n$
(e) $\lim n^n$

9.9 Suppose there exists N_0 such that $s_n \le t_n$ for all $n > N_0$.

(a) Prove that if $\lim s_n = +\infty$, then $\lim t_n = +\infty$.

(b) Prove that if $\lim t_n = -\infty$, then $\lim s_n = -\infty$.

(c) Prove that if $\lim s_n$ and $\lim t_n$ exist, then $\lim s_n \le \lim t_n$.

9.10 (a) Show that if $\lim s_n = +\infty$ and $k > 0$, then $\lim(ks_n) = +\infty$.

(b) Show $\lim s_n = +\infty$ if and only if $\lim(-s_n) = -\infty$.

(c) Show that if $\lim s_n = +\infty$ and $k < 0$, then $\lim(ks_n) = -\infty$.

9.11 (a) Show that if $\lim s_n = +\infty$ and $\inf\{t_n : n \in \mathbb{N}\} > -\infty$, then $\lim(s_n + t_n) = +\infty$.

(b) Show that if $\lim s_n = +\infty$ and $\lim t_n > -\infty$, then $\lim(s_n + t_n) = +\infty$.

(c) Show that if $\lim s_n = +\infty$ and if (t_n) is a bounded sequence, then $\lim(s_n + t_n) = +\infty$.

9.12 ★[3] Assume all $s_n \ne 0$ and that the limit $L = \lim \left|\frac{s_{n+1}}{s_n}\right|$ exists.

(a) Show that if $L < 1$, then $\lim s_n = 0$. *Hint:* Select a so that $L < a < 1$ and obtain N so that $|s_{n+1}| < a|s_n|$ for $n \ge N$. Then show $|s_n| < a^{n-N}|s_N|$ for $n > N$.

(b) Show that if $L > 1$, then $\lim|s_n| = +\infty$. *Hint:* Apply (a) to the sequence $t_n = \frac{1}{|s_n|}$; see Theorem 9.10.

9.13 Show

$$\lim_{n \to \infty} a^n = \begin{cases} 0 & \text{if} \quad |a| < 1 \\ 1 & \text{if} \quad a = 1 \\ +\infty & \text{if} \quad a > 1 \\ \text{does not exist} & \text{if} \quad a \le -1. \end{cases}$$

9.14 Let $p > 0$. Use Exercise 9.12 to show

$$\lim_{n \to \infty} \frac{a^n}{n^p} = \begin{cases} 0 & \text{if} \quad |a| \le 1 \\ +\infty & \text{if} \quad a > 1 \\ \text{does not exist} & \text{if} \quad a < -1. \end{cases}$$

Hint: For the $a > 1$ case, use Exercise 9.12(b).

---

[3]This exercise is referred to in several places.

9.15 Show $\lim_{n\to\infty} \frac{a^n}{n!} = 0$ for all $a \in \mathbb{R}$.

9.16 Use Theorems 9.9 and 9.10 or Exercises 9.9–9.15 to prove the following:

    (a) $\lim \frac{n^4+8n}{n^2+9} = +\infty$

    (b) $\lim[\frac{2^n}{n^2} + (-1)^n] = +\infty$

    (c) $\lim[\frac{3^n}{n^3} - \frac{3^n}{n!}] = +\infty$

9.17 Give a formal proof that $\lim n^2 = +\infty$ using only Definition 9.8.

9.18 (a) Verify $1 + a + a^2 + \cdots + a^n = \frac{1-a^{n+1}}{1-a}$ for $a \neq 1$.

    (b) Find $\lim_{n\to\infty}(1 + a + a^2 + \cdots + a^n)$ for $|a| < 1$.

    (c) Calculate $\lim_{n\to\infty}(1 + \frac{1}{3} + \frac{1}{9} + \frac{1}{27} + \cdots + \frac{1}{3^n})$.

    (d) What is $\lim_{n\to\infty}(1 + a + a^2 + \cdots + a^n)$ for $a \geq 1$?

§10 Monotone Sequences and Cauchy Sequences

In this section we obtain two theorems [Theorems 10.2 and 10.11] that will allow us to conclude certain sequences converge *without* knowing the limit in advance. These theorems are important because in practice the limits are not usually known in advance.

10.1 Definition.
A sequence (s_n) of real numbers is called an *increasing sequence* if $s_n \leq s_{n+1}$ for all n, and (s_n) is called a *decreasing sequence* if $s_n \geq s_{n+1}$ for all n. Note that if (s_n) is increasing, then $s_n \leq s_m$ whenever $n < m$. A sequence that is increasing or decreasing[4] will be called a *monotone sequence* or a *monotonic sequence*.

Example 1
The sequences defined by $a_n = 1 - \frac{1}{n}$, $b_n = n^3$ and $c_n = (1+\frac{1}{n})^n$ are increasing sequences, although this is not obvious for the

---

[4]In the First Edition of this book, increasing and decreasing sequences were referred to as "nondecreasing" and "nonincreasing" sequences, respectively.

sequence (c_n). The sequence $d_n = \frac{1}{n^2}$ is decreasing. The sequences $s_n = (-1)^n$, $t_n = \cos(\frac{n\pi}{3})$, $u_n = (-1)^n n$ and $v_n = \frac{(-1)^n}{n}$ are not monotonic sequences. Also $x_n = n^{1/n}$ is not monotonic, as can be seen by examining the first four values; see Example 1(d) on page 33 in §7.

Of the sequences above, (a_n), (c_n), (d_n), (s_n), (t_n), (v_n) and (x_n) are bounded sequences. The remaining sequences, (b_n) and (u_n), are unbounded sequences. $\qquad\qquad\qquad\qquad\qquad\qquad\qquad\qquad$ $\square$

10.2 Theorem.
All bounded monotone sequences converge.

Proof
Let (s_n) be a bounded increasing sequence. Let S denote the set $\{s_n : n \in \mathbb{N}\}$, and let $u = \sup S$. Since S is bounded, u represents a real number. We show $\lim s_n = u$. Let $\epsilon > 0$. Since $u - \epsilon$ is not an upper bound for S, there exists N such that $s_N > u - \epsilon$. Since (s_n) is increasing, we have $s_N \leq s_n$ for all $n \geq N$. Of course, $s_n \leq u$ for all n, so $n > N$ implies $u - \epsilon < s_n \leq u$, which implies $|s_n - u| < \epsilon$. This shows $\lim s_n = u$.

The proof for bounded decreasing sequences is left to Exercise 10.2. $\qquad\qquad\qquad\qquad\qquad\qquad\qquad\qquad\qquad\qquad$ ∎

Note the Completeness Axiom 4.4 is a vital ingredient in the proof of Theorem 10.2.

Example 2
Consider the sequence (s_n) defined *recursively* by

$$s_1 = 5 \quad \text{and} \quad s_n = \frac{s_{n-1}^2 + 5}{2s_{n-1}} \quad \text{for} \quad n \geq 2. \tag{1}$$

Thus $s_2 = 3$ and $s_3 = \frac{7}{3} \approx 2.333$. First, note a simple induction argument shows $s_n > 0$ for all n. We will show $\lim_n s_n$ exists by showing the sequence is decreasing and bounded; see Theorem 10.2. In fact, we will prove the following by induction:

$$\sqrt{5} < s_{n+1} < s_n \leq 5 \quad \text{for} \quad n \geq 1. \tag{2}$$

Since $\sqrt{5} \approx 2.236$, our computations show (2) holds for $n \leq 2$. For the induction step, assume (2) holds for some $n \geq 2$. To show $s_{n+2} < s_{n+1}$, we need

$$\frac{s_{n+1}^2 + 5}{2s_{n+1}} < s_{n+1} \quad \text{or} \quad s_{n+1}^2 + 5 < 2s_{n+1}^2 \quad \text{or} \quad 5 < s_{n+1}^2,$$

but this holds because $s_{n+1} > \sqrt{5}$ by the assumption (2) for n. To show $s_{n+2} > \sqrt{5}$, we need

$$\frac{s_{n+1}^2 + 5}{2s_{n+1}} > \sqrt{5} \quad \text{or} \quad s_{n+1}^2 + 5 > 2\sqrt{5}s_{n+1}$$

or $s_{n+1}^2 - 2\sqrt{5}s_{n+1} + 5 > 0$, which is true because $s_{n+1}^2 - 2\sqrt{5}s_{n+1} + 5 = (s_{n+1} - \sqrt{5})^2 > 0$. Thus (2) holds for $n+1$ whenever (2) holds for n. Hence (2) holds for all n by induction. Thus $s = \lim_n s_n$ exists.

If one looks at $s_4 = \frac{47}{21} \approx 2.238095$ and compares with $\sqrt{5} \approx 2.236068$, one might suspect $s = \sqrt{5}$. To verify this, we apply the limit Theorems 9.2–9.4 and the fact $s = \lim_n s_{n+1}$ to the equation $2 \cdot s_{n+1}s_n = s_n^2 + 5$ to obtain $2s^2 = s^2 + 5$. Thus $s^2 = 5$ and $s = \sqrt{5}$, since the limit is certainly not $-\sqrt{5}$. ☐

10.3 Discussion of Decimals.

We have not given much attention to the notion that real numbers are simply decimal expansions. This notion is substantially correct, but there are subtleties to be faced. For example, different decimal expansions can represent the same real number. The somewhat more abstract developments of the set $\mathbb{R}$ of real numbers discussed in §6 turn out to be more satisfactory.

We restrict our attention to nonnegative decimal expansions and nonnegative real numbers. From our point of view, every nonnegative decimal expansion is shorthand for the limit of a bounded increasing sequence of real numbers. Suppose we are given a decimal expansion $K.d_1 d_2 d_3 d_4 \cdots$, where K is a nonnegative integer and each d_j belongs to $\{0, 1, 2, 3, 4, 5, 6, 7, 8, 9\}$. Let

$$s_n = K + \frac{d_1}{10} + \frac{d_2}{10^2} + \cdots + \frac{d_n}{10^n}. \tag{1}$$

Then (s_n) is an increasing sequence of real numbers, and (s_n) is bounded [by $K + 1$, in fact]. So by Theorem 10.2, (s_n) converges to a real number we traditionally write as $K.d_1 d_2 d_3 d_4 \cdots$. For example, $3.3333\cdots$ represents

$$\lim_{n\to\infty}\left(3 + \frac{3}{10} + \frac{3}{10^2} + \cdots + \frac{3}{10^n}\right).$$

To calculate this limit, we borrow the following fact about geometric series from Example 1 on page 96 in §14:

$$\lim_{n\to\infty} a(1 + r + r^2 + \cdots + r^n) = \frac{a}{1 - r} \quad \text{for} \quad |r| < 1; \qquad (2)$$

see also Exercise 9.18. In our case, $a = 3$ and $r = \frac{1}{10}$, so $3.3333\cdots$ represents $\frac{3}{1-\frac{1}{10}} = \frac{10}{3}$, as expected. Similarly, $0.9999\cdots$ represents

$$\lim_{n\to\infty}\left(\frac{9}{10} + \frac{9}{10^2} + \cdots + \frac{9}{10^n}\right) = \frac{\frac{9}{10}}{1 - \frac{1}{10}} = 1.$$

Thus $0.9999\cdots$ and $1.0000\cdots$ are different decimal expansions that represent the same real number!

The converse of the preceding discussion also holds. That is, every nonnegative real number x has at least one decimal expansion. This will be proved, along with some related results, in §16. □

Unbounded monotone sequences also have limits.

10.4 Theorem.
(i) If (s_n) is an unbounded increasing sequence, then $\lim s_n = +\infty$.
(ii) If (s_n) is an unbounded decreasing sequence, then $\lim s_n = -\infty$.

Proof
(i) Let (s_n) be an unbounded increasing sequence. Let $M > 0$. Since the set $\{s_n : n \in \mathbb{N}\}$ is unbounded and it is bounded below by s_1, it must be unbounded above. Hence for some N in $\mathbb{N}$ we have $s_N > M$. Clearly $n > N$ implies $s_n \geq s_N > M$, so $\lim s_n = +\infty$.
(ii) The proof is similar and is left to Exercise 10.5. ■

10.5 Corollary.
If (s_n) is a monotone sequence, then the sequence either converges, diverges to $+\infty$, or diverges to $-\infty$. Thus $\lim s_n$ is always meaningful for monotone sequences.

Proof
Apply Theorems 10.2 and 10.4.   ∎

Let (s_n) be a bounded sequence in $\mathbb{R}$; it may or may not converge. It is apparent from the definition of limit in 7.1 that the limiting behavior of (s_n) depends only on sets of the form $\{s_n : n > N\}$. For example, if $\lim s_n$ exists, clearly it lies in the interval $[u_N, v_N]$ where

$$u_N = \inf\{s_n : n > N\} \quad \text{and} \quad v_N = \sup\{s_n : n > N\};$$

see Exercise 8.9. As N increases, the sets $\{s_n : n > N\}$ get smaller, so we have

$$u_1 \leq u_2 \leq u_3 \leq \cdots \quad \text{and} \quad v_1 \geq v_2 \geq v_3 \geq \cdots;$$

see Exercise 4.7(a). By Theorem 10.2 the limits $u = \lim_{N\to\infty} u_N$ and $v = \lim_{N\to\infty} v_N$ both exist, and $u \leq v$ since $u_N \leq v_N$ for all N. If $\lim s_n$ exists then, as noted above, $u_N \leq \lim s_n \leq v_N$ for all N, so we must have $u \leq \lim s_n \leq v$. The numbers u and v are useful whether $\lim s_n$ exists or not and are denoted $\liminf s_n$ and $\limsup s_n$, respectively.

10.6 Definition.
Let (s_n) be a sequence in $\mathbb{R}$. We define

$$\limsup s_n = \lim_{N\to\infty} \sup \{s_n : n > N\} \tag{1}$$

and

$$\liminf s_n = \lim_{N\to\infty} \inf \{s_n : n > N\}. \tag{2}$$

Note that in this definition we do not restrict (s_n) to be bounded. However, we adopt the following conventions. If (s_n) is not bounded above, $\sup\{s_n : n > N\} = +\infty$ for all N and we decree $\limsup s_n = +\infty$. Likewise, if (s_n) is not bounded below, $\inf\{s_n : n > N\} = -\infty$ for all N and we decree $\liminf s_n = -\infty$.

We emphasize $\limsup s_n$ need not equal $\sup\{s_n : n \in \mathbb{N}\}$, but $\limsup s_n \leq \sup\{s_n : n \in \mathbb{N}\}$. Some of the values s_n may be much larger than $\limsup s_n$; $\limsup s_n$ is the largest value that *infinitely many* s_n's can get close to. Similar remarks apply to $\liminf s_n$. These remarks will be clarified in Theorem 11.8 and §12, where we will give a thorough treatment of $\liminf$'s and $\limsup$'s. For now, we need a theorem that shows (s_n) has a limit if and only if $\liminf s_n = \limsup s_n$.

10.7 Theorem.
Let (s_n) be a sequence in $\mathbb{R}$.
  (i) *If $\lim s_n$ is defined [as a real number, $+\infty$ or $-\infty$], then*
      $\liminf s_n = \lim s_n = \limsup s_n$.
  (ii) *If $\liminf s_n = \limsup s_n$, then $\lim s_n$ is defined and $\lim s_n = \liminf s_n = \limsup s_n$.*

Proof
We use the notation $u_N = \inf\{s_n : n > N\}$, $v_N = \sup\{s_n : n > N\}$, $u = \lim u_N = \liminf s_n$ and $v = \lim v_N = \limsup s_n$.
  (i) Suppose $\lim s_n = +\infty$. Let M be a positive real number. Then there is a positive integer N so that

$$n > N \quad \text{implies} \quad s_n > M.$$

Then $u_N = \inf\{s_n : n > N\} \geq M$. It follows that $m > N$ implies $u_m \geq M$. In other words, the sequence (u_N) satisfies the condition defining $\lim u_N = +\infty$, i.e., $\liminf s_n = +\infty$. Likewise $\limsup s_n = +\infty$.
  The case $\lim s_n = -\infty$ is handled in a similar manner.
  Now suppose $\lim s_n = s$ where s is a real number. Consider $\epsilon > 0$. There exists a positive integer N such that $|s_n - s| < \epsilon$ for $n > N$. Thus $s_n < s + \epsilon$ for $n > N$, so

$$v_N = \sup\{s_n : n > N\} \leq s + \epsilon.$$

Also, $m > N$ implies $v_m \leq s+\epsilon$, so $\limsup s_n = \lim v_m \leq s+\epsilon$. Since $\limsup s_n \leq s + \epsilon$ for all $\epsilon > 0$, no matter how small, we conclude $\limsup s_n \leq s = \lim s_n$. A similar argument shows $\lim s_n \leq \liminf s_n$. Since $\liminf s_n \leq \limsup s_n$, we infer all three numbers are equal:

$$\liminf s_n = \lim s_n = \limsup s_n.$$

(ii) If $\liminf s_n = \limsup s_n = +\infty$ it is easy to show $\lim s_n = +\infty$. And if $\liminf s_n = \limsup s_n = -\infty$ it is easy to show $\lim s_n = -\infty$. We leave these two special cases to the reader.

Suppose, finally, that $\liminf s_n = \limsup s_n = s$ where s is a real number. We need to prove $\lim s_n = s$. Let $\epsilon > 0$. Since $s = \lim v_N$ there exists a positive integer N_0 such that

$$|s - \sup\{s_n : n > N_0\}| < \epsilon.$$

Thus $\sup\{s_n : n > N_0\} < s + \epsilon$, so

$$s_n < s + \epsilon \quad \text{for all} \quad n > N_0. \tag{1}$$

Similarly, there exists N_1 such that $|s - \inf\{s_n : n > N_1\}| < \epsilon$, hence $\inf\{s_n : n > N_1\} > s - \epsilon$, hence

$$s_n > s - \epsilon \quad \text{for all} \quad n > N_1. \tag{2}$$

From (1) and (2) we conclude

$$s - \epsilon < s_n < s + \epsilon \quad \text{for} \quad n > \max\{N_0, N_1\},$$

equivalently

$$|s_n - s| < \epsilon \quad \text{for} \quad n > \max\{N_0, N_1\}.$$

This proves $\lim s_n = s$ as desired. ∎

If (s_n) converges, then $\liminf s_n = \limsup s_n$ by the theorem just proved, so for large N the numbers $\sup\{s_n : n > N\}$ and $\inf\{s_n : n > N\}$ are close together. This implies that all the numbers in the set $\{s_n : n > N\}$ are close to each other. This leads us to a concept of great theoretical importance that will be used throughout the book.

10.8 Definition.
A sequence (s_n) of real numbers is called a *Cauchy sequence* if

for each $\epsilon > 0$ there exists a number N such that

$$m, n > N \text{ implies } |s_n - s_m| < \epsilon. \tag{1}$$

Compare this definition with Definition 7.1.

10.9 Lemma.

Convergent sequences are Cauchy sequences.

Proof

Suppose $\lim s_n = s$. The idea is that, since the terms s_n are close to s for large n, they also must be close to each other; indeed

$$|s_n - s_m| = |s_n - s + s - s_m| \le |s_n - s| + |s - s_m|.$$

To be precise, let $\epsilon > 0$. Then there exists N such that

$$n > N \quad \text{implies} \quad |s_n - s| < \frac{\epsilon}{2}.$$

Clearly we may also write

$$m > N \quad \text{implies} \quad |s_m - s| < \frac{\epsilon}{2},$$

so

$$m, n > N \quad \text{implies} \quad |s_n - s_m| \le |s_n - s| + |s - s_m| < \frac{\epsilon}{2} + \frac{\epsilon}{2} = \epsilon.$$

Thus (s_n) is a Cauchy sequence. ∎

10.10 Lemma.

Cauchy sequences are bounded.

Proof

The proof is similar to that of Theorem 9.1. Applying Definition 10.8 with $\epsilon = 1$ we obtain N in $\mathbb{N}$ so that

$$m, n > N \quad \text{implies} \quad |s_n - s_m| < 1.$$

In particular, $|s_n - s_{N+1}| < 1$ for $n > N$, so $|s_n| < |s_{N+1}| + 1$ for $n > N$. If $M = \max\{|s_{N+1}| + 1, |s_1|, |s_2|, \ldots, |s_N|\}$, then $|s_n| \le M$ for all $n \in \mathbb{N}$. ∎

The next theorem is very important because it shows that to verify that a sequence converges it suffices to check it is a Cauchy sequence, a property that does not involve the limit itself.

10.11 Theorem.

A sequence is a convergent sequence if and only if it is a Cauchy sequence.

Proof

The expression "if and only if" indicates that we have two assertions to verify: (i) convergent sequences are Cauchy sequences, and (ii) Cauchy sequences are convergent sequences. We already verified (i) in Lemma 10.9. To check (ii), consider a Cauchy sequence (s_n) and note (s_n) is bounded by Lemma 10.10. By Theorem 10.7 we need only show

$$\liminf s_n = \limsup s_n. \tag{1}$$

Let $\epsilon > 0$. Since (s_n) is a Cauchy sequence, there exists N so that

$$m, n > N \quad \text{implies} \quad |s_n - s_m| < \epsilon.$$

In particular, $s_n < s_m + \epsilon$ for all $m, n > N$. This shows $s_m + \epsilon$ is an upper bound for $\{s_n : n > N\}$, so $v_N = \sup\{s_n : n > N\} \le s_m + \epsilon$ for $m > N$. This, in turn, shows $v_N - \epsilon$ is a lower bound for $\{s_m : m > N\}$, so $v_N - \epsilon \le \inf\{s_m : m > N\} = u_N$. Thus

$$\limsup s_n \le v_N \le u_N + \epsilon \le \liminf s_n + \epsilon.$$

Since this holds for all $\epsilon > 0$, we have $\limsup s_n \le \liminf s_n$. The opposite inequality always holds, so we have established (1). ■

The proof of Theorem 10.11 uses Theorem 10.7, and Theorem 10.7 relies implicitly on the Completeness Axiom 4.4, since without the completeness axiom it is not clear that $\liminf s_n$ and $\limsup s_n$ are meaningful. The completeness axiom assures us that the expressions $\sup\{s_n : n > N\}$ and $\inf\{s_n : n > N\}$ in Definition 10.6 are meaningful, and Theorem 10.2 [which itself relies on the completeness axiom] assures us that the limits in Definition 10.6 also are meaningful.

Exercises on $\limsup$'s and $\liminf$'s appear in §§11 and 12.

Exercises

10.1 Which of the following sequences are increasing? decreasing? bounded?

(a) $\frac{1}{n}$ (b) $\frac{(-1)^n}{n^2}$

(c) n^5 (d) $\sin(\frac{n\pi}{7})$

(e) $(-2)^n$ (f) $\frac{n}{3^n}$

10.2 Prove Theorem 10.2 for bounded decreasing sequences.

10.3 For a decimal expansion $K.d_1d_2d_3d_4\cdots$, let (s_n) be defined as in Discussion 10.3. Prove $s_n < K + 1$ for all $n \in \mathbb{N}$. *Hint:* $\frac{9}{10} + \frac{9}{10^2} + \cdots + \frac{9}{10^n} = 1 - \frac{1}{10^n}$ for all n.

10.4 Discuss why Theorems 10.2 and 10.11 would fail if we restricted our world of numbers to the set $\mathbb{Q}$ of rational numbers.

10.5 Prove Theorem 10.4(ii).

10.6 **(a)** Let (s_n) be a sequence such that

$$|s_{n+1} - s_n| < 2^{-n} \quad \text{for all} \quad n \in \mathbb{N}.$$

Prove (s_n) is a Cauchy sequence and hence a convergent sequence.

(b) Is the result in (a) true if we only assume $|s_{n+1} - s_n| < \frac{1}{n}$ for all $n \in \mathbb{N}$?

10.7 Let S be a bounded nonempty subset of $\mathbb{R}$ such that $\sup S$ is not in S. Prove there is a sequence (s_n) of points in S such that $\lim s_n = \sup S$. See also Exercise 11.11.

10.8 Let (s_n) be an increasing sequence of positive numbers and define $\sigma_n = \frac{1}{n}(s_1 + s_2 + \cdots + s_n)$. Prove (σ_n) is an increasing sequence.

10.9 Let $s_1 = 1$ and $s_{n+1} = (\frac{n}{n+1})s_n^2$ for $n \geq 1$.

   (a) Find s_2, s_3 and s_4.

   (b) Show $\lim s_n$ exists.

   (c) Prove $\lim s_n = 0$.

10.10 Let $s_1 = 1$ and $s_{n+1} = \frac{1}{3}(s_n + 1)$ for $n \geq 1$.

   (a) Find s_2, s_3 and s_4.

   (b) Use induction to show $s_n > \frac{1}{2}$ for all n.

   (c) Show (s_n) is a decreasing sequence.

   (d) Show $\lim s_n$ exists and find $\lim s_n$.

10.11 Let $t_1 = 1$ and $t_{n+1} = [1 - \frac{1}{4n^2}] \cdot t_n$ for $n \geq 1$.

   (a) Show $\lim t_n$ exists.

   (b) What do you think $\lim t_n$ is?

10.12 Let $t_1 = 1$ and $t_{n+1} = [1 - \frac{1}{(n+1)^2}] \cdot t_n$ for $n \geq 1$.

    (a) Show $\lim t_n$ exists.

    (b) What do you think $\lim t_n$ is?

    (c) Use induction to show $t_n = \frac{n+1}{2n}$.

    (d) Repeat part (b).

§11 Subsequences

11.1 Definition.
Suppose $(s_n)_{n\in\mathbb{N}}$ is a sequence. A *subsequence* of this sequence is a sequence of the form $(t_k)_{k\in\mathbb{N}}$ where for each k there is a positive integer n_k such that

$$n_1 < n_2 < \cdots < n_k < n_{k+1} < \cdots \tag{1}$$

and

$$t_k = s_{n_k}. \tag{2}$$

Thus (t_k) is just a selection of some [possibly all] of the s_n's taken in order.

Here are some alternative ways to approach this concept. Note that (1) defines an infinite subset of $\mathbb{N}$, namely $\{n_1, n_2, n_3, \ldots\}$. Conversely, every infinite subset of $\mathbb{N}$ can be described by (1). Thus a subsequence of (s_n) is a sequence obtained by selecting, in order, an infinite subset of the terms.

For a more precise definition, recall we can view the sequence $(s_n)_{n\in\mathbb{N}}$ as a function s with domain $\mathbb{N}$; see §7. For the subset $\{n_1, n_2, n_3, \ldots\}$, there is a natural function σ [lower case Greek sigma] given by $\sigma(k) = n_k$ for $k \in \mathbb{N}$. The function σ "selects" an infinite subset of $\mathbb{N}$, in order. The subsequence of s corresponding to σ is simply the composite function $t = s \circ \sigma$. That is,

$$t_k = t(k) = s \circ \sigma(k) = s(\sigma(k)) = s(n_k) = s_{n_k} \quad \text{for} \quad k \in \mathbb{N}. \tag{3}$$

Thus a sequence t is a subsequence of a sequence s if and only if $t = s \circ \sigma$ for some increasing function σ mapping $\mathbb{N}$ into $\mathbb{N}$. We will usually suppress the notation σ and often suppress the notation t

also. Thus the phrase "a subsequence (s_{n_k}) of (s_n)" will refer to the subsequence defined by (1) and (2) or by (3), depending upon your point of view.

Example 1
Let (s_n) be the sequence defined by $s_n = n^2(-1)^n$. The positive terms of this sequence comprise a subsequence. In this case, the sequence (s_n) is

$$(-1, 4, -9, 16, -25, 36, -49, 64, \ldots)$$

and the subsequence is

$$(4, 16, 36, 64, 100, 144, \ldots).$$

More precisely, the subsequence is $(s_{n_k})_{k \in \mathbb{N}}$ where $n_k = 2k$ so that $s_{n_k} = (2k)^2(-1)^{2k} = 4k^2$. The selection function σ is given by $\sigma(k) = 2k$. □

Example 2
Consider the sequence $a_n = \sin(\frac{n\pi}{3})$ and its subsequence (a_{n_k}) of nonnegative terms. The sequence $(a_n)_{n \in \mathbb{N}}$ is

$$(\frac{1}{2}\sqrt{3}, \frac{1}{2}\sqrt{3}, 0, -\frac{1}{2}\sqrt{3}, -\frac{1}{2}\sqrt{3}, 0, \frac{1}{2}\sqrt{3}, \frac{1}{2}\sqrt{3}, 0, -\frac{1}{2}\sqrt{3}, -\frac{1}{2}\sqrt{3}, 0, \ldots)$$

and the desired subsequence is

$$(\frac{1}{2}\sqrt{3}, \frac{1}{2}\sqrt{3}, 0, 0, \frac{1}{2}\sqrt{3}, \frac{1}{2}\sqrt{3}, 0, 0, \ldots).$$

It is evident that $n_1 = 1$, $n_2 = 2$, $n_3 = 3$, $n_4 = 6$, $n_5 = 7$, $n_6 = 8$, $n_7 = 9$, $n_8 = 12$, $n_9 = 13$, etc. We won't need a formula for n_k, but here is one: $n_k = k + 2\lfloor \frac{k}{4} \rfloor$ for $k \geq 1$, where $\lfloor x \rfloor$ is the "floor function," i.e., $\lfloor x \rfloor$ is the largest integer less than or equal to x, for $x \in \mathbb{R}$. □

After some reflection, the next theorem will seem obvious, but it is good to have a complete proof that covers all situations. The proof is a little bit complicated, but we will apply the theorem several times rather than having to recreate a similar proof several times.[5] Thus it is important to understand the proof.

---

[5]In the first edition of this book, we did create similar proofs instead.

11.2 Theorem.
Let (s_n) be a sequence.

  (i) *If t is in $\mathbb{R}$, then there is a subsequence of (s_n) converging to t if and only if the set $\{n \in \mathbb{N} : |s_n - t| < \epsilon\}$ is infinite for all $\epsilon > 0$.*

  (ii) *If the sequence (s_n) is unbounded above, it has a subsequence with limit $+\infty$.*

  (iii) *Similarly, if (s_n) is unbounded below, a subsequence has limit $-\infty$.*

In each case, the subsequence can be taken to be monotonic.[6]

Proof
The forward implications $\Longrightarrow$ in (i)–(iii) are all easy to check. For example, if $\lim_k s_{n_k} = t$ and $\epsilon > 0$, then all but finitely many of the n_ks are in $\{n \in \mathbb{N} : |s_n - t| < \epsilon\}$. We focus on the other implications.

  (i) First suppose the set $\{n \in \mathbb{N} : s_n = t\}$ is infinite. Then there are subsequences $(s_{n_k})_{k \in \mathbb{N}}$ such that $s_{n_k} = t$ for all k. Such subsequences of (s_n) are boring monotonic sequences converging to t.

Henceforth, we assume $\{n \in \mathbb{N} : s_n = t\}$ is finite. Then

$$\{n \in \mathbb{N} : 0 < |s_n - t| < \epsilon\} \quad \text{is infinite for all} \quad \epsilon > 0.$$

Since these sets equal

$$\{n \in \mathbb{N} : t - \epsilon < s_n < t\} \cup \{n \in \mathbb{N} : t < s_n < t + \epsilon\},$$

and these sets get smaller as $\epsilon \to 0$, we have

$$\{n \in \mathbb{N} : t - \epsilon < s_n < t\} \quad \text{is infinite for all} \quad \epsilon > 0, \qquad (1)$$

or

$$\{n \in \mathbb{N} : t < s_n < t + \epsilon\} \quad \text{is infinite for all} \quad \epsilon > 0; \qquad (2)$$

otherwise, for sufficiently small $\epsilon > 0$, the sets in both (1) and (2) would be finite.

We assume (1) holds, and leave the case that (2) holds to the reader. We will show how to define or construct

---

[6]This will be proved easily here, but is also a consequence of the more general Theorem 11.4.

step-by-step a subsequence $(s_{n_k})_{k\in\mathbb{N}}$ satisfying $t-1 < s_{n_1} < t$ and

$$\max\left\{s_{n_{k-1}}, t-\frac{1}{k}\right\} \le s_{n_k} < t \quad \text{for} \quad k \ge 2. \tag{3}$$

Specifically, we will assume $n_1, n_2, \ldots, n_{k-1}$ have been selected satisfying (3) and show how to select n_k. This will give us an infinite increasing sequence $(n_k)_{k\in\mathbb{N}}$ and hence a subsequence (s_{n_k}) of (s_n) satisfying (3). Since we will have $s_{n_{k-1}} \le s_{n_k}$ for all k, this subsequence will be monotonically increasing. Since (3) also will imply $t-\frac{1}{k} \le s_{n_k} < t$ for all k, we will have $\lim_k s_{n_k} = t$; compare Exercise 8.5(a) on page 44.

A construction like the one described above, and executed below, is called an "inductive definition" or "definition by induction," even though the validity of the process is not a direct consequence of Peano's axiom N5 in §1.[7]

Here is the construction. Select n_1 so that $t-1 < s_{n_1} < t$; this is possible by (1). Suppose $n_1, n_2, \ldots, n_{k-1}$ have been selected so that

$$n_1 < n_2 < \cdots < n_{k-1} \tag{4}$$

and

$$\max\left\{s_{n_{j-1}}, t-\frac{1}{j}\right\} \le s_{n_j} < t \quad \text{for} \quad j = 2, \ldots, k-1. \tag{5}$$

Using (1) with $\epsilon = \max\{s_{n_{k-1}}, t-\frac{1}{k}\}$, we can select $n_k > n_{k-1}$ satisfying (5) for $j = k$, so that (3) holds for k. The procedure defines the sequence $(n_k)_{k\in\mathbb{N}}$. This completes the proof of (i), and is the crux of the full proof.

(ii) Let $n_1 = 1$, say. Given $n_1 < \cdots < n_{k-1}$, select n_k so that $s_{n_k} > \max\{s_{n_{k-1}}, k\}$. This is possible, since (s_n) is unbounded above. The sequence so obtained will be monotonic and have limit $+\infty$. A similar proof verifies (iii). ∎

---

[7]Recursive definitions of sequences, which first appear in Exercises 9.4–9.6, can be viewed as simple examples of definitions by induction.

Example 3

It can be shown that the set $\mathbb{Q}$ of rational numbers can be listed
as a sequence (r_n), though it is tedious to specify an exact formula.
Figure 11.1 suggests such a listing [with repetitions] where $r_1 = 0$,
$r_2 = 1$, $r_3 = \frac{1}{2}$, $r_4 = -\frac{1}{2}$, $r_5 = -1$, $r_6 = -2$, $r_7 = -1$, etc. Readers
familiar with some set theory will recognize this assertion as "$\mathbb{Q}$
is countable." This sequence has an amazing property: given any
real number a there exists a subsequence (r_{n_k}) of (r_n) converging to
a. Since there are infinitely many rational numbers in every interval
$(a-\epsilon, a+\epsilon)$ by Exercise 4.11, Theorem 11.2 shows that a subsequence
of (r_n) converges to a. $\qquad\square$

Example 4

Suppose (s_n) is a sequence of positive numbers such that $\inf\{s_n :
n \in \mathbb{N}\} = 0$. The sequence (s_n) need not converge or even be
bounded, but it has a subsequence converging monotonically to 0. By
Theorem 11.2, it suffices to show $\{n \in \mathbb{N} : s_n < \epsilon\}$ is infinite for each
$\epsilon > 0$. Otherwise, this set would be finite for some $\epsilon_0 > 0$. If the set
is nonempty, then $\inf\{s_n : n \in \mathbb{N}\} = \min\{s_n : s_n < \epsilon_0\} > 0$, because
each s_n is positive and the set $\{s_n : s_n < \epsilon_0\}$ is finite. This contradicts our assumption $\inf\{s_n : n \in \mathbb{N}\} = 0$. If the set is empty, then
$\inf\{s_n : n \in \mathbb{N}\} \geq \epsilon_0 > 0$, again contrary to our assumption. $\qquad\square$

The next theorem is almost obvious.

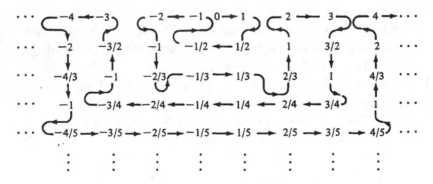

FIGURE 11.1

11.3 Theorem.
If the sequence (s_n) converges, then every subsequence converges to the same limit.

Proof
Let (s_{n_k}) denote a subsequence of (s_n). Note that $n_k \geq k$ for all k. This is easy to prove by induction; in fact, $n_1 \geq 1$ and $n_{k-1} \geq k-1$ implies $n_k > n_{k-1} \geq k-1$ and hence $n_k \geq k$.

Let $s = \lim s_n$ and let $\epsilon > 0$. There exists N so that $n > N$ implies $|s_n - s| < \epsilon$. Now $k > N$ implies $n_k > N$, which implies $|s_{n_k} - s| < \epsilon$. Thus

$$\lim_{k \to \infty} s_{n_k} = s.$$
∎

Our immediate goal is to prove the Bolzano-Weierstrass theorem which asserts that every bounded sequence has a convergent subsequence. First we prove a theorem about monotonic subsequences.

11.4 Theorem.
Every sequence (s_n) has a monotonic subsequence.

Proof
Let's say that the n-th term is *dominant* if it is greater than every term which follows it:

$$s_m < s_n \quad \text{for all} \quad m > n. \tag{1}$$

Case 1. Suppose there are infinitely many dominant terms, and let (s_{n_k}) be any subsequence consisting solely of dominant terms. Then $s_{n_{k+1}} < s_{n_k}$ for all k by (1), so (s_{n_k}) is a decreasing sequence.

Case 2. Suppose there are only finitely many dominant terms. Select n_1 so that s_{n_1} is beyond all the dominant terms of the sequence. Then

$$\text{given } N \geq n_1 \text{ there exists } m > N \text{ such that} \quad s_m \geq s_N. \tag{2}$$

Applying (2) with $N = n_1$ we select $n_2 > n_1$ such that $s_{n_2} \geq s_{n_1}$. Suppose $n_1, n_2, \ldots, n_{k-1}$ have been selected so that

$$n_1 < n_2 < \cdots < n_{k-1} \tag{3}$$

and

$$s_{n_1} \leq s_{n_2} \leq \cdots \leq s_{n_{k-1}}. \tag{4}$$

Applying (2) with $N = n_{k-1}$ we select $n_k > n_{k-1}$ such that $s_{n_k} \geq s_{n_{k-1}}$. Then (3) and (4) hold with k in place of $k-1$, the procedure continues by induction, and we obtain an increasing subsequence (s_{n_k}). ∎

The elegant proof of Theorem 11.4 was brought to our attention by David M. Bloom and is based on a solution in D. J. Newman's beautiful book [48].

11.5 Bolzano-Weierstrass Theorem.
Every bounded sequence has a convergent subsequence.

Proof
If (s_n) is a bounded sequence, it has a monotonic subsequence by Theorem 11.4, which converges by Theorem 10.2. ∎

The Bolzano-Weierstrass theorem is very important and will be used at critical points in Chap. 3. Our proof, based on Theorem 11.4, is somewhat nonstandard for reasons we now discuss. Many of the notions introduced in this chapter make equally good sense in more general settings. For example, the ideas of convergent sequence, Cauchy sequence and bounded sequence all make sense for a sequence (s_n) where each s_n belongs to the plane. But the idea of a monotonic sequence does not carry over. It turns out that the Bolzano-Weierstrass theorem also holds in the plane and in many other settings [see Theorem 13.5], but clearly it would no longer be possible to prove it directly from an analogue of Theorem 11.4. Since the Bolzano-Weierstrass Theorem 11.5 generalizes to settings where Theorem 11.4 makes little sense, in applications we will emphasize the Bolzano-Weierstrass Theorem 11.5 rather than Theorem 11.4.

We need one more notion, and then we will be able to tie our various concepts together in Theorem 11.8.

11.6 Definition.
Let (s_n) be a sequence in $\mathbb{R}$. A *subsequential limit* is any real number or symbol $+\infty$ or $-\infty$ that is the limit of some subsequence of (s_n).

When a sequence has a limit s, then all subsequences have limit s, so $\{s\}$ is the set of subsequential limits. The interesting case is when the original sequence does not have a limit. We return to some of the examples discussed after Definition 11.1.

Example 5
Consider (s_n) where $s_n = n^2(-1)^n$. The subsequence of even terms diverges to $+\infty$, and the subsequence of odd terms diverges to $-\infty$. All subsequences that have a limit diverge to $+\infty$ or $-\infty$, so that $\{-\infty, +\infty\}$ is exactly the set of subsequential limits of (s_n). $\quad\square$

Example 6
Consider the sequence $a_n = \sin(\frac{n\pi}{3})$ in Example 2. This sequence takes each of the values $\frac{1}{2}\sqrt{3}$, 0 and $-\frac{1}{2}\sqrt{3}$ an infinite number of times. The only convergent subsequences are constant from some term on, and $\{-\frac{1}{2}\sqrt{3}, 0, \frac{1}{2}\sqrt{3}\}$ is the set of subsequential limits of (a_n). If $n_k = 3k$, then $a_{n_k} = 0$ for all $k \in \mathbb{N}$ and obviously $\lim_{k\to\infty} a_{n_k} = 0$. If $n_k = 6k + 1$, then $a_{n_k} = \frac{1}{2}\sqrt{3}$ for all k and $\lim_{k\to\infty} a_{n_k} = \frac{1}{2}\sqrt{3}$. And if $n_k = 6k + 4$, then $\lim_{k\to\infty} a_{n_k} = -\frac{1}{2}\sqrt{3}$. $\quad\square$

Example 7
Let (r_n) be a list of all rational numbers. It was shown in Example 3 that every real number is a subsequential limit of (r_n). Also, $+\infty$ and $-\infty$ are subsequential limits; see Exercise 11.7. Consequently, $\mathbb{R} \cup \{-\infty, +\infty\}$ is the set of subsequential limits of (r_n). $\quad\square$

Example 8
Let $b_n = n[1 + (-1)^n]$ for $n \in \mathbb{N}$. Then $b_n = 2n$ for even n and $b_n = 0$ for odd n. Thus the 2-element set $\{0, +\infty\}$ is the set of subsequential limits of (b_n). $\quad\square$

We now turn to the connection between subsequential limits and $\limsup$'s and $\liminf$'s.

11.7 Theorem.
Let (s_n) be any sequence. There exists a monotonic subsequence whose limit is $\limsup s_n$, and there exists a monotonic subsequence whose limit is $\liminf s_n$.

Proof

If (s_n) is not bounded above, then by Theorem 11.2(ii), a monotonic subsequence of (s_n) has limit $+\infty = \limsup s_n$. Similarly, if (s_n) is not bounded below, a monotonic subsequence has limit $-\infty = \liminf s_n$.

The remaining cases are that (s_n) is bounded above or is bounded below. These cases are similar, so we only consider the case that (s_n) is bounded above, so that $\limsup s_n$ is finite. Let $t = \limsup s_n$, and consider $\epsilon > 0$. There exists N_0 so that

$$\sup\{s_n : n > N\} < t + \epsilon \quad \text{for} \quad N \geq N_0.$$

In particular, $s_n < t + \epsilon$ for all $n > N_0$. We now claim

$$\{n \in \mathbb{N} : t - \epsilon < s_n < t + \epsilon\} \quad \text{is infinite.} \tag{1}$$

Otherwise, there exists $N_1 > N_0$ so that $s_n \leq t - \epsilon$ for $n > N_1$. Then $\sup\{s_n : n > N\} \leq t - \epsilon$ for $N \geq N_1$, so that $\limsup s_n < t$, a contradiction. Since (1) holds for each $\epsilon > 0$, Theorem 11.2(i) shows that a monotonic subsequence of (s_n) converges to $t = \limsup s_n$. ∎

11.8 Theorem.

Let (s_n) be any sequence in $\mathbb{R}$, and let S denote the set of subsequential limits of (s_n).

(i) *S is nonempty.*
(ii) *$\sup S = \limsup s_n$ and $\inf S = \liminf s_n$.*
(iii) *$\lim s_n$ exists if and only if S has exactly one element, namely $\lim s_n$.*

Proof

(i) is an immediate consequence of Theorem 11.7.

To prove (ii), consider any limit t of a subsequence (s_{n_k}) of (s_n). By Theorem 10.7 we have $t = \liminf_k s_{n_k} = \limsup_k s_{n_k}$. Since $n_k \geq k$ for all k, we have $\{s_{n_k} : k > N\} \subseteq \{s_n : n > N\}$ for each N in $\mathbb{N}$. Therefore

$$\liminf_n s_n \leq \liminf_k s_{n_k} = t = \limsup_k s_{n_k} \leq \limsup_n s_n.$$

This inequality holds for all t in S; therefore

$$\liminf s_n \leq \inf S \leq \sup S \leq \limsup s_n.$$

Theorem 11.7 shows that $\liminf s_n$ and $\limsup s_n$ both belong to S. Therefore (ii) holds.

Assertion (iii) is simply a reformulation of Theorem 10.7. ∎

Theorems 11.7 and 11.8 show that $\limsup s_n$ is exactly the largest subsequential limit of (s_n), and $\liminf s_n$ is exactly the smallest subsequential limit of (s_n). This makes it easy to calculate $\limsup$'s and $\liminf$'s.

We return to the examples given before Theorem 11.7.

Example 9
If $s_n = n^2(-1)^n$, then $S = \{-\infty, +\infty\}$ as noted in Example 5. Therefore $\limsup s_n = \sup S = +\infty$ and $\liminf s_n = \inf S = -\infty$. □

Example 10
If $a_n = \sin(\frac{n\pi}{3})$, then $S = \{-\frac{1}{2}\sqrt{3}, 0, \frac{1}{2}\sqrt{3}\}$ as observed in Example 6. Hence $\limsup a_n = \sup S = \frac{1}{2}\sqrt{3}$ and $\liminf a_n = \inf S = -\frac{1}{2}\sqrt{3}$. □

Example 11
If (r_n) denotes a list of all rational numbers, then the set $\mathbb{R} \cup \{-\infty, +\infty\}$ is the set of subsequential limits of (r_n). Consequently we have $\limsup r_n = +\infty$ and $\liminf r_n = -\infty$. □

Example 12
If $b_n = n[1 + (-1)^n]$, then $\limsup b_n = +\infty$ and $\liminf b_n = 0$; see Example 8. □

The next result shows that the set S of subsequential limits always contains all limits of sequences *from* S. Such sets are called *closed sets*. Sets of this sort will be discussed further in the enrichment §13.

11.9 Theorem.
Let S denote the set of subsequential limits of a sequence (s_n). Suppose (t_n) is a sequence in $S \cap \mathbb{R}$ and that $t = \lim t_n$. Then t belongs to S.

Proof

Suppose t is finite. Consider the interval $(t - \epsilon, t + \epsilon)$. Then some t_n is in this interval. Let $\delta = \min\{t + \epsilon - t_n, t_n - t + \epsilon\}$, so that

$$(t_n - \delta, t_n + \delta) \subseteq (t - \epsilon, t + \epsilon).$$

Since t_n is a subsequential limit, the set $\{n \in \mathbb{N} : s_n \in (t_n - \delta, t_n + \delta)\}$ is infinite, so the set $\{n \in \mathbb{N} : s_n \in (t - \epsilon, t + \epsilon)\}$ is also infinite. Thus, by Theorem 11.2(i), t itself is a subsequential limit of (s_n).

If $t = +\infty$, then clearly the sequence (s_n) is unbounded above, so a subsequence of (s_n) has limit $+\infty$ by Theorem 11.2(ii). Thus $+\infty$ is also in S. A similar argument applies if $t = -\infty$. ∎

Exercises

11.1 Let $a_n = 3 + 2(-1)^n$ for $n \in \mathbb{N}$.

(a) List the first eight terms of the sequence (a_n).

(b) Give a subsequence that is constant [takes a single value]. Specify the selection function σ.

11.2 Consider the sequences defined as follows:

$$a_n = (-1)^n, \qquad b_n = \frac{1}{n}, \qquad c_n = n^2, \qquad d_n = \frac{6n + 4}{7n - 3}.$$

(a) For each sequence, give an example of a monotone subsequence.

(b) For each sequence, give its set of subsequential limits.

(c) For each sequence, give its lim sup and lim inf.

(d) Which of the sequences converges? diverges to $+\infty$? diverges to $-\infty$?

(e) Which of the sequences is bounded?

11.3 Repeat Exercise 11.2 for the sequences:

$$s_n = \cos\left(\frac{n\pi}{3}\right), \quad t_n = \frac{3}{4n + 1}, \quad u_n = \left(-\frac{1}{2}\right)^n, \quad v_n = (-1)^n + \frac{1}{n}.$$

11.4 Repeat Exercise 11.2 for the sequences:

$$w_n = (-2)^n, \quad x_n = 5^{(-1)^n}, \quad y_n = 1 + (-1)^n, \quad z_n = n \cos\left(\frac{n\pi}{4}\right).$$

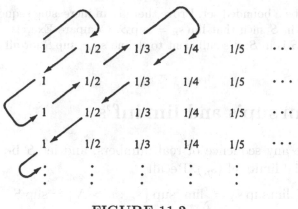

FIGURE 11.2

11.5 Let (q_n) be an enumeration of all the rationals in the interval $(0, 1]$.

(a) Give the set of subsequential limits for (q_n).

(b) Give the values of $\limsup q_n$ and $\liminf q_n$.

11.6 Show every subsequence of a subsequence of a given sequence is itself a subsequence of the given sequence. *Hint*: Define subsequences as in (3) of Definition 11.1.

11.7 Let (r_n) be an enumeration of the set $\mathbb{Q}$ of all rational numbers. Show there exists a subsequence (r_{n_k}) such that $\lim_{k \to \infty} r_{n_k} = +\infty$.

11.8 ★[8] Use Definition 10.6 and Exercise 5.4 to prove $\liminf s_n = -\limsup(-s_n)$ for every sequence (s_n).

11.9 (a) Show the closed interval $[a, b]$ is a closed set.

(b) Is there a sequence (s_n) such that $(0, 1)$ is its set of subsequential limits?

11.10 Let (s_n) be the sequence of numbers in Fig. 11.2 listed in the indicated order.

(a) Find the set S of subsequential limits of (s_n).

(b) Determine $\limsup s_n$ and $\liminf s_n$.

---

[8]This exercise is referred to in several places.

11.11 Let S be a bounded set. Prove there is an increasing sequence (s_n) of points in S such that $\lim s_n = \sup S$. Compare Exercise 10.7. *Note:* If $\sup S$ is in S, it's sufficient to define $s_n = \sup S$ for all n.

§12   lim sup's and lim inf's

Let (s_n) be any sequence of real numbers, and let S be the set of subsequential limits of (s_n). Recall

$$\limsup s_n = \lim_{N\to\infty} \sup\{s_n : n > N\} = \sup S \qquad (*)$$

and

$$\liminf s_n = \lim_{N\to\infty} \inf\{s_n : n > N\} = \inf S. \qquad (**)$$

The first equalities in $(*)$ and $(**)$ are from Definition 10.6, and the second equalities are proved in Theorem 11.8. This section is designed to increase the students' familiarity with these concepts. Most of the material is given in the exercises. We illustrate the techniques by proving some results that will be needed later in the text.

12.1 Theorem.
If (s_n) converges to a positive real number s and (t_n) is any sequence, then

$$\limsup s_n t_n = s \cdot \limsup t_n.$$

Here we allow the conventions $s \cdot (+\infty) = +\infty$ and $s \cdot (-\infty) = -\infty$ for $s > 0$.

Proof
We first show

$$\limsup s_n t_n \geq s \cdot \limsup t_n. \qquad (1)$$

We have three cases. Let $\beta = \limsup t_n$.

   Case 1. Suppose β is finite.

   By Theorem 11.7, there exists a subsequence (t_{n_k}) of (t_n) such that $\lim_{k\to\infty} t_{n_k} = \beta$. We also have $\lim_{k\to\infty} s_{n_k} = s$ [by Theorem 11.3], so $\lim_{k\to\infty} s_{n_k} t_{n_k} = s\beta$. Thus $(s_{n_k} t_{n_k})$ is a subsequence of $(s_n t_n)$ converging to $s\beta$, and therefore $s\beta \leq \limsup s_n t_n$. [Recall that

lim sup $s_n t_n$ is the largest possible limit of a subsequence of $(s_n t_n)$.]
Thus (1) holds.

Case 2. Suppose $\beta = +\infty$.

There exists a subsequence (t_{n_k}) of (t_n) such that $\lim_{k \to \infty} t_{n_k} = +\infty$. Since $\lim_{k \to \infty} s_{n_k} = s > 0$, Theorem 9.9 shows $\lim_{k \to \infty} s_{n_k} t_{n_k} = +\infty$. Hence lim sup $s_n t_n = +\infty$, so (1) clearly holds.

Case 3. Suppose $\beta = -\infty$.

Since $s > 0$, the right-hand side of (1) is equal to $s \cdot (-\infty) = -\infty$. Hence (1) is obvious in this case.

We have now established (1) in all cases. For the reversed inequality, we resort to a little trick. First note that we may ignore the first few terms of (s_n) and assume all $s_n \neq 0$. Then we can write $\lim \frac{1}{s_n} = \frac{1}{s}$ by Lemma 9.5. Now we apply (1) with s_n replaced by $\frac{1}{s_n}$ and t_n replaced by $s_n t_n$:

$$\lim \sup t_n = \lim \sup \left(\frac{1}{s_n}\right)(s_n t_n) \geq \left(\frac{1}{s}\right) \lim \sup s_n t_n,$$

i.e.,

$$\lim \sup s_n t_n \leq s \cdot \lim \sup t_n.$$

This inequality and (1) prove the theorem. ∎

Example 1

The hypothesis s be positive in Theorem 12.1 cannot be relaxed to allow $s = 0$. To see this, consider (s_n) and (t_n), where $s_n = -\frac{1}{n}$ and $t_n = -n^2$ for all n. In this setting, we don't define $0 \cdot (-\infty)$, but even if we did, we wouldn't define this product to be $+\infty$. □

The next theorem will be useful in dealing with infinite series; see the proof of the Ratio Test 14.8.

12.2 Theorem.

Let (s_n) be any sequence of nonzero real numbers. Then we have

$$\lim \inf \left|\frac{s_{n+1}}{s_n}\right| \leq \lim \inf |s_n|^{1/n} \leq \lim \sup |s_n|^{1/n} \leq \lim \sup \left|\frac{s_{n+1}}{s_n}\right|.$$

Proof

The middle inequality is obvious. The first and third inequalities have similar proofs. We will prove the third inequality and leave the first inequality to Exercise 12.11.

Let $\alpha = \limsup |s_n|^{1/n}$ and $L = \limsup |\frac{s_{n+1}}{s_n}|$. We need to prove $\alpha \leq L$. This is obvious if $L = +\infty$, so we assume $L < +\infty$. To prove $\alpha \leq L$ it suffices to show

$$\alpha \leq L_1 \quad \text{for any} \quad L_1 > L. \tag{1}$$

Since

$$L = \limsup \left| \frac{s_{n+1}}{s_n} \right| = \lim_{N \to \infty} \sup \left\{ \left| \frac{s_{n+1}}{s_n} \right| : n > N \right\} < L_1,$$

there exists a positive integer N such that

$$\sup \left\{ \left| \frac{s_{n+1}}{s_n} \right| : n \geq N \right\} < L_1.$$

Thus

$$\left| \frac{s_{n+1}}{s_n} \right| < L_1 \quad \text{for} \quad n \geq N. \tag{2}$$

Now for $n > N$ we can write

$$|s_n| = \left| \frac{s_n}{s_{n-1}} \right| \cdot \left| \frac{s_{n-1}}{s_{n-2}} \right| \cdots \left| \frac{s_{N+1}}{s_N} \right| \cdot |s_N|.$$

There are $n - (N + 1) + 1 = n - N$ fractions here, so applying (2) we see that

$$|s_n| < L_1^{n-N} |s_N| \quad \text{for} \quad n > N.$$

Since L_1 and N are fixed in this argument, $a = L_1^{-N} |s_N|$ is a positive constant and we may write

$$|s_n| < L_1^n a \quad \text{for} \quad n > N.$$

Therefore we have

$$|s_n|^{1/n} < L_1 a^{1/n} \quad \text{for} \quad n > N.$$

Since $\lim_{n \to \infty} a^{1/n} = 1$ by Theorem 9.7(d) on page 48, we conclude $\alpha = \limsup |s_n|^{1/n} \leq L_1$; see Exercise 12.1. Consequently (1) holds as desired. ∎

12.3 Corollary.
If $\lim \left|\frac{s_{n+1}}{s_n}\right|$ *exists [and equals L], then* $\lim |s_n|^{1/n}$ *exists [and equals L].*

Proof
If $\lim \left|\frac{s_{n+1}}{s_n}\right| = L$, then all four values in Theorem 12.2 are equal to L. Hence $\lim |s_n|^{1/n} = L$; see Theorem 10.7. ∎

Example 2
(a) If $L = \limsup s_n \neq \infty$, then for every $\alpha > L$, the set $\{n : s_n > \alpha\}$ is finite. If $L \neq -\infty$, then for every $\beta < L$, the set $\{n : s_n > \beta\}$ is infinite.

   If $L \neq \infty$ and $\alpha > L$, the set $\{n : s_n > \alpha\}$ is finite; otherwise $\sup\{s_n : n > N\} > \alpha$ for all N and hence $L = \limsup s_n \geq \alpha > L$, a contradiction; see Definition 10.6. If $L \neq -\infty$ and $\beta < L$, the set $\{n : s_n > \beta\}$ is infinite; otherwise there exists a positive integer N_0 so that $s_n \leq \beta$ for all $n \geq N_0$, and therefore $\sup\{s_n : n > N\} \leq \beta$ for all $N \geq N_0$. Then $\limsup s_n \leq \beta < L$, a contradiction.

(b) The set $\{n : s_n > \limsup s_n\}$ can be infinite. For example, consider (s_n) where $s_n = \frac{1}{n}$. The set $\{n : s_n < \liminf s_n\}$ can also be infinite; use $s_n = -\frac{1}{n}$.

(c) If $L_0 = \liminf t_n \neq -\infty$, then the set $\{n : t_n < \beta_0\}$ is finite for $\beta_0 < L_0$. If $L_0 \neq \infty$, then the set $\{n : t_n < \alpha_0\}$ is infinite for $\alpha_0 > L_0$.

   This follows from part (a) and Exercise 11.8:

   $$L_0 = \liminf t_n = -\limsup(-t_n) = -L = -\limsup s_n,$$

   where $s_n = -t_n$ and L is as defined in part (a). Now $\beta_0 < L_0$ implies $-\beta_0 > -L_0 = L$, so by part (a),

   $$\{n : t_n < \beta_0\} = \{n : -t_n > -\beta_0\} = \{n : s_n > -\beta_0\} \quad \text{is finite.}$$

   Similarly, $\alpha_0 > L_0$ implies $-\alpha_0 < L$, so

   $$\{n : t_n < \alpha_0\} = \{n : -t_n > -\alpha_0\} = \{n : s_n > -\alpha_0\} \quad \text{is infinite.}$$

(d) If $\liminf s_n < \limsup s_n$, the set

   $$\{n : \liminf s_n \leq s_n \leq \limsup s_n\}$$

   can be empty. Use, for example, $s_n = (-1)^n(1 + \frac{1}{n})$. □

Exercises

12.1 Let (s_n) and (t_n) be sequences and suppose there exists N_0 such that $s_n \leq t_n$ for all $n > N_0$. Show $\liminf s_n \leq \liminf t_n$ and $\limsup s_n \leq \limsup t_n$. *Hint*: Use Definition 10.6 and Exercise 9.9(c).

12.2 Prove $\limsup |s_n| = 0$ if and only if $\lim s_n = 0$.

12.3 Let (s_n) and (t_n) be the following sequences that repeat in cycles of four:

$$(s_n) = (0, 1, 2, 1, 0, 1, 2, 1, 0, 1, 2, 1, 0, 1, 2, 1, 0, \ldots)$$
$$(t_n) = (2, 1, 1, 0, 2, 1, 1, 0, 2, 1, 1, 0, 2, 1, 1, 0, 2, \ldots)$$

Find
(a) $\liminf s_n + \liminf t_n$,   (b) $\liminf(s_n + t_n)$,
(c) $\liminf s_n + \limsup t_n$,   (d) $\limsup(s_n + t_n)$,
(e) $\limsup s_n + \limsup t_n$,   (f) $\liminf(s_n t_n)$,
(g) $\limsup(s_n t_n)$

12.4 Show $\limsup(s_n + t_n) \leq \limsup s_n + \limsup t_n$ for bounded sequences (s_n) and (t_n). *Hint*: First show

$$\sup\{s_n + t_n : n > N\} \leq \sup\{s_n : n > N\} + \sup\{t_n : n > N\}.$$

Then apply Exercise 9.9(c).

12.5 Use Exercises 11.8 and 12.4 to prove

$$\liminf(s_n + t_n) \geq \liminf s_n + \liminf t_n$$

for bounded sequences (s_n) and (t_n).

12.6 Let (s_n) be a bounded sequence, and let k be a nonnegative real number.

   (a) Prove $\limsup(ks_n) = k \cdot \limsup s_n$.

   (b) Do the same for $\liminf$. *Hint*: Use Exercise 11.8.

   (c) What happens in (a) and (b) if $k < 0$?

12.7 Prove if $\limsup s_n = +\infty$ and $k > 0$, then $\limsup(ks_n) = +\infty$.

12.8 Let (s_n) and (t_n) be bounded sequences of nonnegative numbers. Prove $\limsup s_n t_n \leq (\limsup s_n)(\limsup t_n)$.

12.9 **(a)** Prove that if $\lim s_n = +\infty$ and $\liminf t_n > 0$, then $\lim s_n t_n = +\infty$.

   (b) Prove that if $\limsup s_n = +\infty$ and $\liminf t_n > 0$, then $\limsup s_n t_n = +\infty$.

(c) Observe that Exercise 12.7 is the special case of (b) where $t_n = k$ for all $n \in \mathbb{N}$.

12.10 Prove (s_n) is bounded if and only if $\limsup |s_n| < +\infty$.

12.11 Prove the first inequality in Theorem 12.2.

12.12 Let (s_n) be a sequence of nonnegative numbers, and for each n define $\sigma_n = \frac{1}{n}(s_1 + s_2 + \cdots + s_n)$.

(a) Show
$$\liminf s_n \leq \liminf \sigma_n \leq \limsup \sigma_n \leq \limsup s_n.$$

Hint: For the last inequality, show first that $M > N$ implies
$$\sup\{\sigma_n : n > M\} \leq \frac{1}{M}(s_1 + s_2 + \cdots + s_N) + \sup\{s_n : n > N\}.$$

(b) Show that if $\lim s_n$ exists, then $\lim \sigma_n$ exists and $\lim \sigma_n = \lim s_n$.

(c) Give an example where $\lim \sigma_n$ exists, but $\lim s_n$ does not exist.

12.13 Let (s_n) be a bounded sequence in $\mathbb{R}$. Let A be the set of $a \in \mathbb{R}$ such that $\{n \in \mathbb{N} : s_n < a\}$ is finite, i.e., all but finitely many s_n are $\geq a$. Let B be the set of $b \in \mathbb{R}$ such that $\{n \in \mathbb{N} : s_n > b\}$ is finite. Prove $\sup A = \liminf s_n$ and $\inf B = \limsup s_n$.

12.14 Calculate (a) $\lim(n!)^{1/n}$, (b) $\lim \frac{1}{n}(n!)^{1/n}$.

§13 * Some Topological Concepts in Metric Spaces

In this book we are restricting our attention to analysis on $\mathbb{R}$. Accordingly, we have taken full advantage of the order properties of $\mathbb{R}$ and studied such important notions as $\limsup$'s and $\liminf$'s. In §3 we briefly introduced a distance function on $\mathbb{R}$. Most of our analysis could have been based on the notion of distance, in which case it becomes easy and natural to work in a more general setting. For example, analysis on the k-dimensional Euclidean spaces $\mathbb{R}^k$ is important, but these spaces do not have the useful natural ordering that $\mathbb{R}$ has, unless of course $k = 1$.

13.1 Definition.
Let S be a set, and suppose d is a function defined for all pairs (x, y)
of elements from S satisfying
D1. $d(x, x) = 0$ for all $x \in S$ and $d(x, y) > 0$ for distinct x, y in S.
D2. $d(x, y) = d(y, x)$ for all $x, y \in S$.
D3. $d(x, z) \leq d(x, y) + d(y, z)$ for all $x, y, z \in S$.
    Such a function d is called a *distance function* or a *metric* on S.
A *metric space* S is a set S together with a metric on it. Properly
speaking, the metric space is the pair (S, d) since a set S may well
have more than one metric on it; see Exercise 13.1.

Example 1
As in Definition 3.4, let $\operatorname{dist}(a, b) = |a - b|$ for $a, b \in \mathbb{R}$. Then dist is
a metric on $\mathbb{R}$. Note Corollary 3.6 gives D3 in this case. As remarked
there, the inequality

$$\operatorname{dist}(a, c) \leq \operatorname{dist}(a, b) + \operatorname{dist}(b, c)$$

is called the triangle inequality. In fact, for any metric d, property
D3 is called the *triangle inequality*. □

Example 2
The space of all k-tuples

$$\boldsymbol{x} = (x_1, x_2, \ldots, x_k) \quad \text{where} \quad x_j \in \mathbb{R} \quad \text{for} \quad j = 1, 2, \ldots, k,$$

is called k-*dimensional Euclidean space* and written $\mathbb{R}^k$. As noted in
Exercise 13.1, $\mathbb{R}^k$ has several metrics on it. The most familiar metric
is the one that gives the ordinary distance in the plane $\mathbb{R}^2$ or in
3-space $\mathbb{R}^3$:

$$d(\boldsymbol{x}, \boldsymbol{y}) = \left[\sum_{j=1}^{k} (x_j - y_j)^2 \right]^{1/2} \quad \text{for} \quad \boldsymbol{x}, \boldsymbol{y} \in \mathbb{R}^k.$$

[The summation notation $\sum$ is explained in 14.1.] Obviously this
function d satisfies properties D1 and D2. The triangle inequality
D3 is not so obvious. For a proof, see for example [53, §6.1] or
[62, 1.37]. □

13.2 Definition.

A sequence (s_n) in a metric space (S, d) *converges to* s in S if $\lim_{n \to \infty} d(s_n, s) = 0$. A sequence (s_n) in S is a *Cauchy sequence* if for each $\epsilon > 0$ there exists an N such that

$$m, n > N \quad \text{implies} \quad d(s_m, s_n) < \epsilon.$$

The metric space (S, d) is said to be *complete* if every Cauchy sequence in S converges to some element in S.

Since the Completeness Axiom 4.4 deals with least upper bounds, the word "complete" now appears to have two meanings. However, these two uses of the term are very closely related and both reflect the property that the space is complete, i.e., has no gaps. Theorem 10.11 asserts that the metric space $(\mathbb{R}, \text{dist})$ is a complete metric space, and the proof uses the Completeness Axiom 4.4. We could just as well have taken as an axiom the completeness of $(\mathbb{R}, \text{dist})$ as a metric space and proved the least upper bound property in 4.4 as a theorem. We did not do so because the concept of least upper bound in $\mathbb{R}$ seems to us more fundamental than the concept of Cauchy sequence.

We will prove $\mathbb{R}^k$ is complete. But we have a notational problem, since we like subscripts for sequences and for coordinates of points in $\mathbb{R}^k$. When there is a conflict, we will write $(\boldsymbol{x}^{(n)})$ for a sequence instead of $(\boldsymbol{x}_n)$. In this case,

$$\boldsymbol{x}^{(n)} = (x_1^{(n)}, x_2^{(n)}, \ldots, x_k^{(n)}).$$

Unless otherwise specified, *the metric in* $\mathbb{R}^k$ *is always as given in Example 2*.

13.3 Lemma.

A sequence $(\boldsymbol{x}^{(n)})$ in $\mathbb{R}^k$ converges if and only if for each $j = 1, 2, \ldots, k$, the sequence $(x_j^{(n)})$ converges in $\mathbb{R}$. A sequence $(\boldsymbol{x}^{(n)})$ in $\mathbb{R}^k$ is a Cauchy sequence if and only if each sequence $(x_j^{(n)})$ is a Cauchy sequence in $\mathbb{R}$.

Proof

The proof of the first assertion is left to Exercise 13.2(b). For the second assertion, we first observe for $\boldsymbol{x}$, $\boldsymbol{y}$ in $\mathbb{R}^k$ and $j = 1, 2, \ldots, k$,

$$|x_j - y_j| \le d(\boldsymbol{x}, \boldsymbol{y}) \le \sqrt{k} \max\{|x_j - y_j| : j = 1, 2, \ldots, k\}. \qquad (1)$$

Suppose $(\boldsymbol{x}^{(n)})$ is a Cauchy sequence in $\mathbb{R}^k$, and consider fixed j. If $\epsilon > 0$, there exists N such that

$$m, n > N \quad \text{implies} \quad d(\boldsymbol{x}^{(m)}, \boldsymbol{x}^{(n)}) < \epsilon.$$

From the first inequality in (1) we see that

$$m, n > N \quad \text{implies} \quad |x_j^{(m)} - x_j^{(n)}| < \epsilon,$$

so $(x_j^{(n)})$ is a Cauchy sequence in $\mathbb{R}$.

Now suppose each sequence $(x_j^{(n)})$ is a Cauchy sequence in $\mathbb{R}$. Let $\epsilon > 0$. For $j = 1, 2, \ldots, k$, there exist N_j such that

$$m, n > N_j \quad \text{implies} \quad |x_j^{(m)} - x_j^{(n)}| < \frac{\epsilon}{\sqrt{k}}.$$

If $N = \max\{N_1, N_2, \ldots, N_k\}$, then by the second inequality in (1),

$$m, n > N \quad \text{implies} \quad d(\boldsymbol{x}^{(m)}, \boldsymbol{x}^{(n)}) < \epsilon,$$

i.e., $(\boldsymbol{x}^{(n)})$ is a Cauchy sequence in $\mathbb{R}^k$. ■

13.4 Theorem.
Euclidean k-space $\mathbb{R}^k$ is complete.

Proof
Consider a Cauchy sequence $(\boldsymbol{x}^{(n)})$ in $\mathbb{R}^k$. By Lemma 13.3, $(x_j^{(n)})$ is a Cauchy sequence in $\mathbb{R}$ for each j. Hence by Theorem 10.11, $(x_j^{(n)})$ converges to some real number x_j. By Lemma 13.3 again, the sequence $(\boldsymbol{x}^{(n)})$ converges, in fact to $\boldsymbol{x} = (x_1, x_2, \ldots, x_k)$. ■

We now prove the Bolzano-Weierstrass theorem for $\mathbb{R}^k$; compare Theorem 11.5. A set S in $\mathbb{R}^k$ is *bounded* if there exists $M > 0$ such that

$$\max\{|x_j| : j = 1, 2, \ldots, k\} \le M \quad \text{for all} \quad \boldsymbol{x} = (x_1, x_2, \ldots, x_k) \in S.$$

13.5 Bolzano-Weierstrass Theorem.
Every bounded sequence in $\mathbb{R}^k$ has a convergent subsequence.

Proof
Let $(\boldsymbol{x}^{(n)})$ be a bounded sequence in $\mathbb{R}^k$. Then each sequence $(x_j^{(n)})$ is bounded in $\mathbb{R}$. By Theorem 11.5, we may replace $(\boldsymbol{x}^{(n)})$ by a

subsequence such that $(x_1^{(n)})$ converges. By the same theorem, we may replace $(\boldsymbol{x}^{(n)})$ by a subsequence of the subsequence such that $(x_2^{(n)})$ converges. Of course, $(x_1^{(n)})$ still converges by Theorem 11.3. Repeating this argument k times, we obtain a sequence $(\boldsymbol{x}^{(n)})$ so that each sequence $(x_j^{(n)})$ converges, $j = 1, 2, \ldots, k$. This sequence represents a subsequence of the original sequence, and it converges in $\mathbb{R}^k$ by Lemma 13.3. ∎

13.6 Definition.
Let (S, d) be a metric space. Let E be a subset of S. An element $s_0 \in E$ is *interior to* E if for some $r > 0$ we have

$$\{s \in S : d(s, s_0) < r\} \subseteq E.$$

We write E° for the set of points in E that are interior to E. The set E is *open in* S if every point in E is interior to E, i.e., if $E = E^\circ$.

13.7 Discussion.
One can show [Exercise 13.4]
   (i) S is open in S [trivial].
  (ii) The empty set $\varnothing$ is open in S [trivial].
 (iii) The union of *any* collection of open sets is open.
 (iv) The intersection of *finitely many* open sets is again an open set. □

Our study of $\mathbb{R}^k$ and the exercises suggest that metric spaces are fairly general and useful objects. When one is interested in convergence of certain objects [such as points or functions], there is often a metric that assists in the study of the convergence. But sometimes no metric will work and yet there is still some sort of convergence notion. Frequently the appropriate vehicle is what is called a *topology*. This is a set S for which certain subsets are decreed to be *open sets*. In general, all that is required is that the family of open sets satisfies (i)–(iv) above. In particular, the open sets defined by a metric form a topology. We will not pursue this abstract theory. However, because of this abstract theory, concepts that can be defined in terms of open sets [see Definitions 13.8, 13.11, and 22.1] are called *topological*, hence the title of this section.

13.8 Definition.
Let (S, d) be a metric space. A subset E of S is *closed* if its complement $S \setminus E$ is an open set. In other words, E is closed if $E = S \setminus U$ where U is an open set.

Because of (iii) in Discussion 13.7, the intersection of *any* collection of closed sets in closed [Exercise 13.5]. The *closure E^-* of a set E is the intersection of all closed sets containing E. The *boundary* of E is the set $E^- \setminus E^\circ$; points in this set are called *boundary points of E*.

To get a feel for these notions, we state some easy facts and leave the proofs as exercises.

13.9 Proposition.
Let E be a subset of a metric space (S, d).
  (a) *The set E is closed if and only if $E = E^-$.*
  (b) *The set E is closed if and only if it contains the limit of every convergent sequence of points in E.*
  (c) *An element is in E^- if and only if it is the limit of some sequence of points in E.*
  (d) *A point is in the boundary of E if and only if it belongs to the closure of both E and its complement.*

Example 3
In $\mathbb{R}$, open intervals (a, b) are open sets. Closed intervals $[a, b]$ are closed sets. The interior of $[a, b]$ is (a, b). The boundary of both (a, b) and $[a, b]$ is the two-element set $\{a, b\}$.

Every open set in $\mathbb{R}$ is the union of a disjoint sequence of open intervals [Exercise 13.7]. A closed set in $\mathbb{R}$ need not be the union of a disjoint sequence of closed intervals and points; such a set appears in Example 5.

No open interval (a, b) or closed interval $[a, b]$, with $a < b$, can be written as the disjoint union of two or more closed intervals, each having more than one point. This is proved in Theorem 21.11.   □

Example 4
In $\mathbb{R}^k$, open balls $\{x : d(x, x_0) < r\}$ are open sets, and closed balls $\{x : d(x, x_0) \leq r\}$ are closed sets. The boundary of each of these

sets is $\{x : d(x, x_0) = r\}$. In the plane $\mathbb{R}^2$, the sets

$$\{(x_1, x_2) : x_1 > 0\} \quad \text{and} \quad \{(x_1, x_2) : x_1 > 0 \text{ and } x_2 > 0\}$$

are open. If $>$ is replaced by $\geq$, we obtain closed sets. Many sets are neither open nor closed. For example, $[0, 1)$ is neither open nor closed in $\mathbb{R}$, and $\{(x_1, x_2) : x_1 > 0 \text{ and } x_2 \geq 0\}$ is neither open nor closed in $\mathbb{R}^2$. $\qquad\square$

13.10 Theorem.
Let (F_n) be a decreasing sequence $[i.e., F_1 \supseteq F_2 \supseteq \cdots]$ of closed bounded nonempty sets in $\mathbb{R}^k$. Then $F = \cap_{n=1}^{\infty} F_n$ is also closed, bounded and nonempty.

Proof
Clearly F is closed and bounded. It is the nonemptiness that needs proving! For each n, select an element x_n in F_n. By the Bolzano-Weierstrass Theorem 13.5, a subsequence $(x_{n_m})_{m=1}^{\infty}$ of (x_n) converges to some element x_0 in $\mathbb{R}^k$. To show $x_0 \in F$, it suffices to show $x_0 \in F_{n_0}$ with n_0 fixed. If $m \geq n_0$, then $n_m \geq n_0$, so $x_{n_m} \in F_{n_m} \subseteq F_{n_0}$. Hence the sequence $(x_{n_m})_{m=n_0}^{\infty}$ consists of points in F_{n_0} and converges to x_0. Thus x_0 belongs to F_{n_0} by (b) of Proposition 13.9. $\qquad\blacksquare$

Example 5
Here is a famous nonempty closed set in $\mathbb{R}$ called the *Cantor set*. Pictorially, $F = \cap_{n=1}^{\infty} F_n$ where F_n are sketched in Fig. 13.1. The Cantor set has some remarkable properties. The sum of the lengths of the intervals comprising F_n is $(\frac{2}{3})^{n-1}$ and this tends to 0 as $n \to \infty$. Yet the intersection F is so large that it cannot be written as a sequence; in set-theoretic terms it is "uncountable." The interior of F is the empty set, so F is equal to its boundary. For more details, see [62, 2.44], or [31, 6.62]. $\qquad\square$

13.11 Definition.
Let (S, d) be a metric space. A family $\mathcal{U}$ of open sets is said to be an *open cover for a set* E if each point of E belongs to at least one set in $\mathcal{U}$, i.e.,

$$E \subseteq \bigcup \{U : U \in \mathcal{U}\}.$$

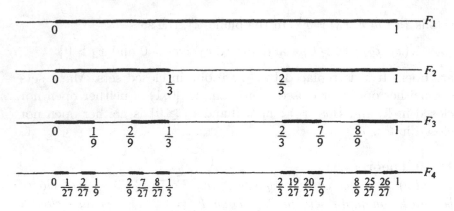

FIGURE 13.1

A *subcover* of $\mathcal{U}$ is any subfamily of $\mathcal{U}$ that also covers E. A cover or subcover is *finite* if it contains only finitely many sets; the sets themselves may be infinite.

A set E is *compact* if every open cover of E has a finite subcover of E.

This rather abstract definition is very important in advanced analysis; see, for example, [30]. In $\mathbb{R}^k$, compact sets are nicely characterized, as follows.

13.12 Heine-Borel Theorem.
A subset E of $\mathbb{R}^k$ is compact if and only if it is closed and bounded.

Proof
Suppose E is compact. For each $m \in \mathbb{N}$, let U_m consist of all $\boldsymbol{x}$ in $\mathbb{R}^k$ such that

$$\max\{|x_j| : j = 1, 2, \ldots, k\} < m.$$

The family $\mathcal{U} = \{U_m : m \in \mathbb{N}\}$ is an open cover of E [it covers $\mathbb{R}^k$!], so a finite subfamily of $\mathcal{U}$ covers E. If U_{m_0} is the largest member of the subfamily, then $E \subseteq U_{m_0}$. It follows that E is bounded. To show E is closed, consider any point $\boldsymbol{x}_0$ in $\mathbb{R}^k \setminus E$. For $m \in \mathbb{N}$, let

$$V_m = \left\{ \boldsymbol{x} \in \mathbb{R}^k : d(\boldsymbol{x}, \boldsymbol{x}_0) > \frac{1}{m} \right\}.$$

Then each V_m is open in $\mathbb{R}^k$ and $\mathcal{V} = \{V_m : m \in \mathbb{N}\}$ covers E since $\cup_{m=1}^{\infty} V_m = \mathbb{R}^k \setminus \{x_0\}$. Since E can be covered by finitely many V_m, for some m_0 we have

$$E \subset \left\{ x \in \mathbb{R}^k : d(x, x_0) > \frac{1}{m_0} \right\}.$$

Thus $\{x \in \mathbb{R}^k : d(x, x_0) < \frac{1}{m_0}\} \subseteq \mathbb{R}^k \setminus E$, so that x_0 is interior to $\mathbb{R}^k \setminus E$. Since x_0 in $\mathbb{R}^k \setminus E$ was arbitrary, $\mathbb{R}^k \setminus E$ is an open set. Hence E is a closed set.

Now suppose E is closed and bounded. Since E is bounded, E is a subset of some set F having the form

$$F = \{x \in \mathbb{R}^k : |x_j| \le m \quad \text{for} \quad j = 1, 2, \ldots, k\}.$$

As noted in Exercise 13.12, it suffices to prove F is compact. We do so in the next proposition after some preparation. ∎

The set F in the last proof is a *k-cell* because it has the following form. There exist closed intervals $[a_1, b_1], [a_2, b_2], \ldots, [a_k, b_k]$ so that

$$F = \{x \in \mathbb{R}^k : x_j \in [a_j, b_j] \quad \text{for} \quad j = 1, 2, \ldots, k\},$$

which is sometimes written as

$$F = [a_1, b_1] \times [a_2, b_2] \times \cdots \times [a_k, b_k],$$

so it is a k-dimensional box in $\mathbb{R}^k$. Thus a 2-cell in $\mathbb{R}^2$ is a closed rectangle. A 3-cell in $\mathbb{R}^3$ is called a "rectangular parallelepiped." The *diameter* of F is

$$\delta = \left[\sum_{j=1}^{k} (b_j - a_j)^2 \right]^{1/2};$$

that is, $\delta = \sup\{d(x, y) : x, y \in F\}$. Using midpoints $c_j = \frac{1}{2}(a_j + b_j)$ of $[a_j, b_j]$, we see that F is a union of 2^k k-cells each having diameter $\frac{\delta}{2}$. If this remark is not clear, consider first the cases $k = 2$ and $k = 3$.

13.13 Proposition.
Every k-cell F in $\mathbb{R}^k$ is compact.

Proof
Assume F is not compact. Then there exists an open cover $\mathcal{U}$ of F, no finite subfamily of which covers F. Let δ denote the diameter of F.

As noted above, F is a union of 2^k k-cells having diameter $\frac{\delta}{2}$. At least one of these 2^k k-cells, which we denote by F_1, cannot be covered by finitely many sets from $\mathcal{U}$. Likewise, F_1 contains a k-cell F_2 of diameter $\frac{\delta}{4}$ which cannot be covered by finitely many sets from $\mathcal{U}$. Continuing in this fashion, we obtain a sequence (F_n) of k-cells such that

$$F_1 \supseteq F_2 \supseteq F_3 \supseteq \cdots; \tag{1}$$

$$F_n \text{ has diameter } \delta \cdot 2^{-n}; \tag{2}$$

$$F_n \text{ cannot be covered by finitely many sets from } \mathcal{U}. \tag{3}$$

By Theorem 13.10, the intersection $\bigcap_{n=1}^{\infty} F_n$ contains a point x_0. This point belongs to some set U_0 in $\mathcal{U}$. Since U_0 is open, there exists $r > 0$ so that

$$\{x \in \mathbb{R}^k : d(x, x_0) < r\} \subseteq U_0.$$

It follows that $F_n \subseteq U_0$ provided $\delta \cdot 2^{-n} < r$, but this contradicts (3) in a dramatic way. ∎

Since $\mathbb{R} = \mathbb{R}^1$, the preceding results apply to $\mathbb{R}$.

Example 6

Let E be a nonempty subset of a metric space (S, d). Consider the function $d(E, x) = \inf\{d(y, x) : y \in E\}$ for $x \in S$. This function satisfies $|d(E, x_1) - d(E, x_2)| \le d(x_1, x_2)$ for x_1, x_2 in S.

We show that if E is compact and if $E \subseteq U$ for some open subset U of S, then for some $\delta > 0$ we have

$$\{x \in S : d(E, x) < \delta\} \subseteq U. \tag{1}$$

For each $x \in E$, we have

$$\{y \in S : d(y, x) < r_x\} \subseteq U \quad \text{for some} \quad r_x > 0. \tag{2}$$

The open balls $\{y \in S : d(y, x) < r_x/2\}$ cover E, so a finite subfamily also covers E. I.e., there are $x_1, \ldots, x_n$ in E so that

$$E \subseteq \bigcup_{k=1}^{n} \left\{y \in S : d(y, x_k) < \frac{r_k}{2}\right\},$$

where we write r_k for r_{x_k}. Now let $\delta = \frac{1}{2}\min\{r_1, \ldots, r_n\}$. To prove (1), consider $x \in S$ and suppose $d(E, x) < \delta$. Then for some

$y \in E$, we have $d(y,x) < \delta$. Moreover, $d(y,x_k) < \frac{r_k}{2}$ for some $k \in \{1,2,\ldots,n\}$. Therefore, for this k we have

$$d(x,x_k) \le d(x,y) + d(y,x_k) < \delta + \frac{r_k}{2} \le \frac{r_k}{2} + \frac{r_k}{2} = r_k.$$

Thus, by (2) applied to $x = x_k$, we see that x belongs to U. Hence (1) holds. $\square$

Exercises

13.1 For points $\boldsymbol{x}$, $\boldsymbol{y}$ in $\mathbb{R}^k$, let

$$d_1(\boldsymbol{x},\boldsymbol{y}) = \max\{|x_j - y_j| : j = 1,2,\ldots,k\}$$

and

$$d_2(\boldsymbol{x},\boldsymbol{y}) = \sum_{j=1}^{k} |x_j - y_j|.$$

(a) Show d_1 and d_2 are metrics for $\mathbb{R}^k$.

(b) Show d_1 and d_2 are complete metrics on $\mathbb{R}^k$.

13.2 (a) Prove (1) in Lemma 13.3.

(b) Prove the first assertion in Lemma 13.3.

13.3 Let B be the set of all bounded sequences $\boldsymbol{x} = (x_1, x_2, \ldots)$, and define $d(\boldsymbol{x},\boldsymbol{y}) = \sup\{|x_j - y_j| : j = 1,2,\ldots\}$.

(a) Show d is a metric for B.

(b) Does $d^*(\boldsymbol{x},\boldsymbol{y}) = \sum_{j=1}^{\infty} |x_j - y_j|$ define a metric for B?

13.4 Prove (iii) and (iv) in Discussion 13.7.

13.5 (a) Verify one of DeMorgan's Laws for sets:

$$\bigcap\{S \setminus U : U \in \mathcal{U}\} = S \setminus \bigcup\{U : U \in \mathcal{U}\}.$$

(b) Show that the intersection of any collection of closed sets is a closed set.

13.6 Prove Proposition 13.9.

13.7 Show that every open set in $\mathbb{R}$ is the disjoint union of a finite or infinite sequence of open intervals.

13.8 **(a)** Verify the assertions in the first paragraph of Example 3.

   (b) Verify the assertions in Example 4.

13.9 Find the closures of the following sets:

   (a) $\{\frac{1}{n} : n \in \mathbb{N}\}$,

   (b) $\mathbb{Q}$, the set of rational numbers,

   (c) $\{r \in \mathbb{Q} : r^2 < 2\}$.

13.10 Show that the interior of each of the following sets is the empty set.

   (a) $\{\frac{1}{n} : n \in \mathbb{N}\}$,

   (b) $\mathbb{Q}$, the set of rational numbers,

   (c) The Cantor set in Example 5.

13.11 Let E be a subset of $\mathbb{R}^k$. Show that E is compact if and only if every sequence in E has a subsequence converging to a point *in* E.

13.12 Let (S, d) be any metric space.

   (a) Show that if E is a closed subset of a compact set F, then E is also compact.

   (b) Show that the finite union of compact sets in S is compact.

13.13 Let E be a compact nonempty subset of $\mathbb{R}$. Show $\sup E$ and $\inf E$ belong to E.

13.14 Let E be a compact nonempty subset of $\mathbb{R}^k$, and let $\delta = \sup\{d(\boldsymbol{x}, \boldsymbol{y}) : \boldsymbol{x}, \boldsymbol{y} \in E\}$. Show E contains points $\boldsymbol{x}_0$, $\boldsymbol{y}_0$ such that $d(\boldsymbol{x}_0, \boldsymbol{y}_0) = \delta$.

13.15 Let (B, d) be as in Exercise 13.3, and let F consist of all $\boldsymbol{x} \in B$ such that $\sup\{|x_j| : j = 1, 2, \ldots\} \leq 1$.

   (a) Show F is closed and bounded. [A set F in a metric space (S, d) is *bounded* if there exist $s_0 \in S$ and $r > 0$ such that $F \subseteq \{s \in S : d(s, s_0) \leq r\}$.]

   (b) Show F is not compact. *Hint:* For each $\boldsymbol{x}$ in F, let $U(\boldsymbol{x}) = \{\boldsymbol{y} \in B : d(\boldsymbol{y}, \boldsymbol{x}) < 1\}$, and consider the cover $\mathcal{U}$ of F consisting of all $U(\boldsymbol{x})$. For each $n \in \mathbb{N}$, let $\boldsymbol{x}^{(n)}$ be defined so that $x_n^{(n)} = -1$ and $x_j^{(n)} = 1$ for $j \neq n$. Show that distinct $\boldsymbol{x}^{(n)}$ cannot belong to the same member of $\mathcal{U}$.

§14 Series

Our thorough treatment of sequences allows us to now quickly obtain the basic properties of infinite series.

14.1 Summation Notation.

The notation $\sum_{k=m}^{n} a_k$ is shorthand for the sum $a_m + a_{m+1} + \cdots + a_n$. The symbol "$\sum$" instructs us to sum and the decorations "$k = m$" and "n" tell us to sum the summands obtained by successively substituting $m, m+1, \ldots, n$ for k. For example, $\sum_{k=2}^{5} \frac{1}{k^2+k}$ is shorthand for

$$\frac{1}{2^2+2} + \frac{1}{3^2+3} + \frac{1}{4^2+4} + \frac{1}{5^2+5} = \frac{1}{6} + \frac{1}{12} + \frac{1}{20} + \frac{1}{30}$$

and $\sum_{k=0}^{n} 2^{-k}$ is shorthand for $1 + \frac{1}{2} + \frac{1}{4} + \cdots + \frac{1}{2^n}$.

The symbol $\sum_{n=m}^{\infty} a_n$ is shorthand for $a_m + a_{m+1} + a_{m+2} + \cdots$, although we have not yet assigned meaning to such an infinite sum. We now do so. $\qquad \square$

14.2 Infinite Series.

To assign meaning to $\sum_{n=m}^{\infty} a_n$, we consider the sequences $(s_n)_{n=m}^{\infty}$ of *partial sums*:

$$s_n = a_m + a_{m+1} + \cdots + a_n = \sum_{k=m}^{n} a_k.$$

The infinite series $\sum_{n=m}^{\infty} a_n$ is said to *converge* provided the sequence (s_n) of partial sums converges to a real number S, in which case we define $\sum_{n=m}^{\infty} a_n = S$. Thus

$$\sum_{n=m}^{\infty} a_n = S \quad \text{means} \quad \lim s_n = S \quad \text{or} \quad \lim_{n \to \infty} \left(\sum_{k=m}^{n} a_k \right) = S.$$

A series that does not converge is said to *diverge*. We say that $\sum_{n=m}^{\infty} a_n$ *diverges to* $+\infty$ and we write $\sum_{n=m}^{\infty} a_n = +\infty$ provided $\lim s_n = +\infty$; a similar remark applies to $-\infty$. The symbol $\sum_{n=m}^{\infty} a_n$ has no meaning unless the series converges, or diverges to $+\infty$ or $-\infty$. Often we will be concerned with properties of infinite series but not their exact values or precisely where the summation begins, in which case we may write $\sum a_n$ rather than $\sum_{n=m}^{\infty} a_n$.

If the terms a_n of an infinite series $\sum a_n$ are all nonnegative, then the partial sums (s_n) form an increasing sequence, so Theorems 10.2 and 10.4 show that $\sum a_n$ either converges, or diverges to $+\infty$. In particular, $\sum |a_n|$ is meaningful for any sequence (a_n) whatever. The series $\sum a_n$ is said to *converge absolutely* or to be *absolutely convergent* if $\sum |a_n|$ converges. Absolutely convergent series are convergent, as we shall see in Corollary 14.7. □

Example 1

A series of the form $\sum_{n=0}^{\infty} ar^n$ for constants a and r is called a *geometric series*. These are the easiest series to sum. For $r \neq 1$, the partial sums s_n are given by

$$\sum_{k=0}^{n} ar^k = a\frac{1 - r^{n+1}}{1 - r}. \tag{1}$$

This identity can be verified by mathematical induction or by multiplying both sides by $1 - r$, in which case the right hand side equals $a - ar^{n+1}$ and the left side becomes

$$(1 - r) \sum_{k=0}^{n} ar^k = \sum_{k=0}^{n} ar^k - \sum_{k=0}^{n} ar^{k+1}$$
$$= a + ar + ar^2 + \cdots + ar^n$$
$$\quad -(ar + ar^2 + \cdots + ar^n + ar^{n+1})$$
$$= a - ar^{n+1}.$$

For $|r| < 1$, we have $\lim_{n \to \infty} r^{n+1} = 0$ by Theorem 9.7(b) on page 48, so from (1) we have $\lim_{n \to \infty} s_n = \frac{a}{1-r}$. This proves

$$\sum_{n=0}^{\infty} ar^n = \frac{a}{1 - r} \quad \text{if} \quad |r| < 1. \tag{2}$$

If $a \neq 0$ and $|r| \geq 1$, then the sequence (ar^n) does not converge to 0, so the series $\sum ar^n$ diverges by Corollary 14.5 below. □

Example 2

Formula (2) of Example 1 and the next result are very important and both should be used whenever possible, even though we will not prove (1) below until the next section. Consider a fixed positive real number p. Then

$$\sum_{n=1}^{\infty} \frac{1}{n^p} \quad \text{converges if and only if} \quad p > 1. \tag{1}$$

In particular, for $p \le 1$, we can write $\sum 1/n^p = +\infty$. The exact values of the series for $p > 1$ are not easy to determine. Here are some remarkable formulas that can be shown by techniques [Fourier series or complex variables, to name two possibilities] that will not be covered in this text.

$$\sum_{n=1}^{\infty} \frac{1}{n^2} = \frac{\pi^2}{6} = 1.6449\cdots, \tag{2}$$

$$\sum_{n=1}^{\infty} \frac{1}{n^4} = \frac{\pi^4}{90} = 1.0823\cdots. \tag{3}$$

Similar formulas hold for $\sum_{n=1}^{\infty} \frac{1}{n^p}$ when p is any even integer, but no such elegant formulas are known for p odd. In particular, no such formula is known for $\sum_{n=1}^{\infty} \frac{1}{n^3}$ though of course this series converges and can be approximated as closely as desired. $\qquad\square$

It is worth emphasizing that it is often easier to prove limits exist or series converge than to determine their exact values. In the next section we will show without much difficulty that $\sum \frac{1}{n^p}$ converges for all $p > 1$, but it is a lot harder to show the sum is $\frac{\pi^2}{6}$ when $p = 2$ and no one knows exactly what the sum is for $p = 3$.

14.3 Definition.
We say a series $\sum a_n$ satisfies the *Cauchy criterion* if its sequence (s_n) of partial sums is a Cauchy sequence [see Definition 10.8]:

for each $\epsilon > 0$ there exists a number N such that
$$m, n > N \quad \text{implies} \quad |s_n - s_m| < \epsilon. \tag{1}$$

Nothing is lost in this definition if we impose the restriction $n > m$. Moreover, it is only a notational matter to work with $m - 1$ where $m \le n$ instead of m where $m < n$. Therefore (1) is equivalent to

for each $\epsilon > 0$ there exists a number N such that
$$n \ge m > N \quad \text{implies} \quad |s_n - s_{m-1}| < \epsilon. \tag{2}$$

Since $s_n - s_{m-1} = \sum_{k=m}^{n} a_k$, condition (2) can be rewritten

for each $\epsilon > 0$ there exists a number N such that

$$n \geq m > N \quad \text{implies} \quad \left| \sum_{k=m}^{n} a_k \right| < \epsilon. \tag{3}$$

We will usually use version (3) of the *Cauchy criterion*. Theorem 10.11 implies the following.

14.4 Theorem.
A series converges if and only if it satisfies the Cauchy criterion.

14.5 Corollary.
If a series $\sum a_n$ converges, then $\lim a_n = 0$.

Proof
Since the series converges, (3) in Definition 14.3 holds. In particular, (3) in 14.3 holds for $n = m$; i.e., for each $\epsilon > 0$ there exists a number N such that $n > N$ implies $|a_n| < \epsilon$. Thus $\lim a_n = 0$. ∎

The converse of Corollary 14.5 does not hold as the example $\sum 1/n = +\infty$ shows.

We next give several tests to assist us in determining whether a series converges. The first test is elementary but useful.

14.6 Comparison Test.
Let $\sum a_n$ be a series where $a_n \geq 0$ for all n.
(i) If $\sum a_n$ converges and $|b_n| \leq a_n$ for all n, then $\sum b_n$ converges.
(ii) If $\sum a_n = +\infty$ and $b_n \geq a_n$ for all n, then $\sum b_n = +\infty$.

Proof
(i) For $n \geq m$ we have

$$\left| \sum_{k=m}^{n} b_k \right| \leq \sum_{k=m}^{n} |b_k| \leq \sum_{k=m}^{n} a_k; \tag{1}$$

the first inequality follows from the triangle inequality [Exercise 3.6(b)]. Since $\sum a_n$ converges, it satisfies the Cauchy criterion 14.3(1). It follows from (1) that $\sum b_n$ also satisfies the Cauchy criterion, and hence $\sum b_n$ converges.

(ii) Let (s_n) and (t_n) be the sequences of partial sums for $\sum a_n$ and $\sum b_n$, respectively. Since $b_n \geq a_n$ for all n, we obviously have $t_n \geq s_n$ for all n. Since $\lim s_n = +\infty$, we conclude $\lim t_n = +\infty$, i.e., $\sum b_n = +\infty$. ∎

14.7 Corollary.
Absolutely convergent series are convergent.[9]

Proof
Suppose $\sum b_n$ is absolutely convergent. This means $\sum a_n$ converges where $a_n = |b_n|$ for all n. Then $|b_n| \leq a_n$ trivially, so $\sum b_n$ converges by 14.6(i). ∎

We next state the Ratio Test which is popular because it is often easy to use. But it has defects: It isn't as general as the Root Test. Moreover, an important result concerning the radius of convergence of a power series uses the Root Test. Finally, the Ratio Test is worthless if some of the a_n's equal 0. To review lim sup's and lim inf's, see Definition 10.6, Theorems 10.7 and 11.8, and §12.

14.8 Ratio Test.
A series $\sum a_n$ of nonzero terms
   (i) *converges absolutely if* $\limsup |a_{n+1}/a_n| < 1$,
  (ii) *diverges if* $\liminf |a_{n+1}/a_n| > 1$.
 (iii) *Otherwise* $\liminf |a_{n+1}/a_n| \leq 1 \leq \limsup |a_{n+1}/a_n|$ *and the test gives no information.*

We give the proof after the proof of the Root Test.

Remember that if $\lim |a_{n+1}/a_n|$ exists, then it is equal to both $\limsup |a_{n+1}/a_n|$ and $\liminf |a_{n+1}/a_n|$ and hence the Ratio Test will give information unless, of course, the limit $\lim |a_{n+1}/a_n|$ equals 1.

14.9 Root Test.
Let $\sum a_n$ be a series and let $\alpha = \limsup |a_n|^{1/n}$. The series $\sum a_n$
   (i) *converges absolutely if* $\alpha < 1$,
  (ii) *diverges if* $\alpha > 1$.

---

[9]As noted in [35], the proofs of this corollary and the Alternating Series Theorem 15.3 use the completeness of $\mathbb{R}$.

(iii) *Otherwise $\alpha = 1$ and the test gives no information.*

Proof

   (i) Suppose $\alpha < 1$, and select $\epsilon > 0$ so that $\alpha + \epsilon < 1$. Then by Definition 10.6 there is a positive integer N such that

$$\alpha - \epsilon < \sup\{|a_n|^{1/n} : n > N\} < \alpha + \epsilon.$$

In particular, we have $|a_n|^{1/n} < \alpha + \epsilon$ for $n > N$, so

$$|a_n| < (\alpha + \epsilon)^n \quad \text{for} \quad n > N.$$

Since $0 < \alpha + \epsilon < 1$, the geometric series $\sum_{n=N+1}^{\infty}(\alpha + \epsilon)^n$ converges and the Comparison Test shows the series $\sum_{n=N+1}^{\infty} a_n$ also converges. Then clearly $\sum a_n$ converges; see Exercise 14.9.

   (ii) If $\alpha > 1$, then by Theorem 11.7 a subsequence of $|a_n|^{1/n}$ has limit $\alpha > 1$. It follows that $|a_n| > 1$ for infinitely many choices of n. In particular, the sequence (a_n) cannot possibly converge to 0, so the series $\sum a_n$ cannot converge by Corollary 14.5.

   (iii) For each of the series $\sum \frac{1}{n}$ and $\sum \frac{1}{n^2}$, α turns out to equal 1 as can be seen by applying Theorem 9.7(c) on page 48. Since $\sum \frac{1}{n}$ diverges and $\sum \frac{1}{n^2}$ converges, the equality $\alpha = 1$ does not guarantee either convergence or divergence of the series. ∎

Proof of the Ratio Test

Let $\alpha = \limsup |a_n|^{1/n}$. By Theorem 12.2 we have

$$\liminf \left|\frac{a_{n+1}}{a_n}\right| \leq \alpha \leq \limsup \left|\frac{a_{n+1}}{a_n}\right|. \tag{1}$$

If $\limsup |a_{n+1}/a_n| < 1$, then $\alpha < 1$ and the series converges by the Root Test. If $\liminf |a_{n+1}/a_n| > 1$, then $\alpha > 1$ and the series diverges by the Root Test. Assertion 14.8(iii) is verified by again examining the series $\sum 1/n$ and $\sum 1/n^2$. ∎

Inequality (1) in the proof of the Ratio Test shows that the Root Test is superior to the Ratio Test in the following sense: Whenever the Root Test gives no information [i.e., $\alpha = 1$] the Ratio Test will surely also give no information. On the other hand, Example 8 below gives a series for which the Ratio Test gives no information but which converges by the Root Test. Nevertheless, the tests usually fail together as the next remark shows.

14.10 Remark.
If the terms a_n are nonzero and if $\lim |a_{n+1}/a_n| = 1$, then $\alpha = \limsup |a_n|^{1/n} = 1$ by Corollary 12.3, so neither the Ratio Test nor the Root Test gives information concerning the convergence of $\sum a_n$.

We have three tests for convergence of a series [Comparison, Ratio, Root], and we will obtain two more in the next section. There is no clearcut strategy advising us which test to try first. However, if the form of a given series $\sum a_n$ does not suggest a particular strategy, and if the ratios a_{n+1}/a_n are easy to calculate, one may as well try the Ratio Test first.

Example 3
Consider the series

$$\sum_{n=2}^{\infty} \left(-\frac{1}{3}\right)^n = \frac{1}{9} - \frac{1}{27} + \frac{1}{81} - \frac{1}{243} + \cdots. \tag{1}$$

This is a geometric series and has the form $\sum_{n=0}^{\infty} ar^n$ if we write it as $(1/9) \sum_{n=0}^{\infty} (-1/3)^n$. Here $a = 1/9$ and $r = -1/3$, so by (2) of Example 1 the sum is $(1/9)/[1 - (-1/3)] = 1/12$.

The series (1) can also be shown to converge by the Comparison Test, since $\sum 1/3^n$ converges by the Ratio Test or by the Root Test. In fact, if $a_n = (-1/3)^n$, then $\lim |a_{n+1}/a_n| = \limsup |a_n|^{1/n} = 1/3$. Of course, none of these tests will give us the exact value of the series (1). □

Example 4
Consider the series

$$\sum \frac{n}{n^2 + 3}. \tag{1}$$

If $a_n = \dfrac{n}{n^2 + 3}$, then

$$\frac{a_{n+1}}{a_n} = \frac{n+1}{(n+1)^2 + 3} \cdot \frac{n^2 + 3}{n} = \frac{n+1}{n} \cdot \frac{n^2 + 3}{n^2 + 2n + 4},$$

so $\lim |a_{n+1}/a_n| = 1$. As noted in 14.10, neither the Ratio Test nor the Root Test gives any information in this case. Before trying the Comparison Test we need to decide whether we *believe* the series converges or not. Since a_n is approximately $1/n$ for large n and since $\sum(1/n)$ diverges, we expect the series (1) to diverge. Now

$$\frac{n}{n^2 + 3} \geq \frac{n}{n^2 + 3n^2} = \frac{n}{4n^2} = \frac{1}{4n}.$$

Since $\sum(1/n)$ diverges, $\sum(1/4n)$ also diverges [its partial sums are $s_n/4$ where $s_n = \sum_{k=1}^{n}(1/k)$], so (1) diverges by the Comparison Test. □

Example 5
Consider the series

$$\sum \frac{1}{n^2 + 1}. \tag{1}$$

As the reader should check, neither the Ratio Rest nor the Root Test gives any information. The nth term is approximately $\frac{1}{n^2}$ and in fact $\frac{1}{n^2+1} \leq \frac{1}{n^2}$. Since $\sum \frac{1}{n^2}$ converges, the series (1) converges by the Comparison Test. □

Example 6
Consider the series

$$\sum \frac{n}{3^n}. \tag{1}$$

If $a_n = n/3^n$, then $a_{n+1}/a_n = (n+1)/(3n)$, so $\lim |a_{n+1}/a_n| = 1/3$. Hence the series (1) converges by the Ratio Test. In this case, applying the Root Test is not much more difficult provided we recall $\lim n^{1/n} = 1$. It is also possible to show (1) converges by comparing it with a suitable geometric series. □

Example 7
Consider the series

$$\sum a_n \quad \text{where} \quad a_n = \left[\frac{2}{(-1)^n - 3}\right]^n. \tag{1}$$

The form of a_n suggests the Root Test. Since $|a_n|^{1/n} = 1$ for even n and $|a_n|^{1/n} = 1/2$ for odd n, we have $\alpha = \limsup |a_n|^{1/n} = 1$.

So the Root Test gives no information, and the Ratio Test cannot help either. On the other hand, if we had been alert, we would have observed $a_n = 1$ for even n, so (a_n) cannot converge to 0. Therefore the series (1) diverges by Corollary 14.5. $\square$

Example 8
Consider the series

$$\sum_{n=0}^{\infty} 2^{(-1)^n - n} = 2 + \frac{1}{4} + \frac{1}{2} + \frac{1}{16} + \frac{1}{8} + \frac{1}{64} + \cdots. \qquad (1)$$

Let $a_n = 2^{(-1)^n - n}$. Since $a_n \leq \frac{1}{2^{n-1}}$ for all n, we can quickly conclude the series converges by the Comparison Test. But our real interest in this series is that it illustrates the difference between the Ratio Test and the Root Test. Since $a_{n+1}/a_n = 1/8$ for even n and $a_{n+1}/a_n = 2$ for odd n, we have

$$\frac{1}{8} = \liminf \left| \frac{a_{n+1}}{a_n} \right| < 1 < \limsup \left| \frac{a_{n+1}}{a_n} \right| = 2.$$

Hence the Ratio Test gives no information.

Note that $(a_n)^{1/n} = 2^{\frac{1}{n} - 1}$ for even n and $(a_n)^{1/n} = 2^{-\frac{1}{n} - 1}$ for odd n. Since $\lim 2^{\frac{1}{n}} = \lim 2^{-\frac{1}{n}} = 1$ by Theorem 9.7(d) on page 48, we conclude $\lim(a_n)^{1/n} = \frac{1}{2}$. Therefore $\alpha = \limsup(a_n)^{1/n} = \frac{1}{2} < 1$ and the series (1) converges by the Root Test. $\square$

Example 9
Consider the series

$$\sum_{n=1}^{\infty} \frac{(-1)^{n+1}}{\sqrt{n}} = 1 - \frac{1}{\sqrt{2}} + \frac{1}{\sqrt{3}} - \frac{1}{2} + \frac{1}{\sqrt{5}} - \cdots. \qquad (1)$$

Since $\lim \sqrt{n/(n+1)} = 1$, neither the Ratio Test nor the Root Test gives any information. Since $\sum \frac{1}{\sqrt{n}}$ diverges, we will not be able to use the Comparison Test 14.6(i) to show (1) converges. Since the terms of the series (1) are not all nonnegative, we will not be able to use the Comparison Test 14.6(ii) to show (1) diverges. It turns out that this series converges by the Alternating Series Test 15.3 which we have deferred to the next section. $\square$

Exercises

14.1 Determine which of the following series converge. Justify your answers.

(a) $\sum \frac{n^4}{2^n}$

(b) $\sum \frac{2^n}{n!}$

(c) $\sum \frac{n^2}{3^n}$

(d) $\sum \frac{n!}{n^4+3}$

(e) $\sum \frac{\cos^2 n}{n^2}$

(f) $\sum_{n=2}^{\infty} \frac{1}{\log n}$

14.2 Repeat Exercise 14.1 for the following.

(a) $\sum \frac{n-1}{n^2}$

(b) $\sum (-1)^n$

(c) $\sum \frac{3n}{n^3}$

(d) $\sum \frac{n^3}{3^n}$

(e) $\sum \frac{n^2}{n!}$

(f) $\sum \frac{1}{n^n}$

(g) $\sum \frac{n}{2^n}$

14.3 Repeat Exercise 14.1 for the following.

(a) $\sum \frac{1}{\sqrt{n!}}$

(b) $\sum \frac{2+\cos n}{3^n}$

(c) $\sum \frac{1}{2^n+n}$

(d) $\sum (\frac{1}{2})^n (50 + \frac{2}{n})$

(e) $\sum \sin(\frac{n\pi}{9})$

(f) $\sum \frac{(100)^n}{n!}$

14.4 Repeat Exercise 14.1 for the following.

(a) $\sum_{n=2}^{\infty} \frac{1}{[n+(-1)^n]^2}$

(b) $\sum [\sqrt{n+1} - \sqrt{n}]$

(c) $\sum \frac{n!}{n^n}$

14.5 Suppose $\sum a_n = A$ and $\sum b_n = B$ where A and B are real numbers. Use limit theorems from §9 to quickly prove the following.

(a) $\sum (a_n + b_n) = A + B$.

(b) $\sum k a_n = kA$ for $k \in \mathbb{R}$.

(c) Is $\sum a_n b_n = AB$ a reasonable conjecture? Discuss.

14.6 (a) Prove that if $\sum |a_n|$ converges and (b_n) is a bounded sequence, then $\sum a_n b_n$ converges. *Hint*: Use Theorem 14.4.

(b) Observe that Corollary 14.7 is a special case of part (a).

14.7 Prove that if $\sum a_n$ is a convergent series of nonnegative numbers and $p > 1$, then $\sum a_n^p$ converges.

14.8 Show that if $\sum a_n$ and $\sum b_n$ are convergent series of nonnegative numbers, then $\sum \sqrt{a_n b_n}$ converges. *Hint*: Show $\sqrt{a_n b_n} \leq a_n + b_n$ for all n.

14.9 The convergence of a series does not depend on any finite number of the terms, though of course the value of the limit does. More precisely, consider series $\sum a_n$ and $\sum b_n$ and suppose the set $\{n \in \mathbb{N} : a_n \neq b_n\}$

is finite. Then the series both converge or else they both diverge. Prove this. *Hint*: This is almost obvious from Theorem 14.4.

14.10 Find a series $\sum a_n$ which diverges by the Root Test but for which the Ratio Test gives no information. Compare Example 8.

14.11 Let (a_n) be a sequence of nonzero real numbers such that the sequence $\left(\frac{a_{n+1}}{a_n}\right)$ of ratios is a constant sequence. Show $\sum a_n$ is a geometric series.

14.12 Let $(a_n)_{n\in\mathbb{N}}$ be a sequence such that $\liminf |a_n| = 0$. Prove there is a subsequence $(a_{n_k})_{k\in\mathbb{N}}$ such that $\sum_{k=1}^{\infty} a_{n_k}$ converges.

14.13 We have seen that it is often a lot harder to find the value of an infinite sum than to show it exists. Here are some sums that can be handled.

(a) Calculate $\sum_{n=1}^{\infty}\left(\frac{2}{3}\right)^n$ and $\sum_{n=1}^{\infty}\left(-\frac{2}{3}\right)^n$.

(b) Prove $\sum_{n=1}^{\infty} \frac{1}{n(n+1)} = 1$. *Hint*: Note that $\sum_{k=1}^{n} \frac{1}{k(k+1)} = \sum_{k=1}^{n}\left[\frac{1}{k} - \frac{1}{k+1}\right]$.

(c) Prove $\sum_{n=1}^{\infty} \frac{n-1}{2^{n+1}} = \frac{1}{2}$. *Hint*: Note $\frac{k-1}{2^{k+1}} = \frac{k}{2^k} - \frac{k+1}{2^{k+1}}$.

(d) Use (c) to calculate $\sum_{n=1}^{\infty} \frac{n}{2^n}$.

14.14 Prove $\sum_{n=1}^{\infty} \frac{1}{n}$ diverges by comparing with the series $\sum_{n=2}^{\infty} a_n$ where (a_n) is the sequence

$$\left(\frac{1}{2}, \frac{1}{4}, \frac{1}{4}, \frac{1}{8}, \frac{1}{8}, \frac{1}{8}, \frac{1}{8}, \frac{1}{16}, \frac{1}{16}, \frac{1}{16}, \frac{1}{16}, \frac{1}{16}, \frac{1}{16}, \frac{1}{16}, \frac{1}{16}, \frac{1}{32}, \frac{1}{32}, \ldots\right).$$

§15   Alternating Series and Integral Tests

Sometimes one can check convergence or divergence of series by comparing the partial sums with familiar integrals. We illustrate.

Example 1
We show $\sum \frac{1}{n} = +\infty$.

Consider the picture of the function $f(x) = \frac{1}{x}$ in Fig. 15.1. For $n \geq 1$ it is evident that

$$\sum_{k=1}^{n} \frac{1}{k} = \text{Sum of the areas of the first } n \text{ rectangles in Fig. 15.1}$$

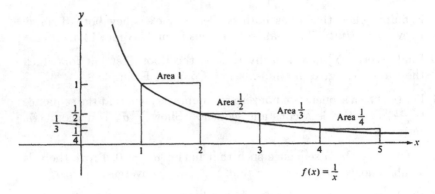

FIGURE 15.1

$$\geq \text{Area under the curve } \frac{1}{x} \text{ between 1 and } n+1$$

$$= \int_1^{n+1} \frac{1}{x}\, dx = \log_e(n+1).$$

Since $\lim_{n\to\infty} \log_e(n+1) = +\infty$, we conclude $\sum_{n=1}^{\infty} \frac{1}{n} = +\infty$.

The series $\sum \frac{1}{n}$ diverges very slowly. In Example 7 on page 120, we observe $\sum_{n=1}^{N} \frac{1}{n}$ is approximately $\log_e N + 0.5772$. Thus for $N = 1,000$ the sum is approximately 7.485, and for $N = 1,000,000$ the sum is approximately 14.393. □

Another proof that $\sum \frac{1}{n}$ diverges was indicated in Exercise 14.14. However, an integral test is useful to establish the next result.

Example 2

We show $\sum \frac{1}{n^2}$ converges.

Consider the graph of $f(x) = \frac{1}{x^2}$ in Fig. 15.2. Then we have

$$\sum_{k=1}^{n} \frac{1}{k^2} = \text{Sum of the areas of the first } n \text{ rectangles}$$

$$\leq 1 + \int_1^n \frac{1}{x^2}\, dx = 2 - \frac{1}{n} < 2$$

for all $n \geq 1$. Thus the partial sums form an increasing sequence that is bounded above by 2. Therefore $\sum_{n=1}^{\infty} \frac{1}{n^2}$ converges and its

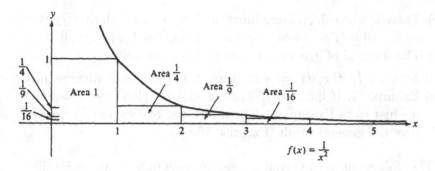

FIGURE 15.2

sum is less than or equal to 2. Actually, we have already mentioned [without proof!] that the sum is $\frac{\pi^2}{6} = 1.6449\cdots$.

Note that in estimating $\sum_{k=1}^{n} \frac{1}{k^2}$ we did not simply write $\sum_{k=1}^{n} \frac{1}{k^2} \leq \int_{0}^{n} \frac{1}{x^2}\, dx$, even though this is true, because this integral is infinite. We were after a *finite* upper bound for the partial sums. □

The techniques just illustrated can be used to prove the following theorem.

15.1 Theorem.
$\sum \frac{1}{n^p}$ *converges if and only if* $p > 1$.

Proof
Supply your own picture and observe that if $p > 1$, then

$$\sum_{k=1}^{n} \frac{1}{k^p} \leq 1 + \int_{1}^{n} \frac{1}{x^p}\, dx = 1 + \frac{1}{p-1}\left(1 - \frac{1}{n^{p-1}}\right) < 1 + \frac{1}{p-1} = \frac{p}{p-1}.$$

Consequently $\sum_{n=1}^{\infty} \frac{1}{n^p} \leq \frac{p}{p-1} < +\infty$.

Suppose $0 < p \leq 1$. Then $\frac{1}{n} \leq \frac{1}{n^p}$ for all n. Since $\sum \frac{1}{n}$ diverges, we see that $\sum \frac{1}{n^p}$ diverges by the Comparison Test. ∎

15.2 Integral Tests.
Here are the conditions under which an integral test is advisable:
  (a) The tests in §14 do not seem to apply.
  (b) The terms a_n of the series $\sum a_n$ are nonnegative.

(c) There is a nice decreasing function f on $[1, \infty)$ such that $f(n) = a_n$ for all n [f is *decreasing* if $x < y$ implies $f(x) \geq f(y)$].

(d) The integral of f is easy to calculate or estimate.

If $\lim_{n \to \infty} \int_1^n f(x)\,dx = +\infty$, then the series will diverge just as in Example 1. If $\lim_{n \to \infty} \int_1^n f(x)\,dx < +\infty$, then the series will converge just as in Example 2. The interested reader may formulate and prove the general result [Exercise 15.8].　　　　□

The following result enables us to conclude that series like $\sum \frac{(-1)^n}{\sqrt{n}}$ converge even though they do not converge absolutely. See Example 9 in §14.

15.3 Alternating Series Theorem.

If $a_1 \geq a_2 \geq \cdots \geq a_n \geq \cdots \geq 0$ and $\lim a_n = 0$, then the alternating series $\sum (-1)^{n+1} a_n$ converges. Moreover, the partial sums $s_n = \sum_{k=1}^n (-1)^{k+1} a_k$ satisfy $|s - s_n| \leq a_n$ for all n.

The series $\sum (-1)^n a_n$ is called an *alternating* series because the signs of the terms alternate between $+$ and $-$.

Proof

We need to show that the sequence (s_n) converges. Note that the subsequence (s_{2n}) is increasing because $s_{2n+2} - s_{2n} = -a_{2n+2} + a_{2n+1} \geq 0$. Similarly, the subsequence (s_{2n-1}) is decreasing since $s_{2n+1} - s_{2n-1} = a_{2n+1} - a_{2n} \leq 0$. We claim

$$s_{2m} \leq s_{2n+1} \quad \text{for all} \quad m, n \in \mathbb{N}. \tag{1}$$

First note that $s_{2n} \leq s_{2n+1}$ for all n, because $s_{2n+1} - s_{2n} = a_{2n+1} \geq 0$. If $m \leq n$, then (1) holds because $s_{2m} \leq s_{2n} \leq s_{2n+1}$. If $m \geq n$, then (1) holds because $s_{2n+1} \geq s_{2m+1} \geq s_{2m}$. Thanks to (1), we see that (s_{2n}) is an increasing subsequence of (s_n) bounded above by each odd partial sum, and (s_{2n+1}) is a decreasing subsequence of (s_n) bounded below by each even partial sum. By Theorem 10.2, these subsequences converge, say to s and t. Now

$$t - s = \lim_{n \to \infty} s_{2n+1} - \lim_{n \to \infty} s_{2n} = \lim_{n \to \infty} (s_{2n+1} - s_{2n}) = \lim_{n \to \infty} a_{2n+1} = 0,$$

so $s = t$. It follows that $\lim_n s_n = s$.

To check the last claim, note that $s_{2k} \leq s \leq s_{2k+1}$, so both $s_{2k+1} - s$ and $s - s_{2k}$ are clearly bounded by $s_{2k+1} - s_{2k} = a_{2k+1} \leq a_{2k}$. So, whether n is even or odd, we have $|s - s_n| \leq a_n$. ∎

Exercises

15.1 Determine which of the following series converge. Justify your answers.

(a) $\sum \frac{(-1)^n}{n}$
(b) $\sum \frac{(-1)^n n!}{2^n}$

15.2 Repeat Exercise 15.1 for the following.

(a) $\sum [\sin(\frac{n\pi}{6})]^n$
(b) $\sum [\sin(\frac{n\pi}{7})]^n$

15.3 Show $\sum_{n=2}^{\infty} \frac{1}{n(\log n)^p}$ converges if and only if $p > 1$.

15.4 Determine which of the following series converge. Justify your answers.

(a) $\sum_{n=2}^{\infty} \frac{1}{\sqrt{n} \log n}$
(b) $\sum_{n=2}^{\infty} \frac{\log n}{n}$
(c) $\sum_{n=4}^{\infty} \frac{1}{n(\log n)(\log \log n)}$
(d) $\sum_{n=2}^{\infty} \frac{\log n}{n^2}$

15.5 Why didn't we use the Comparison Test to prove Theorem 15.1 for $p > 1$?

15.6 (a) Give an example of a divergent series $\sum a_n$ for which $\sum a_n^2$ converges.

(b) Observe that if $\sum a_n$ is a convergent series of nonnegative terms, then $\sum a_n^2$ also converges. See Exercise 14.7.

(c) Give an example of a convergent series $\sum a_n$ for which $\sum a_n^2$ diverges.

15.7 (a) Prove if (a_n) is a decreasing sequence of real numbers and if $\sum a_n$ converges, then $\lim n a_n = 0$. Hint: Consider $|a_{N+1} + a_{N+2} + \cdots + a_n|$ for suitable N.

(b) Use (a) to give another proof that $\sum \frac{1}{n}$ diverges.

15.8 Formulate and prove a general integral test as advised in 15.2.

§16 * Decimal Expansions of Real Numbers

We begin by recalling the brief discussion of decimals in Discussion 10.3. There we considered a decimal expansion $K.d_1 d_2 d_3 \cdots$, where K is a nonnegative integer and each digit d_j belongs to $\{0,1,2,3,4,5,6,7,8,9\}$. This expansion represents the real number

$$K + \sum_{j=1}^{\infty} \frac{d_j}{10^j} = K + \sum_{j=1}^{\infty} d_j \cdot 10^{-j}$$

which we also can write as

$$\lim_{n \to \infty} s_n \quad \text{where} \quad s_n = K + \sum_{j=1}^{n} d_j \cdot 10^{-j}.$$

Thus *every such decimal expansion represents a nonnegative real number.* We will prove the converse after we formalize the process of long division. The development here is based on some suggestions by Karl Stromberg.

16.1 Long Division.

Consider a/b where a and b are positive integers. Since the K term above causes no difficulty, we assume $a < b$. We analyze the familiar long division process which gives a decimal expansion for a/b. Figure 16.1 shows the first few steps where $a = 3$ and $b = 7$. If we name the digits $d_1, d_2, d_3, \ldots$ and the remainders $r_1, r_2, r_3, \ldots$, then so far $d_1 = 4$, $d_2 = 2$ and $r_1 = 2$, $r_2 = 6$. At the next step we divide 7 into $60 = 10 \cdot r_2$ and obtain $60 = 7 \cdot 8 + 4$. The quotient 8 becomes the third digit d_3, we place the product 56 under 60, subtract and obtain a new remainder $4 = r_3$. That is, we are calculating the remainder

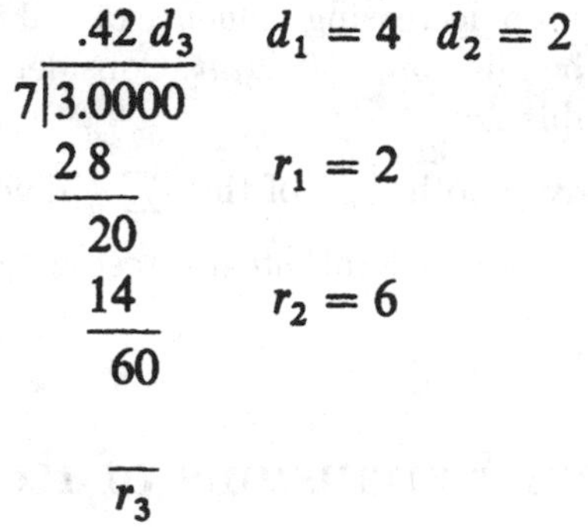

$$
\begin{array}{r}
.42\,d_3 \qquad d_1 = 4 \quad d_2 = 2 \\
7\,\overline{)3.0000} \\
2\,8 \qquad\qquad r_1 = 2 \\
\overline{20} \\
14 \qquad\qquad r_2 = 6 \\
\overline{60} \\
\\
\overline{r_3}
\end{array}
$$

FIGURE 16.1

obtained by dividing 60 by 7. Next we multiply the remainder $r_3 = 4$ by 10 and repeat the process. At each stage

$$d_n \in \{0, 1, 2, 3, 4, 5, 6, 7, 8, 9\}$$
$$r_n = 10 \cdot r_{n-1} - 7 \cdot d_n$$
$$0 \le r_n < 7.$$

These results hold for $n = 1, 2, \ldots$ if we set $r_0 = 3$. In general, we set $r_0 = a$ and obtain

$$d_n \in \{0, 1, 2, 3, 4, 5, 6, 7, 8, 9\} \tag{1}$$
$$r_n = 10 \cdot r_{n-1} - b \cdot d_n \tag{2}$$
$$0 \le r_n < b. \tag{3}$$

As we will show, this construction yields:

$$\frac{a}{b} = \frac{d_1}{10} + \frac{1}{10}\frac{r_1}{b} \tag{4}$$

and

$$\frac{a}{b} = \frac{d_1}{10} + \cdots + \frac{d_n}{10^n} + \frac{1}{10^n}\frac{r_n}{b}. \tag{5}$$

for $n \ge 2$. Since each r_n is less than b, we have $\lim_n \frac{1}{10^n}\frac{r_n}{b} = 0$, so (5) shows

$$\frac{a}{b} = \sum_{j=1}^{\infty} \frac{d_j}{10^j};$$

thus $.d_1 d_2 d_3 \cdots$ is a decimal expansion for $\frac{a}{b}$.

Now Eq. (4) follows from $r_1 = 10a - bd_1$, and we will verify (5) by mathematical induction. From (2), we have

$$r_{n+1} = 10r_n - bd_{n+1}, \quad \text{so} \quad \frac{r_n}{b} = \frac{1}{10}d_{n+1} + \frac{1}{10}\frac{r_{n+1}}{b}.$$

Substituting this into (5) gives

$$\frac{a}{b} = \frac{d_1}{10} + \cdots + \frac{d_n}{10^n} + \frac{d_{n+1}}{10^{n+1}} + \frac{1}{10^{n+1}}\frac{r_{n+1}}{b}, \tag{6}$$

i.e., (5) holds with n replaced by $n+1$. Thus by induction, (5) holds for all n. $\qquad\square$

16.2 Theorem.
Every nonnegative real number x has at least one decimal expansion.

Proof

It suffices to consider x in $[0, 1)$. The proof will be similar to that for results in 16.1, starting with Eqs. (4) and (5) in its proof. We will use the "floor function" $\lfloor \ \rfloor$ on $\mathbb{R}$, where $\lfloor y \rfloor$ is defined to be the largest integer less than or equal to y, for each $y \in \mathbb{R}$.

Since $10x < 10$, we can write

$$10x = d_1 + x_1 \quad \text{where} \quad d_1 = \lfloor 10x \rfloor \quad \text{and} \quad x_1 \in [0, 1). \quad (1)$$

Note d_1 is in $\{0, 1, 2, 3, \ldots, 9\}$ since $10x < 10$. Therefore

$$x = \frac{d_1}{10} + \frac{1}{10} x_1. \quad (2)$$

Suppose, for some $n \geq 1$, we have chosen $d_1, \ldots, d_n$ in $\{0, 1, 2, 3, \ldots, 9\}$ and $x_1, \ldots, x_n$ in $[0, 1)$ so that

$$x = \frac{d_1}{10} + \cdots + \frac{d_n}{10^n} + \frac{1}{10^n} x_n. \quad (3)$$

Since $10x_n < 10$, we can write

$$10x_n = d_{n+1} + x_{n+1} \quad \text{where} \quad d_{n+1} = \lfloor 10x_n \rfloor, x_{n+1} \in [0, 1). \quad (4)$$

Solving for x_n in (4) and substituting the value in (3), we get

$$x = \frac{d_1}{10} + \cdots + \frac{d_{n+1}}{10^{n+1}} + \frac{1}{10^{n+1}} x_{n+1}.$$

This completes the induction step. Since $\lim_n \frac{1}{10^{n+1}} x_{n+1} = 0$, we conclude

$$x = \lim_n \sum_{j=1}^{n+1} \frac{d_j}{10^j} = \sum_{j=1}^{\infty} \frac{d_j}{10^j},$$

so that $.d_1 d_2 d_3 \cdots$ is a decimal expansion for x. ∎

As noted in Discussion 10.3, $1.000 \cdots$ and $0.999 \cdots$ are decimal expansions for the same real number. That is, the series

$$1 + \sum_{j=1}^{\infty} 0 \cdot 10^{-j} \quad \text{and} \quad \sum_{j=1}^{\infty} 9 \cdot 10^{-j}$$

have the same value, namely 1. Similarly, $2.75000\cdots$ and $2.74999\cdots$ are both decimal expansions for $\frac{11}{4}$ [Exercise 16.1]. The next theorem shows this is essentially the only way a number can have distinct decimal expansions.

16.3 Theorem.

A real number x has exactly one decimal expansion or else x has two decimal expansions, one ending in a sequence of all 0's and the other ending in a sequence of all 9's.

Proof

We assume $x \geq 0$. If x has decimal expansion $K.000\cdots$ with $K > 0$, then it has one other decimal expansion, namely $(K-1).999\cdots$. If x has decimal expansion $K.d_1 d_2 d_3 \cdots d_r 000 \cdots$ where $d_r \neq 0$, then it has one other decimal expansion $K.d_1 d_2 d_3 \cdots (d_r - 1)9999\cdots$. The reader can easily check these claims [Exercise 16.2].

Now suppose x has two distinct decimal expansions $K.d_1 d_2 d_3 \cdots$ and $L.e_1 e_2 e_3 \cdots$. Suppose $K < L$. If any $d_j < 9$, then by Exercise 16.3 we have

$$x < K + \sum_{j=1}^{\infty} 9 \cdot 10^{-j} = K + 1 \leq L \leq x,$$

a contradiction. It follows that $x = K + 1 = L$ and its decimal expansions are $K.999\cdots$ and $(K+1).000\cdots$. In the remaining case, we have $K = L$. Let

$$m = \min\{j : d_j \neq e_j\}.$$

We may assume $d_m < e_m$. If $d_j < 9$ for any $j > m$, then by Exercise 16.3,

$$x < K + \sum_{j=1}^{m} d_j \cdot 10^{-j} + \sum_{j=m+1}^{\infty} 9 \cdot 10^{-j} = K + \sum_{j=1}^{m} d_j \cdot 10^{-j} + 10^{-m}$$

$$= K + \sum_{j=1}^{m-1} e_j \cdot 10^{-j} + d_m \cdot 10^{-m} + 10^{-m} \leq K + \sum_{j=1}^{m} e_j \cdot 10^{-j} \leq x,$$

a contradiction. Thus $d_j = 9$ for $j > m$. Likewise, if $e_j > 0$ for any $j > m$, then

$$x > K + \sum_{j=1}^{m} e_j \cdot 10^{-j} = K + \sum_{j=1}^{m-1} d_j \cdot 10^{-j} + e_m \cdot 10^{-m}$$

$$\geq K + \sum_{j=1}^{m-1} d_j \cdot 10^{-j} + d_m \cdot 10^{-m} + 10^{-m}$$

$$= K + \sum_{j=1}^{m} d_j \cdot 10^{-j} + \sum_{j=m+1}^{\infty} 9 \cdot 10^{-j} \geq x,$$

a contradiction. So in this case, $d_j = 9$ for $j > m$, $e_m = d_m + 1$ and $e_j = 0$ for $j > m$. ∎

16.4 Definition.
An expression of the form

$$K.d_1 d_2 \cdots d_\ell \overline{d_{\ell+1} \cdots d_{\ell+r}}$$

represents the decimal expansion in which the block $d_{\ell+1} \cdots d_{\ell+r}$ is repeated indefinitely:

$$K.d_1 d_2 \cdots d_\ell d_{\ell+1} \cdots d_{\ell+r} d_{\ell+1} \cdots d_{\ell+r} d_{\ell+1} \cdots d_{\ell+r} d_{\ell+1} \cdots d_{\ell+r} \cdots .$$

We call such an expansion a *repeating decimal*.

Example 1
Every integer is a repeating decimal. For example, $17 = 17.\overline{0} = 17.000\cdots$. Another simple example is

$$.\overline{8} = .888\cdots = \sum_{j=1}^{\infty} 8 \cdot 10^{-j} = \frac{8}{10} \sum_{j=0}^{\infty} 10^{-j} = \frac{8}{10} \cdot \frac{10}{9} = \frac{8}{9}.$$

$\square$

Example 2
The expression $3.9\overline{67}$ represents the repeating decimal $3.9676767\cdots$. We evaluate this as follows:

$$3.9\overline{67} = 3 + 9 \cdot 10^{-1} + 6 \cdot 10^{-2} + 7 \cdot 10^{-3} + 6 \cdot 10^{-4} + 7 \cdot 10^{-5} + \cdots$$

$$= 3 + 9 \cdot 10^{-1} + 67 \cdot 10^{-3} \sum_{j=0}^{\infty} (10^{-2})^j$$

$$= 3 + 9 \cdot 10^{-1} + 67 \cdot 10^{-3} \left(\frac{100}{99} \right) = 3 + \frac{9}{10} + \frac{67}{990}$$

$$= \frac{3,928}{990} = \frac{1,964}{495}.$$

Thus the repeating decimal $3.9\overline{67}$ represents the rational number $\frac{1,964}{495}$. Any repeating decimal can be evaluated as a rational number in this way, as we'll show in the next theorem. □

Example 3
We find the decimal expansion for $\frac{11}{7}$. By the usual long division process in 16.1, we find

$$\frac{11}{7} = 1.5714285714285714285714285714285714285714571 \cdots,$$

i.e., $\frac{11}{7} = 1.\overline{571428}$. To check this, observe

$$1.\overline{571428} = 1 + 571,428 \cdot 10^{-6} \sum_{j=0}^{\infty} (10^{-6})^j = 1 + \frac{571,428}{999,999}$$

$$= 1 + \frac{4}{7} = \frac{11}{7}.$$

□

Many books give the next theorem as an exercise, probably to avoid the complicated notation. If the details seem too complicated to you, move on to Examples 4–7.

16.5 Theorem.
A real number x is rational if and only if its decimal expansion is repeating. [*Theorem 16.3 shows that if x has two decimal expansions, they are both repeating.*]

Proof
First assume $x \geq 0$ has a repeating decimal expansion $x = K.d_1 d_2 \cdots d_\ell \overline{d_{\ell+1} \cdots d_{\ell+r}}$. Then

$$x = K + \sum_{j=1}^{\ell} d_j \cdot 10^{-j} + 10^{-\ell} y$$

where

$$y = .\overline{d_{\ell+1} \cdots d_{\ell+r}},$$

so it suffices to show y is rational. To simplify the notation, we write

$$y = .\overline{e_1 e_2 \cdots e_r}.$$

A little computation shows

$$y = \sum_{j=1}^{r} e_j \cdot 10^{-j} \left[\sum_{j=0}^{\infty} (10^{-r})^j \right] = \sum_{j=1}^{r} e_j \cdot 10^{-j} \frac{10^r}{10^r - 1}.$$

Thus y is rational. In fact, if we write $e_1 e_2 \cdots e_r$ for the usual *decimal* $\sum_{j=0}^{r-1} e_j \cdot 10^{r-1-j}$ *not the product*, then $y = \frac{e_1 e_2 \cdots e_r}{10^r - 1}$; see Example 3.

Next consider any positive rational, say $\frac{a}{b}$ where $a, b \in \mathbb{N}$. We may assume $a < b$. As we saw in 16.1, $\frac{a}{b}$ is given by the decimal expansion $.d_1 d_2 d_3 \cdots$ where $r_0 = a$,

$$d_k \in \{0, 1, 2, 3, 4, 5, 6, 7, 8, 9\} \tag{1}$$
$$r_k = 10 \cdot r_{k-1} - d_k b \tag{2}$$
$$0 \le r_k < b, \tag{3}$$

for $k \ge 1$. Since a and b are integers, each r_k is an integer. Thus (3) can be written

$$r_k \in \{0, 1, 2, \ldots, b-1\} \quad \text{for} \quad k \ge 0. \tag{4}$$

This set has b elements, so the first $b + 1$ remainders r_k cannot all be distinct. That is, there exist integers $m \ge 0$ and $p > 0$ so that

$$0 \le m < m + p \le b \quad \text{and} \quad r_m = r_{m+p}.$$

From the construction giving (1)–(3) it is clear that given r_{k-1}, the integers r_k and d_k are uniquely determined. Thus

$$r_j = r_k \quad \text{implies} \quad r_{j+1} = r_{k+1} \quad \text{and} \quad d_{j+1} = d_{k+1}.$$

Since $r_m = r_{m+p}$, we conclude $r_{m+1} = r_{m+1+p}$ and $d_{m+1} = d_{m+1+p}$. A simple induction shows that the statement

$$\text{``}r_k = r_{k+p} \quad \text{and} \quad d_k = d_{k+p}\text{''}$$

holds for all integers $k \ge m + 1$. Thus the decimal expansion of $\frac{a}{b}$ is periodic with period p after the first m digits. That is,

$$\frac{a}{b} = .d_1 d_2 \cdots d_m \overline{d_{m+1} \cdots d_{m+p}}. \qquad \blacksquare$$

Remark. Given r_{k-1}, the uniqueness of r_k and d_k follows from the so-called "division algorithm," which is actually a theorem that shows the algorithm for division never breaks down. It says that if b is a positive integer and $m \in \mathbb{Z}$, then there are unique integers q and r so that

$$m = bq + r \quad \text{and} \quad 0 \le r < b;$$

q is called the quotient and r is called the remainder. With $m = 10 \cdot r_{k-1}$ in Theorem 16.5, this yields

$$10 \cdot r_{k-1} = bq + r \quad \text{where} \quad 0 \le r < b.$$

If we name $q = d_k$ and $r = r_k$, then we obtain formula (2) in Theorem 16.5. For more details, see for example, [60, §3.5].

Example 4
An expansion such as

.10100100010000100000100000010000000100000000010000000000100···

represents an irrational number, since it cannot be a repeating decimal: we've arranged for arbitrarily long blocks of 0's. □

Example 5
We do not know the complete decimal expansions of $\sqrt{2}$, $\sqrt{3}$ and many other familiar irrational numbers, but we know that they cannot be repeating by virtue of the last theorem. □

Example 6
We have claimed π and e are irrational. These facts and many others are proved in a fascinating book by Ivan Niven [49].

(a) Here is a proof that

$$e = \sum_{k=0}^{\infty} \frac{1}{k!}$$

is irrational. Assume $e = \frac{a}{b}$ where $a, b \in \mathbb{N}$. Then both $(b+1)! \cdot e$ and $(b+1)! \cdot \sum_{k=0}^{b+1} \frac{1}{k!}$ must be integers, so the difference

$$(b+1)! \sum_{k=b+2}^{\infty} \frac{1}{k!}$$

must be a positive integer. On the other hand, this last number is less than

$$\frac{1}{b+2} + \frac{1}{(b+2)^2} + \frac{1}{(b+2)^3} + \cdots = \frac{\frac{1}{b+2}}{1 - \frac{1}{b+2}} = \frac{1}{b+1} < 1,$$

a contradiction.

(b) We will prove π^2 is irrational, from which the irrationality of π follows; see Exercise 16.10.

The key to the proof that π^2 is irrational is the sequence of integrals $I_n = \int_0^\pi \frac{(x(\pi-x))^n}{n!} \sin x \, dx$. Thus

$$I_n = \int_0^\pi P_n(x) \sin x \, dx \quad \text{where} \quad P_n(x) = \frac{(x(\pi - x))^n}{n!}.$$

Claim 1. There is a sequence $(Q_n)_{n=0}^\infty$ of polynomials with *integer* coefficients, of degree at most n, satisfying $I_n = Q_n(\pi^2)$ for all n.

Proof. First, we obtain a recursive relation for I_n; see (3). We use integration by parts (Theorem 34.2) twice to show

$$I_n = \int_0^\pi P_n(x) \sin x \, dx = -\int_0^\pi P_n''(x) \sin x \, dx \quad \text{for} \quad n \geq 1.$$
$$(1)$$

In fact,

$$\int_0^\pi P_n(x) \sin x \, dx = [-P_n(\pi)\cos(\pi) + P_n(0)\cos(0)] + \int_0^\pi P_n'(x) \cos x \, dx$$

$$= \int_0^\pi P_n'(x) \cos x \, dx,$$

since $P_n(x)$ contains a factor of $x(\pi - x)$ which is 0 at both π and 0. Therefore

$$I_n = \int_0^\pi P_n'(x) \cos x \, dx$$

$$= [P_n'(\pi)\sin(\pi) - P_n'(0)\sin(0)] - \int_0^\pi P_n''(x) \sin x \, dx$$

$$= -\int_0^\pi P_n''(x) \sin x \, dx,$$

and (1) holds. For $n \geq 2$, we have

$$P_n'(x) = \frac{(x(\pi - x))^{n-1}}{(n-1)!}(\pi - 2x) = P_{n-1}(x)(\pi - 2x),$$

so, using this for $n - 1$ and the product rule, we have

$$P_n''(x) = P_{n-2}(x)(\pi - 2x)^2 + P_{n-1}(x)(-2)$$
$$= \pi^2 P_{n-2}(x) + P_{n-2}(x)(-4\pi x + 4x^2) - 2P_{n-1}(x).$$

Since

$$P_{n-2}(x)(-4\pi x + 4x^2) = -4 \cdot \frac{(x(\pi - x))^{n-1}}{(n-2)!} = -4 \cdot P_{n-1}(x)(n-1),$$

we conclude

$$P_n''(x) = \pi^2 P_{n-2}(x) - [4(n-1) + 2]P_{n-1}(x).$$

Therefore

$$P_n''(x) = \pi^2 P_{n-2}(x) - (4n - 2)P_{n-1}(x) \quad \text{for} \quad n \geq 2. \quad (2)$$

To prove Claim 1, note that $I_0 = 2$ is clear. Using (1) and $P_1''(x) = -2$, we see that $I_1 = 4$. And from (1) and (2), we see that

$$I_n = -\pi^2 I_{n-2} + (4n - 2)I_{n-1} \quad \text{for} \quad n \geq 2. \quad (3)$$

Now Claim 1 holds by a simple induction argument, where $Q_0 = 2$, $Q_1 = 4$, and

$$Q_n(x) = -xQ_{n-2}(x) + (4n - 2)Q_{n-1}(x) \quad \text{for} \quad n \geq 2.$$

Claim 2. π^2 is irrational.

Proof. Suppose $\pi^2 = a/b$. Using Claim 1, we see that each $b^n I_n = b^n Q_n(a/b)$ is an integer. Since $x(\pi - x)$ takes its maximum at $\pi/2$, we can write

$$0 < b^n I_n = b^n \int_0^\pi \frac{(x(\pi - x))^n}{n!} \sin x \, dx$$
$$< b^n \int_0^\pi \frac{(\frac{\pi}{2} \cdot \frac{\pi}{2})^n}{n!} \, dx = \frac{\left(\frac{b\pi^2}{4}\right)^n}{n!} \pi.$$

As noted in Exercise 9.15, the right-hand side converges to 0. So, for large n, the integer $b^n I_n$ lies in the interval $(0,1)$, a contradiction.

This simplification of Ivan Niven's famous short proof (1947) is due to Zhou and Markov [72]. Zhou and Markov use a similar technique to prove $\tan r$ is irrational for nonzero rational r and $\cos r$ is irrational if r^2 is a nonzero rational. Compare with results in Niven's book [49, Chap. 2].

(c) It is even more difficult to prove π and e are not algebraic numbers; see Definition 2.1. These results are proved in Niven's book [49, Theorems 2.12 and 9.11]. □

Example 7

There is a famous number introduced by Euler over 200 years ago that arises in the study of the gamma function. It is known as *Euler's constant* and is defined by

$$\gamma = \lim_{n \to \infty} \left[\sum_{k=1}^{n} \frac{1}{k} - \log_e n \right].$$

Even though

$$\lim_{n \to \infty} \sum_{k=1}^{n} \frac{1}{k} = +\infty \quad \text{and} \quad \lim_{n \to \infty} \log_e n = +\infty,$$

the limit defining γ exists and is finite [Exercise 16.9]. In fact, γ is approximately 0.577216. The amazing fact is that no one knows whether γ is rational or not. Most mathematicians believe γ is irrational. This is because it is "easier" for a number to be irrational, since repeating decimal expansions are regular. The remark in Exercise 16.8 hints at another reason it is easier for a number to be irrational. □

Exercises

16.1 (a) Show $2.74\overline{9}$ and $2.75\overline{0}$ are both decimal expansions for $\frac{11}{4}$.

   (b) Which of these expansions arises from the long division process described in 16.1?

16.2 Verify the claims in the first paragraph of the proof of Theorem 16.3.

16.3 Suppose $\sum a_n$ and $\sum b_n$ are convergent series of nonnegative numbers. Show that if $a_n \leq b_n$ for all n and if $a_n < b_n$ for at least one n, then $\sum a_n < \sum b_n$.

16.4 Write the following repeating decimals as rationals, i.e., as fractions of integers.

(a) $.2$ (b) $.0\overline{2}$
(c) $.\overline{02}$ (d) $3.\overline{14}$
(e) $.\overline{10}$ (f) $.1\overline{492}$

16.5 Find the decimal expansions of the following rational numbers.

(a) $1/8$ (b) $1/16$
(c) $2/3$ (d) $7/9$
(e) $6/11$ (f) $22/7$

16.6 Find the decimal expansions of $\frac{1}{7}$, $\frac{2}{7}$, $\frac{3}{7}$, $\frac{4}{7}$, $\frac{5}{7}$ and $\frac{6}{7}$. Note the interesting pattern.

16.7 Is $.12345678910111213141516171819202122232425 26\cdots$ rational?

16.8 Let (s_n) be a sequence of numbers in $(0,1)$. Each s_n has a decimal expansion $0.d_1^{(n)} d_2^{(n)} d_3^{(n)} \cdots$. For each n, let $e_n = 6$ if $d_n^{(n)} \neq 6$ and $e_n = 7$ if $d_n^{(n)} = 6$. Show $.e_1 e_2 e_3 \cdots$ is the decimal expansion for some number y in $(0,1)$ and $y \neq s_n$ for all n. *Remark*: This shows the elements of $(0,1)$ cannot be listed as a sequence. In set-theoretic parlance, $(0,1)$ is "uncountable." Since the set $\mathbb{Q} \cap (0,1)$ can be listed as a sequence, there are a lot of irrational numbers in $(0,1)$!

16.9 Let $\gamma_n = (\sum_{k=1}^{n} \frac{1}{k}) - \log_e n = \sum_{k=1}^{n} \frac{1}{k} - \int_1^n \frac{1}{t} dt$.

  (a) Show (γ_n) is a decreasing sequence. *Hint*: Look at $\gamma_n - \gamma_{n+1}$.

  (b) Show $0 < \gamma_n \leq 1$ for all n.

  (c) Observe that $\gamma = \lim_n \gamma_n$ exists and is finite.

16.10 In Example 6(b), we showed π^2 is irrational. Use this to show π is irrational. What can you say about $\sqrt{\pi}$ and $\sqrt[3]{\pi}$? π^4?

3

CHAPTER

Continuity

Most of the calculus involves the study of continuous functions. In this chapter we study continuous and uniformly continuous functions.

§17 Continuous Functions

Recall that the salient features of a function f are:

(a) The set on which f is defined, called the *domain of f* and written $\mathrm{dom}(f)$;

(b) The assignment, rule or formula specifying the value $f(x)$ of f at each x in $\mathrm{dom}(f)$.

We will be concerned with functions f such that $\mathrm{dom}(f) \subseteq \mathbb{R}$ and such that f is a *real-valued function*, i.e., $f(x) \in \mathbb{R}$ for all $x \in \mathrm{dom}(f)$. Properly speaking, the symbol f represents the function while $f(x)$ represents the value of the function at x. However, a function is often given by specifying its values and without mentioning its domain. In this case, the domain is understood to be the *natural domain*: the largest subset of $\mathbb{R}$ on which the function is a well defined

K.A. Ross, *Elementary Analysis: The Theory of Calculus*,
Undergraduate Texts in Mathematics, DOI 10.1007/978-1-4614-6271-2_3,
© Springer Science+Business Media New York 2013

real-valued function. Thus "the function $f(x) = \frac{1}{x}$" is shorthand for
"the function f given by $f(x) = \frac{1}{x}$ with natural domain $\{x \in \mathbb{R} :$
$x \neq 0\}$." Similarly, the natural domain of $g(x) = \sqrt{4 - x^2}$ is $[-2, 2]$
and the natural domain of $\csc x = \frac{1}{\sin x}$ is the set of real numbers x
not of the form $n\pi, n \in \mathbb{Z}$.

In keeping with the approach in this book, we will define continuity in terms of sequences. We then show our definition is equivalent
to the usual ϵ–δ definition.

17.1 Definition.
Let f be a real-valued function whose domain is a subset of $\mathbb{R}$. The
function f is *continuous at x_0* in $\mathrm{dom}(f)$ if, for every sequence (x_n)
in $\mathrm{dom}(f)$ converging to x_0, we have $\lim_n f(x_n) = f(x_0)$. If f is
continuous at each point of a set $S \subseteq \mathrm{dom}(f)$, then f is said to
be *continuous on S*. The function f is said to be *continuous* if it is
continuous on $\mathrm{dom}(f)$.

Our definition implies that the values $f(x)$ are close to $f(x_0)$
when the values x are close to x_0. The next theorem says this in
another way. In fact, condition (1) of the next theorem is the ϵ–δ
definition of continuity given in many calculus books.

17.2 Theorem.
*Let f be a real-valued function whose domain is a subset of $\mathbb{R}$. Then
f is continuous at x_0 in $\mathrm{dom}(f)$ if and only if*

$$\text{for each} \quad \epsilon > 0 \quad \text{there exists} \quad \delta > 0 \quad \text{such that}$$
$$x \in \mathrm{dom}(f) \quad \text{and} \quad |x - x_0| < \delta \quad \text{imply} \quad |f(x) - f(x_0)| < \epsilon.$$
$$\tag{1}$$

Proof
Suppose (1) holds, and consider a sequence (x_n) in $\mathrm{dom}(f)$ such that
$\lim x_n = x_0$. We need to prove $\lim f(x_n) = f(x_0)$. Let $\epsilon > 0$. By (1),
there exists $\delta > 0$ such that

$$x \in \mathrm{dom}(f) \quad \text{and} \quad |x - x_0| < \delta \quad \text{imply} \quad |f(x) - f(x_0)| < \epsilon.$$

Since $\lim x_n = x_0$, there exists N so that

$$n > N \quad \text{implies} \quad |x_n - x_0| < \delta.$$

It follows that

$$n > N \quad \text{implies} \quad |f(x_n) - f(x_0)| < \epsilon.$$

This proves $\lim f(x_n) = f(x_0)$.

Now assume f is continuous at x_0, but (1) fails. Then there exists $\epsilon > 0$ so that the implication

"$x \in \text{dom}(f) \quad \text{and} \quad |x - x_0| < \delta \quad \text{imply} \quad |f(x) - f(x_0)| < \epsilon$"

fails for each $\delta > 0$. In particular, the implication

"$x \in \text{dom}(f) \quad \text{and} \quad |x - x_0| < \dfrac{1}{n} \quad \text{imply} \quad |f(x) - f(x_0)| < \epsilon$"

fails for each $n \in \mathbb{N}$. So for each $n \in \mathbb{N}$ there exists x_n in $\text{dom}(f)$ such that $|x_n - x_0| < \frac{1}{n}$ and yet $|f(x_0) - f(x_n)| \geq \epsilon$. Thus we have $\lim x_n = x_0$, but we cannot have $\lim f(x_n) = f(x_0)$ since $|f(x_0) - f(x_n)| \geq \epsilon$ for all n. This shows f cannot be continuous at x_0, contrary to our assumption. ■

As the next example illustrates, it is sometimes easier to work with the sequential definition of continuity in Definition 17.1 than the ϵ–δ property in Theorem 17.2. However, it is important to get comfortable with the ϵ–δ property, partly because the definition of uniform continuity is more closely related to the ϵ–δ property than the sequential definition.

Example 1
Let $f(x) = 2x^2 + 1$ for $x \in \mathbb{R}$. Prove f is continuous on $\mathbb{R}$ by
(a) Using the definition,
(b) Using the ϵ–δ property of Theorem 17.2.

Solution
(a) Suppose $\lim x_n = x_0$. Then we have

$$\lim f(x_n) = \lim[2x_n^2 + 1] = 2[\lim x_n]^2 + 1 = 2x_0^2 + 1 = f(x_0)$$

where the second equality is an application of the limit Theorems 9.2–9.4. Hence f is continuous at each x_0 in $\mathbb{R}$.

(b) Let x_0 be in $\mathbb{R}$ and let $\epsilon > 0$. We want to show $|f(x) - f(x_0)| < \epsilon$ provided $|x - x_0|$ is sufficiently small, i.e., less than some δ. We observe

$$|f(x) - f(x_0)| = |2x^2 + 1 - (2x_0^2 + 1)| = |2x^2 - 2x_0^2|$$
$$= 2|x - x_0| \cdot |x + x_0|.$$

We need to get a bound for $|x + x_0|$ that does not depend on x. We notice that if $|x - x_0| < 1$, say, then $|x| < |x_0| + 1$ and hence $|x + x_0| \le |x| + |x_0| < 2|x_0| + 1$. Thus we have

$$|f(x) - f(x_0)| \le 2|x - x_0|(2|x_0| + 1)$$

provided $|x - x_0| < 1$. To arrange for $2|x - x_0|(2|x_0| + 1) < \epsilon$, it suffices to have $|x - x_0| < \frac{\epsilon}{2(2|x_0|+1)}$ and also $|x - x_0| < 1$. So we put

$$\delta = \min\left\{1, \frac{\epsilon}{2(2|x_0| + 1)}\right\}.$$

The work above shows $|x - x_0| < \delta$ implies $|f(x) - f(x_0)| < \epsilon$, as desired. □

The reason that solution (a) in Example 1 is so much easier is that the careful analysis was done in proving the limit theorems in §9.

Example 2
Let $f(x) = x^2 \sin(\frac{1}{x})$ for $x \ne 0$ and $f(0) = 0$. The graph of f in Fig. 17.1 *looks* continuous. Prove f is continuous at 0.

Solution
Let $\epsilon > 0$. Clearly $|f(x) - f(0)| = |f(x)| \le x^2$ for all x. Since we want this to be less than ϵ, we set $\delta = \sqrt{\epsilon}$. Then $|x - 0| < \delta$ implies $x^2 < \delta^2 = \epsilon$, so

$$|x - 0| < \delta \quad \text{implies} \quad |f(x) - f(0)| < \epsilon.$$

Hence the ϵ–δ property holds and f is continuous at 0. □

In Example 2 it would have been equally easy to show that if $\lim x_n = 0$ then $\lim f(x_n) = 0$. The function f in Example 2 is also continuous at the other points of $\mathbb{R}$; see Example 4.

Example 3
Let $f(x) = \frac{1}{x} \sin(\frac{1}{x^2})$ for $x \ne 0$ and $f(0) = 0$. Show f is *discontinuous*, i.e., not continuous, at 0.

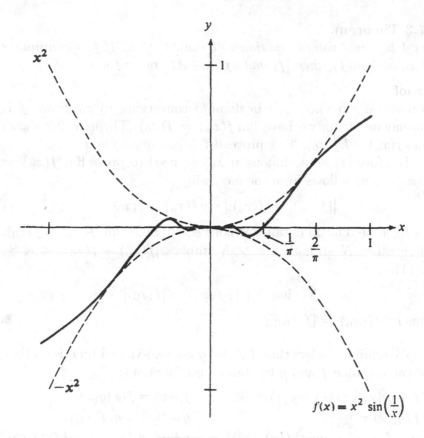

$$f(x) = x^2 \sin\left(\frac{1}{x}\right)$$

FIGURE 17.1

Solution

It suffices for us to find a sequence (x_n) converging to 0 such that $f(x_n)$ does not converge to $f(0) = 0$. So we will arrange for $\frac{1}{x_n}\sin(\frac{1}{x_n^2}) = \frac{1}{x_n}$ where $x_n \to 0$. Thus we want $\sin(\frac{1}{x_n^2}) = 1$, $\frac{1}{x_n^2} = 2\pi n + \frac{\pi}{2}$, $x_n^2 = \frac{1}{2\pi n + \frac{\pi}{2}}$ or $x_n = \frac{1}{\sqrt{2\pi n + \frac{\pi}{2}}}$. Then $\lim x_n = 0$ while $\lim f(x_n) = \lim \frac{1}{x_n} = +\infty$. $\qquad\square$

Let f be a real-valued function. For k in $\mathbb{R}$, kf signifies the function defined by $(kf)(x) = kf(x)$ for $x \in \text{dom}(f)$. Also $|f|$ is the function defined by $|f|(x) = |f(x)|$ for $x \in \text{dom}(f)$. Thus if f is given by $f(x) = \sqrt{x} - 4$ for $x \geq 0$, then $3f$ is given by $(3f)(x) = 3\sqrt{x} - 12$ for $x \geq 0$, and $|f|$ is given by $|f|(x) = |\sqrt{x} - 4|$ for $x \geq 0$. Here is an easy theorem.

17.3 Theorem.
Let f be a real-valued function with $\text{dom}(f) \subseteq \mathbb{R}$. If f is continuous at x_0 in $\text{dom}(f)$, then $|f|$ and kf, $k \in \mathbb{R}$, are continuous at x_0.

Proof
Consider a sequence (x_n) in $\text{dom}(f)$ converging to x_0. Since f is continuous at x_0, we have $\lim f(x_n) = f(x_0)$. Theorem 9.2 shows $\lim kf(x_n) = kf(x_0)$. This proves kf is continuous at x_0.

To prove $|f|$ is continuous at x_0, we need to prove $\lim |f(x_n)| = |f(x_0)|$. This follows from the inequality

$$||f(x_n)| - |f(x_0)|| \leq |f(x_n) - f(x_0)|; \tag{1}$$

see Exercise 3.5. [In detail, consider $\epsilon > 0$. Since $\lim f(x_n) = f(x_0)$, there exists N such that $n > N$ implies $|f(x_n) - f(x_0)| < \epsilon$. So by (1),

$$n > N \quad \text{implies} \quad ||f(x_n)| - |f(x_0)|| < \epsilon;$$

thus $\lim |f(x_n)| = |f(x_0)|$.] ∎

We remind readers that if f and g are real-valued functions, then we can combine f and g to obtain new functions:

$$(f + g)(x) = f(x) + g(x); \qquad fg(x) = f(x)g(x);$$
$$(f/g)(x) = \frac{f(x)}{g(x)}; \qquad g \circ f(x) = g(f(x));$$
$$\max(f,g)(x) = \max\{f(x), g(x)\}; \qquad \min(f,g)(x) = \min\{f(x), g(x)\}.$$

The function $g \circ f$ is called the *composition of g and f*. Each of these new functions is defined exactly where they make sense. Thus the domains of $f + g$, fg, $\max(f,g)$ and $\min(f,g)$ are $\text{dom}(f) \cap \text{dom}(g)$, the domain of f/g is the set $\text{dom}(f) \cap \{x \in \text{dom}(g) : g(x) \neq 0\}$, and the domain of $g \circ f$ is $\{x \in \text{dom}(f) : f(x) \in \text{dom}(g)\}$. Note $f + g = g + f$ and $fg = gf$ but that in general $f \circ g \neq g \circ f$.

These new functions are continuous if f and g are continuous.

17.4 Theorem.
Let f and g be real-valued functions that are continuous at x_0 in $\mathbb{R}$. Then

  (i) $f + g$ *is continuous at x_0;*
  (ii) fg *is continuous at x_0;*
  (iii) f/g *is continuous at x_0 if $g(x_0) \neq 0$.*

Proof
We are given that $x_0 \in \mathrm{dom}(f) \cap \mathrm{dom}(g)$. Let (x_n) be a sequence in $\mathrm{dom}(f) \cap \mathrm{dom}(g)$ converging to x_0. Then we have $\lim f(x_n) = f(x_0)$ and $\lim g(x_n) = g(x_0)$. Theorem 9.3 shows

$$\lim(f+g)(x_n) = \lim[f(x_n) + g(x_n)] = \lim f(x_n) + \lim g(x_n)$$
$$= f(x_0) + g(x_0) = (f+g)(x_0),$$

so $f+g$ is continuous at x_0. Likewise, Theorem 9.4 implies fg is continuous at x_0.

To handle f/g we assume $g(x_0) \neq 0$ and consider a sequence (x_n) in $\mathrm{dom}(f) \cap \{x \in \mathrm{dom}(g) : g(x) \neq 0\}$ converging to x_0. Then Theorem 9.6 shows

$$\lim\left(\frac{f}{g}\right)(x_n) = \lim \frac{f(x_n)}{g(x_n)} = \frac{f(x_0)}{g(x_0}) = \left(\frac{f}{g}\right)(x_0);$$

so f/g is continuous at x_0. ∎

17.5 Theorem.
If f is continuous at x_0 and g is continuous at $f(x_0)$, then the composite function $g \circ f$ is continuous at x_0.

Proof
We are given that $x_0 \in \mathrm{dom}(f)$ and $f(x_0) \in \mathrm{dom}(g)$. Let (x_n) be a sequence in $\{x \in \mathrm{dom}(f) : f(x) \in \mathrm{dom}(g)\}$ converging to x_0. Since f is continuous at x_0, we have $\lim f(x_n) = f(x_0)$. Since the sequence $(f(x_n))$ converges to $f(x_0)$ and g is continuous at $f(x_0)$, we also have $\lim g(f(x_n)) = g(f(x_0))$; that is, $\lim g \circ f(x_n) = g \circ f(x_0)$. Hence $g \circ f$ is continuous at x_0. ∎

Example 4
For this example, let us accept as known that polynomial functions and the functions $\sin x$, $\cos x$ and e^x are continuous on $\mathbb{R}$. Then $4e^x$ and $|\sin x|$ are continuous on $\mathbb{R}$ by Theorem 17.3. The function $\sin x + 4e^x + x^3$ is continuous on $\mathbb{R}$ by (i) of Theorem 17.4. The function $x^4 \sin x$ is continuous on $\mathbb{R}$ by (ii) of Theorem 17.4, and (iii) of Theorem 17.4 shows $\tan x = \frac{\sin x}{\cos x}$ is continuous wherever $\cos x \neq 0$, i.e., at all x not of the form $n\pi + \frac{\pi}{2}$, $n \in \mathbb{Z}$. Theorem 17.5 tells us $e^{\sin x}$ and $\cos(e^x)$ are continuous on $\mathbb{R}$; for example, $\cos(e^x) =$

$g \circ f(x)$ where $f(x) = e^x$ and $g(x) = \cos x$. Several applications of Theorems 17.3–17.5 will show $x^2 \sin(\frac{1}{x})$ and $\frac{1}{x}\sin(\frac{1}{x^2})$ are continuous at all nonzero x in $\mathbb{R}$. □

Example 5
Let f and g be continuous at x_0 in $\mathbb{R}$. Prove $\max(f,g)$ is continuous at x_0. □

Solution
We could prove this using Definition 17.1 or the ϵ–δ condition in Theorem 17.2. However, we illustrate a useful technique by reducing the problem to results we have already established. First, observe

$$\max(f,g) = \frac{1}{2}(f+g) + \frac{1}{2}|f-g|,$$

which shows that the function $\max(f,g)$ is a combination of functions to which our theorems apply. This equation holds because $\max\{a,b\} = \frac{1}{2}(a+b) + \frac{1}{2}|a-b|$ is true for all $a,b \in \mathbb{R}$, a fact which is easily checked by considering the cases $a \geq b$ and $a < b$. By Theorem 17.4(i), $f+g$ and $f-g$ are continuous at x_0. Hence $|f-g|$ is continuous at x_0 by Theorem 17.3. Then $\frac{1}{2}(f+g)$ and $\frac{1}{2}|f-g|$ are continuous at x_0 by Theorem 17.3, and another application of Theorem 17.4(i) shows $\max(f,g)$ is continuous at x_0. □

Exercises

17.1 Let $f(x) = \sqrt{4-x}$ for $x \leq 4$ and $g(x) = x^2$ for all $x \in \mathbb{R}$.

(a) Give the domains of $f+g$, fg, $f \circ g$ and $g \circ f$.

(b) Find the values $f \circ g(0)$, $g \circ f(0)$, $f \circ g(1)$, $g \circ f(1)$, $f \circ g(2)$ and $g \circ f(2)$.

(c) Are the functions $f \circ g$ and $g \circ f$ equal?

(d) Are $f \circ g(3)$ and $g \circ f(3)$ meaningful?

17.2 Let $f(x) = 4$ for $x \geq 0$, $f(x) = 0$ for $x < 0$, and $g(x) = x^2$ for all x. Thus $\operatorname{dom}(f) = \operatorname{dom}(g) = \mathbb{R}$.

(a) Determine the following functions: $f+g$, fg, $f \circ g$, $g \circ f$. Be sure to specify their domains.

(b) Which of the functions f, g, $f+g$, fg, $f \circ g$, $g \circ f$ is continuous?

17.3 Accept on faith that the following familiar functions are continuous on their domains: $\sin x$, $\cos x$, e^x, 2^x, $\log_e x$ for $x > 0$, x^p for $x > 0$ [p any real number]. Use these facts and theorems in this section to prove the following functions are also continuous.

(a) $\log_e(1 + \cos^4 x)$

(b) $[\sin^2 x + \cos^6 x]^\pi$

(c) 2^{x^2}

(d) 8^x

(e) $\tan x$ for $x \neq$ odd multiple of $\frac{\pi}{2}$

(f) $x \sin(\frac{1}{x})$ for $x \neq 0$

(g) $x^2 \sin(\frac{1}{x})$ for $x \neq 0$

(h) $\frac{1}{x} \sin(\frac{1}{x^2})$ for $x \neq 0$

17.4 Prove the function $\sqrt{x}$ is continuous on its domain $[0, \infty)$. *Hint*: Apply Example 5 in §8.

17.5 (a) Prove that if $m \in \mathbb{N}$, then the function $f(x) = x^m$ is continuous on $\mathbb{R}$.

(b) Prove every *polynomial function* $p(x) = a_0 + a_1 x + \cdots + a_n x^n$ is continuous on $\mathbb{R}$.

17.6 A *rational function* is a function f of the form p/q where p and q are polynomial functions. The domain of f is $\{x \in \mathbb{R} : q(x) \neq 0\}$. Prove every rational function is continuous. *Hint*: Use Exercise 17.5.

17.7 (a) Observe that if $k \in \mathbb{R}$, then the function $g(x) = kx$ is continuous by Exercise 17.5.

(b) Prove $|x|$ is a continuous function on $\mathbb{R}$.

(c) Use (a) and (b) and Theorem 17.5 to give another proof of Theorem 17.3.

17.8 Let f and g be real-valued functions.

(a) Show $\min(f, g) = \frac{1}{2}(f + g) - \frac{1}{2}|f - g|$.

(b) Show $\min(f, g) = -\max(-f, -g)$.

(c) Use (a) or (b) to prove that if f and g are continuous at x_0 in $\mathbb{R}$, then $\min(f, g)$ is continuous at x_0.

17.9 Prove each of the following functions in continuous at x_0 by verifying the ϵ–δ property of Theorem 17.2.

    (a) $f(x) = x^2$, $x_0 = 2$;

    (b) $f(x) = \sqrt{x}$, $x_0 = 0$;

    (c) $f(x) = x\sin(\frac{1}{x})$ for $x \neq 0$ and $f(0) = 0$, $x_0 = 0$;

    (d) $g(x) = x^3$, x_0 arbitrary.
       Hint for (d): $x^3 - x_0^3 = (x - x_0)(x^2 + x_0 x + x_0^2)$.

17.10 Prove the following functions are discontinuous at the indicated points. You may use either Definition 17.1 or the ϵ–δ property in Theorem 17.2.

    (a) $f(x) = 1$ for $x > 0$ and $f(x) = 0$ for $x \leq 0$, $x_0 = 0$;

    (b) $g(x) = \sin(\frac{1}{x})$ for $x \neq 0$ and $g(0) = 0$, $x_0 = 0$;

    (c) $\mathrm{sgn}(x) = -1$ for $x < 0$, $\mathrm{sgn}(x) = 1$ for $x > 0$, and $\mathrm{sgn}(0) = 0$, $x_0 = 0$. The function sgn is called the *signum function*; note $\mathrm{sgn}(x) = \frac{x}{|x|}$ for $x \neq 0$.

17.11 Let f be a real-valued function with $\mathrm{dom}(f) \subseteq \mathbb{R}$. Prove f is continuous at x_0 if and only if, for every *monotonic* sequence (x_n) in $\mathrm{dom}(f)$ converging to x_0, we have $\lim f(x_n) = f(x_0)$. *Hint*: Don't forget Theorem 11.4.

17.12 **(a)** Let f be a continuous real-valued function with domain (a, b). Show that if $f(r) = 0$ for each rational number r in (a, b), then $f(x) = 0$ for all $x \in (a, b)$.

    (b) Let f and g be continuous real-valued functions on (a, b) such that $f(r) = g(r)$ for each rational number r in (a, b). Prove $f(x) = g(x)$ for all $x \in (a, b)$. *Hint*: Use part (a).

17.13 **(a)** Let $f(x) = 1$ for rational numbers x and $f(x) = 0$ for irrational numbers. Show f is discontinuous at every x in $\mathbb{R}$.

    (b) Let $h(x) = x$ for rational numbers x and $h(x) = 0$ for irrational numbers. Show h is continuous at $x = 0$ and at no other point.

17.14 For each nonzero rational number x, write x as $\frac{p}{q}$ where p, q are integers with no common factors and $q > 0$, and then define $f(x) = \frac{1}{q}$. Also define $f(0) = 1$ and $f(x) = 0$ for all $x \in \mathbb{R} \backslash \mathbb{Q}$. Thus $f(x) = 1$ for each integer, $f(\frac{1}{2}) = f(-\frac{1}{2}) = f(\frac{3}{2}) = \cdots = \frac{1}{2}$, etc. Show f is continuous at each point of $\mathbb{R} \setminus \mathbb{Q}$ and discontinuous at each point of $\mathbb{Q}$.

17.15 Let f be a real-valued function whose domain is a subset of $\mathbb{R}$. Show f is continuous at x_0 in dom(f) if and only if, for every sequence (x_n) in dom$(f) \setminus \{x_0\}$ converging to x_0, we have $\lim f(x_n) = f(x_0)$.

17.16 The postage-stamp function P is defined by $P(x) = A$ for $0 \leq x < 1$ and $P(x) = A + Bn$ for $n \leq x < n+1$. The definition of P means that P takes the value A on the interval $[0,1)$, the value $A + B$ on the interval $[1,2)$, the value $A + 2B$ on the interval $[2,3)$, etc. Here A is postage needed for the first ounce, and B is the postage needed for each additional ounce. Show P is discontinuous at every positive integer. Because postage rates tend to increase over time, A and B are actually functions.

§18 Properties of Continuous Functions

A real-valued function f is said to be *bounded* if $\{f(x) : x \in \text{dom}(f)\}$ is a bounded set, i.e., if there exists a real number M such that $|f(x)| \leq M$ for all $x \in \text{dom}(f)$.

18.1 Theorem.
Let f be a continuous real-valued function on a closed interval $[a, b]$. Then f is a bounded function. Moreover, f assumes its maximum and minimum values on $[a, b]$; that is, there exist x_0, y_0 in $[a, b]$ such that $f(x_0) \leq f(x) \leq f(y_0)$ for all $x \in [a, b]$.

Proof
Assume f is not bounded on $[a, b]$. Then to each $n \in \mathbb{N}$ there corresponds an $x_n \in [a, b]$ such that $|f(x_n)| > n$. By the Bolzano-Weierstrass Theorem 11.5, (x_n) has a subsequence (x_{n_k}) that converges to some real number x_0. The number x_0 also must belong to the closed interval $[a, b]$, as noted in Exercise 8.9. Since f is continuous at x_0, we have $\lim_{k \to \infty} f(x_{n_k}) = f(x_0)$, but we also have $\lim_{k \to \infty} |f(x_{n_k})| = +\infty$, which is a contradiction. It follows that f is bounded.

Now let $M = \sup\{f(x) : x \in [a, b]\}$; M is finite by the preceding paragraph. For each $n \in \mathbb{N}$ there exists $y_n \in [a, b]$ such that $M - \frac{1}{n} < f(y_n) \leq M$. Hence we have $\lim f(y_n) = M$. By the Bolzano-Weierstrass theorem, there is a subsequence (y_{n_k}) of (y_n) converging to a limit y_0 in $[a, b]$. Since f is continuous at y_0, we have $f(y_0) = \lim_{k \to \infty} f(y_{n_k})$. Since $(f(y_{n_k}))_{k \in \mathbb{N}}$ is a subsequence of $(f(y_n))_{n \in \mathbb{N}}$, Theorem 11.3 shows $\lim_{k \to \infty} f(y_{n_k}) = \lim_{n \to \infty} f(y_n) = M$ and therefore $f(y_0) = M$. Thus f assumes its maximum at y_0.

The last paragraph applies to the function $-f$, so $-f$ assumes its maximum at some $x_0 \in [a, b]$. It follows easily that f assumes its minimum at x_0; see Exercise 18.1. ■

Theorem 18.1 is used all the time, at least implicitly, in solving maximum-minimum problems in calculus because it is taken for granted that the problems have solutions, namely that a continuous function on a closed interval actually takes on a maximum and a minimum. If the domain is not a closed interval, one needs to be careful; see Exercise 18.3.

Theorem 18.1 is false if the closed interval $[a, b]$ is replaced by an open interval. For example, $f(x) = \frac{1}{x}$ is continuous but unbounded on $(0, 1)$. The function x^2 is continuous and bounded on $(-1, 1)$, but it does not have a maximum value on $(-1, 1)$.

18.2 Intermediate Value Theorem.
If f is a continuous real-valued function on an interval I, then f has the intermediate value property on I: Whenever $a, b \in I$, $a < b$ and y lies between $f(a)$ and $f(b)$ [i.e., $f(a) < y < f(b)$ or $f(b) < y < f(a)$], there exists at least one x in (a, b) such that $f(x) = y$.

Proof
We assume $f(a) < y < f(b)$; the other case is similar. Let $S = \{x \in [a, b] : f(x) < y\}$; see Fig. 18.1. Since a belongs to S, the set S is nonempty, so $x_0 = \sup S$ represents a number in $[a, b]$. For each $n \in \mathbb{N}$, $x_0 - \frac{1}{n}$ is not an upper bound for S, so there exists $s_n \in S$ such that $x_0 - \frac{1}{n} < s_n \le x_0$. Thus $\lim s_n = x_0$ and, since $f(s_n) < y$ for all n, we have

$$f(x_0) = \lim f(s_n) \le y. \tag{1}$$

Let $t_n = \min\{b, x_0 + \frac{1}{n}\}$. Since $x_0 \le t_n \le x_0 + \frac{1}{n}$ we have $\lim t_n = x_0$. Each t_n belongs to $[a, b]$ but not to S, so $f(t_n) \ge y$ for all n. Therefore we have

$$f(x_0) = \lim f(t_n) \ge y;$$

this and (1) imply $f(x_0) = y$. ■

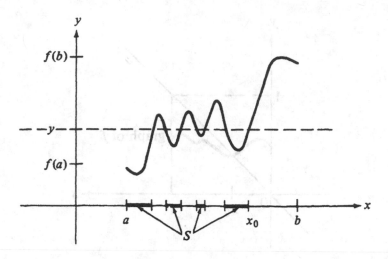

FIGURE 18.1

18.3 Corollary.
If f is a continuous real-valued function on an interval I, then the set $f(I) = \{f(x) : x \in I\}$ is also an interval or a single point.

Proof
The set $J = f(I)$ has the property:

$$y_0, y_1 \in J \quad \text{and} \quad y_0 < y < y_1 \quad \text{imply} \quad y \in J. \qquad (1)$$

If $\inf J < \sup J$, then such a set J will be an interval. In fact, we will show

$$\inf J < y < \sup J \quad \text{implies} \quad y \in J, \qquad (2)$$

so J is an interval with endpoints $\inf J$ and $\sup J$; $\inf J$ and $\sup J$ may or may not belong to J and they may or may not be finite.

To prove (2) from (1), consider $\inf J < y < \sup J$. Then there exist y_0, y_1 in J so that $y_0 < y < y_1$. Thus $y \in J$ by (1). ∎

Example 1
Let f be a continuous function mapping $[0, 1]$ into $[0, 1]$. In other words, $\text{dom}(f) = [0, 1]$ and $f(x) \in [0, 1]$ for all $x \in [0, 1]$. Show f has a *fixed point*, i.e., a point $x_0 \in [0, 1]$ such that $f(x_0) = x_0$; x_0 is left "fixed" by f. □

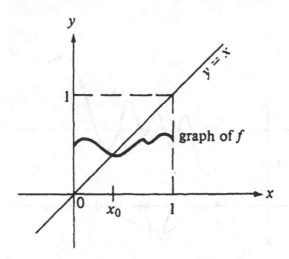

FIGURE 18.2

Solution

The graph of f lies in the unit square; see Fig. 18.2. Our assertion is equivalent to the assertion that the graph of f crosses the $y = x$ line, which is almost obvious.

A rigorous proof involves a little trick. Consider $g(x) = f(x) - x$ which is also a continuous function on $[0, 1]$. Since $g(0) = f(0) - 0 = f(0) \geq 0$ and $g(1) = f(1) - 1 \leq 1 - 1 = 0$, the Intermediate Value theorem shows $g(x_0) = 0$ for some $x_0 \in [0, 1]$. Then obviously we have $f(x_0) = x_0$.   □

Example 2

Show that if $y > 0$ and $m \in \mathbb{N}$, then y has a positive mth root.   □

Solution

The function $f(x) = x^m$ is continuous [Exercise 17.5]. There exists $b > 0$ so that $y \leq b^m$; in fact, if $y \leq 1$ let $b = 1$ and if $y > 1$ let $b = y$. Thus $f(0) < y \leq f(b)$ and the Intermediate Value theorem implies $f(x) = y$ for some x in $(0, b]$. So $y = x^m$ and x is an mth root of y.   □

Let us analyze the function $f(x) = x^m$ in Example 2 more closely. It is a *strictly increasing function* on $[0, \infty)$:

$$0 \leq x_1 < x_2 \quad \text{implies} \quad x_1^m < x_2^m.$$

Therefore f is one-to-one and each nonnegative y has exactly one nonnegative mth root. This assures us that the notation $y^{1/m}$ is unambiguous. In fact, $f^{-1}(y) = y^{1/m}$ is the *inverse function* of f since $f^{-1} \circ f(x) = x$ for $x \in \text{dom}(f)$ and $f \circ f^{-1}(y) = y$ for $y \in \text{dom}(f^{-1})$. Since $f(x) = x^m$ is continuous, the function $f^{-1}(y) = y^{1/m}$ is continuous on $[0, \infty)$ by the next theorem. Note that for $m = 2$ this result appears in Exercise 17.4.

18.4 Theorem.
Let f be a continuous strictly increasing function on some interval I. Then $f(I)$ is an interval J by Corollary 18.3 and f^{-1} represents a function with domain J. The function f^{-1} is a continuous strictly increasing function on J.

Proof
The function f^{-1} is easily shown to be strictly increasing. Since f^{-1} maps J onto I, the next theorem shows f^{-1} is continuous. ∎

18.5 Theorem.
Let g be a strictly increasing function on an interval J such that $g(J)$ is an interval I. Then g is continuous on J.

Proof
Consider x_0 in J. We assume x_0 is not an endpoint of J; tiny changes in the proof are needed otherwise. Then $g(x_0)$ is not an endpoint of I, so there exists $\epsilon_0 > 0$ such that $(g(x_0) - \epsilon_0, g(x_0) + \epsilon_0) \subseteq I$.

Let $\epsilon > 0$. Since we only need to verify the ϵ–δ property of Theorem 17.2 for small ϵ, we may assume $\epsilon < \epsilon_0$. Then there exist $x_1, x_2 \in J$ such that $g(x_1) = g(x_0) - \epsilon$ and $g(x_2) = g(x_0) + \epsilon$. Clearly we have $x_1 < x_0 < x_2$. Also, if $x_1 < x < x_2$, then $g(x_1) < g(x) < g(x_2)$, hence $g(x_0) - \epsilon < g(x) < g(x_0) + \epsilon$, and hence $|g(x) - g(x_0)| < \epsilon$. Now if we put $\delta = \min\{x_2 - x_0, x_0 - x_1\}$, then $|x - x_0| < \delta$ implies $x_1 < x < x_2$ and hence $|g(x) - g(x_0)| < \epsilon$. ∎

Theorem 18.5 provides a partial converse to the Intermediate Value theorem, since it tells us that a strictly increasing function with the intermediate value property is continuous. However, Exercise 18.12 shows that a function can have the intermediate value property without being continuous.

18.6 Theorem.
Let f be a one-to-one continuous function on an interval I. Then f is strictly increasing $[x_1 < x_2$ implies $f(x_1) < f(x_2)]$ or strictly decreasing $[x_1 < x_2$ implies $f(x_1) > f(x_2)]$.

Proof
First we show

$$\text{if } a < b < c \text{ in } I, \text{ then } f(b) \text{ lies between } f(a) \text{ and } f(c). \quad (1)$$

If not, then $f(b) > \max\{f(a), f(c)\}$, say. Select y so that $f(b) > y > \max\{f(a), f(c)\}$. By the Intermediate Value Theorem 18.2 applied to $[a, b]$ and to $[b, c]$, there exist $x_1 \in (a, b)$ and $x_2 \in (b, c)$ such that $f(x_1) = f(x_2) = y$. This contradicts the one-to-one property of f.

Now select any $a_0 < b_0$ in I and suppose, say, that $f(a_0) < f(b_0)$. We will show f is strictly increasing on I. By (1) we have

$$
\begin{array}{llll}
f(x) < f(a_0) & \text{for} & x < a_0 & [\text{since} \quad x < a_0 < b_0], \\
f(a_0) < f(x) < f(b_0) & \text{for} & a_0 < x < b_0, & \\
f(b_0) < f(x) & \text{for} & x > b_0 & [\text{since} \quad a_0 < b_0 < x].
\end{array}
$$

In particular,

$$f(x) < f(a_0) \quad \text{for all} \quad x < a_0, \quad (2)$$

$$f(a_0) < f(x) \quad \text{for all} \quad x > a_0. \quad (3)$$

Consider any $x_1 < x_2$ in I. If $x_1 \leq a_0 \leq x_2$, then $f(x_1) < f(x_2)$ by (2) and (3). If $x_1 < x_2 < a_0$, then $f(x_1) < f(a_0)$ by (2), so by (1) we have $f(x_1) < f(x_2)$. Finally, if $a_0 < x_1 < x_2$, then $f(a_0) < f(x_2)$, so that $f(x_1) < f(x_2)$. ∎

Exercises

18.1 Let f be as in Theorem 18.1. Show that if $-f$ assumes its maximum at $x_0 \in [a, b]$, then f assumes its minimum at x_0.

18.2 Reread the proof of Theorem 18.1 with $[a, b]$ replaced by (a, b). Where does it break down? Discuss.

18.3 Use calculus to find the maximum and minimum of $f(x) = x^3 - 6x^2 + 9x + 1$ on $[0, 5)$.

18.4 Let $S \subseteq \mathbb{R}$ and suppose there exists a sequence (x_n) in S converging to a number $x_0 \notin S$. Show there exists an unbounded continuous function on S.

18.5 **(a)** Let f and g be continuous functions on $[a, b]$ such that $f(a) \geq g(a)$ and $f(b) \leq g(b)$. Prove $f(x_0) = g(x_0)$ for at least one x_0 in $[a, b]$.

(b) Show Example 1 can be viewed as a special case of part (a).

18.6 Prove $x = \cos x$ for some x in $(0, \frac{\pi}{2})$.

18.7 Prove $xe^x = 2$ for some x in $(0, 1)$.

18.8 Suppose f is a real-valued continuous function on $\mathbb{R}$ and $f(a)f(b) < 0$ for some $a, b \in \mathbb{R}$. Prove there exists x between a and b such that $f(x) = 0$.

18.9 Prove that a polynomial function f of odd degree has at least one real root. *Hint*: It may help to consider first the case of a cubic, i.e., $f(x) = a_0 + a_1 x + a_2 x^2 + a_3 x^3$ where $a_3 \neq 0$.

18.10 Suppose f is continuous on $[0, 2]$ and $f(0) = f(2)$. Prove there exist x, y in $[0, 2]$ such that $|y - x| = 1$ and $f(x) = f(y)$. *Hint*: Consider $g(x) = f(x + 1) - f(x)$ on $[0, 1]$.

18.11 **(a)** Show that if f is strictly increasing on an interval I, then $-f$ is strictly decreasing on I.

(b) State and prove Theorems 18.4 and 18.5 for strictly decreasing functions.

18.12 Let $f(x) = \sin(\frac{1}{x})$ for $x \neq 0$ and let $f(0) = 0$.

(a) Observe that f is discontinuous at 0 by Exercise 17.10(b).

(b) Show f has the intermediate value property on $\mathbb{R}$.

§19 Uniform Continuity

Let f be a real-valued function whose domain is a subset of $\mathbb{R}$. Theorem 17.2 tells us that f is continuous on a set $S \subseteq \text{dom}(f)$ if and only if

$$\text{for each } x_0 \in S \text{ and } \epsilon > 0 \text{ there is } \delta > 0 \text{ so that} \atop x \in \text{dom}(f) \text{ and } |x - x_0| < \delta \text{ imply } |f(x) - f(x_0)| < \epsilon. \tag{*}$$

The choice of δ depends on $\epsilon > 0$ and on the point x_0 in S.

Example 1

We verify (*) for the function $f(x) = \frac{1}{x^2}$ on $(0, \infty)$. Let $x_0 > 0$ and $\epsilon > 0$. We need to show $|f(x) - f(x_0)| < \epsilon$ for $|x - x_0|$ sufficiently small. Note that

$$f(x) - f(x_0) = \frac{1}{x^2} - \frac{1}{x_0^2} = \frac{x_0^2 - x^2}{x^2 x_0^2} = \frac{(x_0 - x)(x_0 + x)}{x^2 x_0^2}. \quad (1)$$

If $|x - x_0| < \frac{x_0}{2}$, then we have $|x| > \frac{x_0}{2}$, $|x| < \frac{3x_0}{2}$ and $|x_0 + x| < \frac{5x_0}{2}$. These observations and (1) show that if $|x - x_0| < \frac{x_0}{2}$, then

$$|f(x) - f(x_0)| < \frac{|x_0 - x| \cdot \frac{5x_0}{2}}{(\frac{x_0}{2})^2 x_0^2} = \frac{10|x_0 - x|}{x_0^3}.$$

Thus if we set $\delta = \min\{\frac{x_0}{2}, \frac{x_0^3 \epsilon}{10}\}$, then

$$|x - x_0| < \delta \quad \text{implies} \quad |f(x) - f(x_0)| < \epsilon.$$

This establishes (*) for f on $(0, \infty)$. Note that δ depends on both ϵ and x_0. Even if ϵ is fixed, δ gets small when x_0 is small. This shows that *our* choice of δ definitely depends on x_0 as well as ϵ, though this may be because we obtained δ via sloppy estimates. As a matter of fact, in this case δ *must* depend on x_0 as well as ϵ; see Example 3. Figure 19.1 shows how a fixed ϵ requires smaller and smaller δ as x_0 approaches 0. [In the figure, δ_1 signifies a δ that works for x_1 and ϵ, δ_2 signifies a δ that works for x_2 and ϵ, etc.] □

It turns out to be very useful to know when the δ in condition (*) can be chosen to depend only on $\epsilon > 0$ and S, so that δ does not depend on the particular point x_0. Such functions are said to be uniformly continuous on S. In the definition, the points x and x_0 play a symmetric role, so we will call them x and y.

19.1 Definition.

Let f be a real-valued function defined on a set $S \subseteq \mathbb{R}$. Then f is *uniformly continuous on S* if

$$\text{for each } \epsilon > 0 \text{ there exists } \delta > 0 \text{ such that}$$
$$x, y \in S \text{ and } |x - y| < \delta \text{ imply } |f(x) - f(y)| < \epsilon. \quad (1)$$

We will say f is *uniformly continuous* if f is uniformly continuous on dom(f).

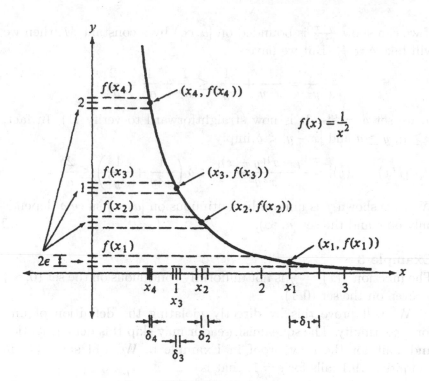

FIGURE 19.1

Note that if a function is uniformly continuous on its domain, then it is continuous on its domain. This should be obvious; if it isn't, Theorem 17.2 and Definition 19.1 should be carefully scrutinized. Note also that uniform continuity is a property concerning a function *and a set* [on which it is defined]. It makes no sense to speak of a function being uniformly continuous at each point.

Example 2
We show $f(x) = \frac{1}{x^2}$ is uniformly continuous on any set of the form $[a, \infty)$ where $a > 0$. Here a is fixed. Let $\epsilon > 0$. We need to show there exists $\delta > 0$ such that

$$x \geq a, \quad y \geq a \quad \text{and} \quad |x - y| < \delta \quad \text{imply} \quad |f(x) - f(y)| < \epsilon. \quad (1)$$

As in formula (1) of Example 1, we have

$$f(x) - f(y) = \frac{(y - x)(y + x)}{x^2 y^2}.$$

If we can show $\frac{y+x}{x^2y^2}$ is bounded on $[a, \infty)$ by a constant M, then we will take $\delta = \frac{\epsilon}{M}$. But we have

$$\frac{y+x}{x^2y^2} = \frac{1}{x^2y} + \frac{1}{xy^2} \leq \frac{1}{a^3} + \frac{1}{a^3} = \frac{2}{a^3},$$

so we set $\delta = \frac{\epsilon a^3}{2}$. It is now straightforward to verify (1). In fact, $x \geq a$, $y \geq a$ and $|x - y| < \delta$ imply

$$|f(x) - f(y)| = \frac{|y - x|(y + x)|}{x^2y^2} < \delta\left(\frac{1}{x^2y} + \frac{1}{xy^2}\right) \leq \frac{2\delta}{a^3} = \epsilon.$$

We have shown f is uniformly continuous on $[a, \infty)$ since δ depends only on ϵ and the set $[a, \infty)$. $\qquad \square$

Example 3
The function $f(x) = \frac{1}{x^2}$ is not uniformly continuous on the set $(0, \infty)$ or even on the set $(0, 1)$.

We will prove this by directly violating the definition of uniform continuity. The squeamish reader may skip this demonstration and wait for the easy proof in Example 6. We will show (1) in Definition 19.1 fails for $\epsilon = 1$; that is

$$\text{for each } \delta > 0 \text{ there exist } x, y \text{ in } (0, 1) \text{ such that} \tag{1}$$
$$|x - y| < \delta \text{ and yet } |f(x) - f(y)| \geq 1.$$

[Actually, for this function, (1) in Definition 19.1 fails for all $\epsilon > 0$.] To show (1) it suffices to take $y = x + \frac{\delta}{2}$ and arrange for

$$\left|f(x) - f\left(x + \frac{\delta}{2}\right)\right| \geq 1. \tag{2}$$

[The motivation for this maneuver is to go from two variables, x and y, in (1) to one variable, x, in (2).] By (1) in Example 1, (2) is equivalent to

$$1 \leq \frac{(x + \frac{\delta}{2} - x)(x + \frac{\delta}{2} + x)}{x^2(x + \frac{\delta}{2})^2} = \frac{\delta(2x + \frac{\delta}{2})}{2x^2(x + \frac{\delta}{2})^2}. \tag{3}$$

It suffices to prove (1) for $\delta < \frac{1}{2}$. To apply (3), let us try $x = \delta$ for no particular reason. Then

$$\frac{\delta(2\delta + \frac{\delta}{2})}{2\delta^2(\delta + \frac{\delta}{2})^2} = \frac{5\delta^2/2}{9\delta^4/2} = \frac{5}{9\delta^2} > \frac{5}{9(\frac{1}{2})^2} = \frac{20}{9} > 1.$$

We were lucky! To summarize, we have shown that if $0 < \delta < \frac{1}{2}$, then $|f(\delta) - f(\delta + \frac{\delta}{2})| > 1$, so (1) holds with $x = \delta$ and $y = \delta + \frac{\delta}{2}$. □

Example 4
Is the function $f(x) = x^2$ uniformly continuous on $[-7, 7]$? To see if it is, consider $\epsilon > 0$. Note that $|f(x) - f(y)| = |x^2 - y^2| = |x - y| \cdot |x + y|$. Since $|x + y| \le 14$ for x, y in $[-7, 7]$, we have

$$|f(x) - f(y)| \le 14|x - y| \quad \text{for} \quad x, y \in [-7, 7].$$

Thus if $\delta = \frac{\epsilon}{14}$, then

$$x, y \in [-7, 7] \quad \text{and} \quad |x - y| < \delta \quad \text{imply} \quad |f(x) - f(y)| < \epsilon.$$

We have shown that f is uniformly continuous on $[-7, 7]$. A similar proof would work for $f(x) = x^2$ on any closed interval. However, these results are not accidents as the next important theorem shows. □

19.2 Theorem.
If f is continuous on a closed interval $[a, b]$, then f is uniformly continuous on $[a, b]$.

Proof
Assume f is not uniformly continuous on $[a, b]$. Then there exists $\epsilon > 0$ such that for each $\delta > 0$ the implication

$$\text{"}|x - y| < \delta \quad \text{implies} \quad |f(x) - f(y)| < \epsilon\text{"}$$

fails. That is, for each $\delta > 0$ there exist $x, y \in [a, b]$ such that $|x - y| < \delta$ and yet $|f(x) - f(y)| \ge \epsilon$. Thus for each $n \in \mathbb{N}$ there exist x_n, y_n in $[a, b]$ such that $|x_n - y_n| < \frac{1}{n}$ and yet $|f(x_n) - f(y_n)| \ge \epsilon$. By the Bolzano-Weierstrass Theorem 11.5, a subsequence (x_{n_k}) of (x_n) converges. Moreover, if $x_0 = \lim_{k \to \infty} x_{n_k}$, then x_0 belongs to $[a, b]$; see Exercise 8.9. Clearly we also have $x_0 = \lim_{k \to \infty} y_{n_k}$. Since f is continuous at x_0, we have

$$f(x_0) = \lim_{k \to \infty} f(x_{n_k}) = \lim_{k \to \infty} f(y_{n_k}),$$

so

$$\lim_{k \to \infty} [f(x_{n_k}) - f(y_{n_k})] = 0.$$

Since $|f(x_{n_k}) - f(y_{n_k})| \ge \epsilon$ for all k, we have a contradiction. We conclude f is uniformly continuous on $[a, b]$. ∎

The preceding proof used only two properties of $[a, b]$:

(a) Boundedness, so the Bolzano-Weierstrass theorem applies;

(b) A convergent sequence in $[a, b]$ converges to an element in $[a, b]$.

As noted prior to Theorem 11.9, sets with property (b) are called *closed sets*. Hence Theorem 19.2 has the following generalization. *If f is continuous on a closed and bounded set S, then f is uniformly continuous on S.* See also Theorems 21.4 and 13.12 that appear in enrichment sections.

Example 5

In view of Theorem 19.2, the following functions are uniformly continuous on the indicated sets: x^{73} on $[-13, 13]$, $\sqrt{x}$ on $[0, 400]$, $x^{17} \sin(e^x) - e^{4x} \cos 2x$ on $[-8\pi, 8\pi]$, and $\frac{1}{x^6}$ on $[\frac{1}{4}, 44]$. □

19.3 Discussion.

Example 5 illustrates the power of Theorem 19.2, but it still may not be clear why uniform continuity is worth studying. One of the important applications of uniform continuity concerns the integrability of continuous functions on closed intervals. To see the relevance of uniform continuity, consider a continuous nonnegative real-valued function f on $[0, 1]$. For $n \in \mathbb{N}$ and $i = 0, 1, 2, \ldots, n - 1$, let

$$M_{i,n} = \sup\left\{f(x) : x \in \left[\tfrac{i}{n}, \tfrac{i+1}{n}\right]\right\} \quad \text{and} \quad m_{i,n} = \inf\left\{f(x) : x \in \left[\tfrac{i}{n}, \tfrac{i+1}{n}\right]\right\}.$$

Then the sum of the areas of the rectangles in Fig. 19.2a equals

$$U_n = \frac{1}{n} \sum_{i=0}^{n-1} M_{i,n}$$

and the sum of the areas of the rectangles in Fig. 19.2b equals

$$L_n = \frac{1}{n} \sum_{i=0}^{n-1} m_{i,n}.$$

The function f would turn out to be Riemann integrable provided the numbers U_n and L_n are close together for large n, i.e., if

$$\lim_{n \to \infty} (U_n - L_n) = 0; \tag{1}$$

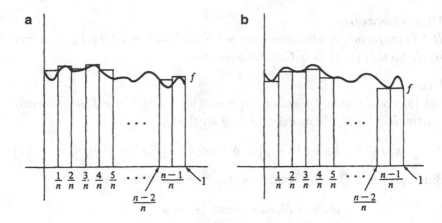

FIGURE 19.2

see Exercise 32.6. Moreover, we would have $\int_0^1 f(x)dx = \lim U_n = \lim L_n$. Relation (1) may appear obvious from Fig. 19.2, but uniform continuity is needed to prove it. First note

$$0 \le U_n - L_n = \frac{1}{n}\sum_{i=0}^{n-1}(M_{i,n} - m_{i,n})$$

for all n. Let $\epsilon > 0$. By Theorem 19.2, f is uniformly continuous on $[0,1]$, so there exists $\delta > 0$ such that

$$x, y \in [0,1] \quad \text{and} \quad |x - y| < \delta \quad \text{imply} \quad |f(x) - f(y)| < \epsilon. \qquad (2)$$

Select N so that $\frac{1}{N} < \delta$. Consider $n > N$; for $i = 0, 1, 2, \ldots, n-1$, Theorem 18.1 shows there exist x_i, y_i in $[\frac{i}{n}, \frac{i+1}{n}]$ satisfying $f(x_i) = m_{i,n}$ and $f(y_i) = M_{i,n}$. Since $|x_i - y_i| \le \frac{1}{n} < \frac{1}{N} < \delta$, (2) shows

$$M_{i,n} - m_{i,n} = f(y_i) - f(x_i) < \epsilon,$$

so

$$0 \le U_n - L_n = \frac{1}{n}\sum_{i=0}^{n-1}(M_{i,n} - m_{i,n}) < \frac{1}{n}\sum_{i=0}^{n-1}\epsilon = \epsilon.$$

This proves (1) as desired. $\qquad\qquad\qquad\qquad\qquad\qquad\qquad$ $\square$

The next two theorems show uniformly continuous functions have nice properties.

19.4 Theorem.
If f is uniformly continuous on a set S and (s_n) is a Cauchy sequence in S, then $(f(s_n))$ is a Cauchy sequence.

Proof
Let (s_n) be a Cauchy sequence in S and let $\epsilon > 0$. Since f is uniformly continuous on S, there exists $\delta > 0$ so that

$$x, y \in S \quad \text{and} \quad |x - y| < \delta \quad \text{imply} \quad |f(x) - f(y)| < \epsilon. \qquad (1)$$

Since (s_n) is a Cauchy sequence, there exists N so that

$$m, n > N \quad \text{implies} \quad |s_n - s_m| < \delta.$$

From (1) we see that

$$m, n > N \quad \text{implies} \quad |f(s_n) - f(s_m)| < \epsilon.$$

This proves $(f(s_n))$ is also a Cauchy sequence. ∎

Example 6
We show $f(x) = \frac{1}{x^2}$ is not uniformly continuous on $(0, 1)$. Let $s_n = \frac{1}{n}$ for $n \in \mathbb{N}$. Then (s_n) is obviously a Cauchy sequence in $(0, 1)$. Since $f(s_n) = n^2$, $(f(s_n))$ is not a Cauchy sequence. Therefore f cannot be uniformly continuous on $(0, 1)$ by Theorem 19.4. □

The next theorem involves extensions of functions. We say a function $\tilde{f}$ is an *extension of a function f* if

$$\text{dom}(f) \subseteq \text{dom}(\tilde{f}) \quad \text{and} \quad f(x) = \tilde{f}(x) \quad \text{for all} \quad x \in \text{dom}(f).$$

Example 7
Let $f(x) = x \sin(\frac{1}{x})$ for $x \in (0, \frac{1}{\pi}]$. The function defined by

$$\tilde{f}(x) = \begin{cases} x \sin(\frac{1}{x}) & \text{for} \quad 0 < x \leq \frac{1}{\pi} \\ 0 & \text{for} \quad x = 0 \end{cases}$$

is an extension of f. Note that $\text{dom}(f) = (0, \frac{1}{\pi}]$ and $\text{dom}(\tilde{f}) = [0, \frac{1}{\pi}]$. In this case, $\tilde{f}$ is a continuous extension of f. See Fig. 19.3 as well as Exercises 17.3(f) and 17.9(c). □

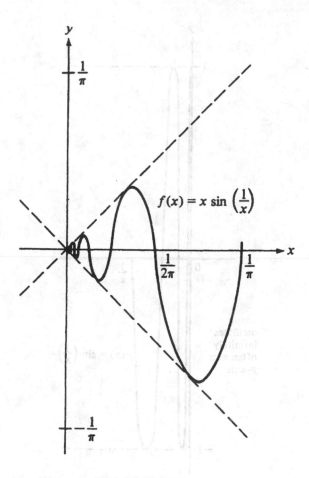

$$f(x) = x \sin\left(\frac{1}{x}\right)$$

FIGURE 19.3

Example 8

Let $g(x) = \sin(\frac{1}{x})$ for $x \in (0, \frac{1}{\pi}]$. The function g can be extended to a function $\tilde{g}$ with domain $[0, \frac{1}{\pi}]$ in many ways, but $\tilde{g}$ will not be continuous. See Fig. 19.4. ☐

The function f in Example 7 is uniformly continuous [since $\tilde{f}$ is], and f extends to a continuous function on the closed interval. The function g in Example 8 does not extend to a continuous function on the closed interval, and it turns out that g is not uniformly continuous. These examples illustrate the next theorem.

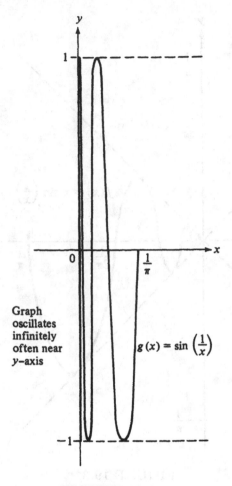

FIGURE 19.4

19.5 Theorem.
A real-valued function f on (a, b) is uniformly continuous on (a, b) if and only if it can be extended to a continuous function $\tilde{f}$ on $[a, b]$.

Proof
First suppose f can be extended to a continuous function $\tilde{f}$ on $[a, b]$. Then $\tilde{f}$ is uniformly continuous on $[a, b]$ by Theorem 19.2, so clearly f is uniformly continuous on (a, b).

Suppose now that f is uniformly continuous on (a, b). We need to define $\tilde{f}(a)$ and $\tilde{f}(b)$ so that the extended function will be continuous.

It suffices for us to deal with $\tilde{f}(a)$. We make two claims:

$$\text{if } (s_n) \text{ is a sequence in } (a,b) \text{ converging} \atop \text{to } a, \text{ then } (f(s_n)) \text{ converges,}} \tag{1}$$

and

$$\text{if } (s_n) \text{ and } (t_n) \text{ are sequences in } (a,b) \text{ converging} \atop \text{to } a, \text{ then } \lim f(s_n) = \lim f(t_n).} \tag{2}$$

Momentarily accepting (1) and (2) as valid, we define

$$\tilde{f}(a) = \lim f(s_n) \text{ for any sequence} \atop (s_n) \text{ in } (a,b) \text{ converging to } a.} \tag{3}$$

Assertion (1) guarantees the limit exists, and assertion (2) guarantees this definition is unambiguous. The continuity of $\tilde{f}$ at a follows directly from (3); see Exercise 17.15.

To prove (1), note that (s_n) is a Cauchy sequence, so $(f(s_n))$ is also a Cauchy sequence by Theorem 19.4. Hence $(f(s_n))$ converges by Theorem 10.11. To prove (2) we create a third sequence (u_n) such that (s_n) and (t_n) are both subsequences of (u_n). In fact, we simply interleaf (s_n) and (t_n):

$$(u_n)_{n=1}^{\infty} = (s_1, t_1, s_2, t_2, s_3, t_3, s_4, t_4, s_5, t_5, \ldots).$$

It is evident that $\lim u_n = a$, so $\lim f(u_n)$ exists by (1). Theorem 11.3 shows the subsequences $(f(s_n))$ and $(f(t_n))$ of $(f(u_n))$ both converge to $\lim f(u_n)$, so $\lim f(s_n) = \lim f(t_n)$. ∎

Example 9

Let $h(x) = \frac{\sin x}{x}$ for $x \neq 0$. The function $\tilde{h}$ defined on $\mathbb{R}$ by

$$\tilde{h}(x) = \begin{cases} \frac{\sin x}{x} & \text{for} \quad x \neq 0 \\ 1 & \text{for} \quad x = 0 \end{cases}$$

is an extension of h. Clearly h and $\tilde{h}$ are continuous at all $x \neq 0$. It turns out that $\tilde{h}$ is continuous at $x = 0$ [see below], so h is uniformly continuous on $(a, 0)$ and $(0, b)$ for any $a < 0 < b$ by Theorem 19.5. In fact, $\tilde{h}$ is uniformly continuous on $\mathbb{R}$ [Exercise 19.11].

We cannot prove the continuity of $\tilde{h}$ at 0 in this book because we do not give a definition of $\sin x$. The continuity of $\tilde{h}$ at 0 reflects

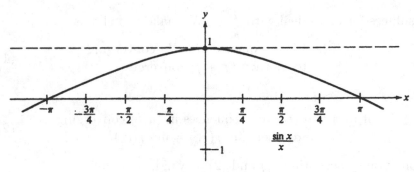

FIGURE 19.5

the fact that $\sin x$ is differentiable at 0 and its derivative there is $\cos(0) = 1$, i.e.,

$$1 = \lim_{x \to 0} \frac{\sin x - \sin 0}{x - 0} = \lim_{x \to 0} \frac{\sin x}{x};$$

see Fig. 19.5. The proof of this depends on how $\sin x$ is defined; see the brief discussion in 37.12. For a discussion of this limit and L'Hospital's rule, see Example 1 in §30. □

Here is another useful criterion that implies uniform continuity.

19.6 Theorem.
Let f be a continuous function on an interval I $[I$ may be bounded or unbounded]. Let I° be the interval obtained by removing from I any endpoints that happen to be in I. If f is differentiable on I° and if f' is bounded on I°, then f is uniformly continuous on I.

Proof
For this proof we need the Mean Value theorem, which can be found in most calculus texts or later in this book [Theorem 29.3].

Let M be a bound for f' on I so that $|f'(x)| \le M$ for all x. Let $\epsilon > 0$ and let $\delta = \frac{\epsilon}{M}$. Consider $a, b \in I$ where $a < b$ and $|b - a| < \delta$. By the Mean Value theorem, there exists $x \in (a, b)$ such that $f'(x) = \frac{f(b) - f(a)}{b - a}$, so

$$|f(b) - f(a)| = |f'(x)| \cdot |b - a| \le M|b - a| < M\delta = \epsilon.$$

This proves the uniform continuity of f on I. ■

Example 10

Let $a > 0$ and consider $f(x) = \frac{1}{x^2}$. Since $f'(x) = -\frac{2}{x^3}$ we have $|f'(x)| \le \frac{2}{a^3}$ on $[a, \infty)$. Hence f is uniformly continuous on $[a, \infty)$ by Theorem 19.6. For a direct proof of this fact, see Example 2. $\square$

Exercises

19.1 Which of the following continuous functions are uniformly continuous on the specified set? Justify your answers. Use any theorems you wish.

(a) $f(x) = x^{17} \sin x - e^x \cos 3x$ on $[0, \pi]$,

(b) $f(x) = x^3$ on $[0, 1]$,

(c) $f(x) = x^3$ on $(0, 1)$,

(d) $f(x) = x^3$ on $\mathbb{R}$,

(e) $f(x) = \frac{1}{x^3}$ on $(0, 1]$,

(f) $f(x) = \sin \frac{1}{x^2}$ on $(0, 1]$,

(g) $f(x) = x^2 \sin \frac{1}{x}$ on $(0, 1]$.

19.2 Prove each of the following functions is uniformly continuous on the indicated set by directly verifying the ϵ-δ property in Definition 19.1.

(a) $f(x) = 3x + 11$ on $\mathbb{R}$,

(b) $f(x) = x^2$ on $[0, 3]$,

(c) $f(x) = \frac{1}{x}$ on $[\frac{1}{2}, \infty)$.

19.3 Repeat Exercise 19.2 for the following.

(a) $f(x) = \frac{x}{x+1}$ on $[0, 2]$,

(b) $f(x) = \frac{5x}{2x-1}$ on $[1, \infty)$.

19.4 (a) Prove that if f is uniformly continuous on a bounded set S, then f is a bounded function on S. *Hint*: Assume not. Use Theorems 11.5 and 19.4.

(b) Use (a) to give yet another proof that $\frac{1}{x^2}$ is not uniformly continuous on $(0, 1)$.

19.5 Which of the following continuous functions is uniformly continuous on the specified set? Justify your answers, using appropriate theorems or Exercise 19.4(a).

(a) $\tan x$ on $[0, \frac{\pi}{4}]$,

(b) $\tan x$ on $[0, \frac{\pi}{2})$,

(c) $\frac{1}{x} \sin^2 x$ on $(0, \pi]$,

(d) $\frac{1}{x-3}$ on $(0, 3)$,

(e) $\frac{1}{x-3}$ on $(3, \infty)$,

(f) $\frac{1}{x-3}$ on $(4, \infty)$.

19.6 (a) Let $f(x) = \sqrt{x}$ for $x \geq 0$. Show f' is unbounded on $(0, 1]$ but f is nevertheless uniformly continuous on $(0, 1]$. Compare with Theorem 19.6.

(b) Show f is uniformly continuous on $[1, \infty)$.

19.7 (a) Let f be a continuous function on $[0, \infty)$. Prove that if f is uniformly continuous on $[k, \infty)$ for some k, then f is uniformly continuous on $[0, \infty)$.

(b) Use (a) and Exercise 19.6(b) to prove $\sqrt{x}$ is uniformly continuous on $[0, \infty)$.

19.8 (a) Use the Mean Value theorem to prove

$$|\sin x - \sin y| \leq |x - y|$$

for all x, y in $\mathbb{R}$; see the proof of Theorem 19.6.

(b) Show $\sin x$ is uniformly continuous on $\mathbb{R}$.

19.9 Let $f(x) = x \sin(\frac{1}{x})$ for $x \neq 0$ and $f(0) = 0$.

(a) Observe f is continuous on $\mathbb{R}$; see Exercises 17.3(f) and 17.9(c).

(b) Why is f uniformly continuous on any bounded subset of $\mathbb{R}$?

(c) Is f uniformly continuous on $\mathbb{R}$?

19.10 Repeat Exercise 19.9 for the function g where $g(x) = x^2 \sin(\frac{1}{x})$ for $x \neq 0$ and $g(0) = 0$.

19.11 Accept the fact that the function $\tilde{h}$ in Example 9 is continuous on $\mathbb{R}$; prove it is uniformly continuous on $\mathbb{R}$.

19.12 Let f be a continuous function on $[a, b]$. Show the function f^* defined by $f^*(x) = \sup\{f(y) : a \le y \le x\}$, for $x \in [a, b]$, is an increasing continuous function on $[a, b]$.

§20 Limits of Functions

A function f is continuous at a point a provided the values $f(x)$ are near the value $f(a)$ for x near a [and $x \in \text{dom}(f)$]. See Definition 17.1 and Theorem 17.2. It would be reasonable to view $f(a)$ as the limit of the values $f(x)$, for x near a, and to write $\lim_{x \to a} f(x) = f(a)$. In this section we formalize this notion. This section is needed for our careful study of derivatives in Chap. 5, but it may be deferred until then.

We will be interested in ordinary limits, left-handed and right-handed limits and infinite limits. In order to handle these various concepts efficiently and also to emphasize their common features, we begin with a very general definition, which is not a standard definition.

20.1 Definition.
Let S be a subset of $\mathbb{R}$, let a be a real number or symbol ∞ or $-\infty$ that is the limit of some sequence in S, and let L be a real number or symbol $+\infty$ or $-\infty$. We write $\lim_{x \to a^S} f(x) = L$ if

$$f \text{ is a function defined on } S, \tag{1}$$

and

$$\begin{aligned} &\text{for every sequence } (x_n) \text{ in } S \text{ with limit } a, \\ &\text{we have } \lim_{n \to \infty} f(x_n) = L. \end{aligned} \tag{2}$$

The expression "$\lim_{x \to a^S} f(x)$" is read "limit, as x tends to a along S, of $f(x)$."

20.2 Remarks.
(a) From Definition 17.1 we see that a function f is continuous at a in $\text{dom}(f) = S$ if and only if $\lim_{x \to a^S} f(x) = f(a)$.

(b) Observe that limits, when they exist, are unique. This follows from (2) of Definition 20.1, since limits of sequences are unique, a fact that is verified at the end of §7.   □

We now define the various standard limit concepts for functions.

20.3 Definition.

(a) For $a \in \mathbb{R}$ and a function f we write $\lim_{x \to a} f(x) = L$ provided $\lim_{x \to a} S f(x) = L$ for some set $S = J \setminus \{a\}$ where J is an open interval containing a. $\lim_{x \to a} f(x)$ is called the [*two-sided*] *limit of f at a*. Note f need not be defined at a and, even if f is defined at a, the value $f(a)$ need not equal $\lim_{x \to a} f(x)$. In fact, $f(a) = \lim_{x \to a} f(x)$ if and only if f is defined on an open interval containing a and f is continuous at a.

(b) For $a \in \mathbb{R}$ and a function f we write $\lim_{x \to a^+} f(x) = L$ provided $\lim_{x \to a} S f(x) = L$ for some open interval $S = (a, b)$. $\lim_{x \to a^+} f(x)$ is the *right-hand limit of f at a*. Again f need not be defined at a.

(c) For $a \in \mathbb{R}$ and a function f we write $\lim_{x \to a^-} f(x) = L$ provided $\lim_{x \to a} S f(x) = L$ for some open interval $S = (c, a)$. $\lim_{x \to a^-} f(x)$ is the *left-hand limit of f at a*.

(d) For a function f we write $\lim_{x \to \infty} f(x) = L$ provided $\lim_{x \to \infty} S f(x) = L$ for some interval $S = (c, \infty)$. Likewise, we write $\lim_{x \to -\infty} f(x) = L$ provided $\lim_{x \to -\infty} S f(x) = L$ for some interval $S = (-\infty, b)$.

The limits defined above are unique; i.e., they do not depend on the exact choice of the set S [Exercise 20.19].

Example 1
We have $\lim_{x \to 4} x^3 = 64$ and $\lim_{x \to 2} \frac{1}{x} = \frac{1}{2}$ because the functions x^3 and $\frac{1}{x}$ are continuous at 4 and 2, respectively. It is easy to show $\lim_{x \to 0^+} \frac{1}{x} = +\infty$ and $\lim_{x \to 0^-} \frac{1}{x} = -\infty$; see Exercise 20.14. It follows that $\lim_{x \to 0} \frac{1}{x}$ does *not* exist; see Theorem 20.10.   □

Example 2
Consider $\lim_{x \to 2} \frac{x^2 - 4}{x - 2}$. This is not like Example 1, because the function under the limit is not even defined at $x = 2$. However, we can

rewrite the function as

$$\frac{x^2 - 4}{x - 2} = \frac{(x - 2)(x + 2)}{x - 2} = x + 2 \quad \text{for} \quad x \neq 2.$$

Now it is clear that $\lim_{x \to 2} \frac{x^2-4}{x-2} = \lim_{x \to 2}(x + 2) = 4$. We should emphasize that the functions $\frac{x^2-4}{x-2}$ and $x + 2$ are *not* identical. The domain of $f(x) = \frac{x^2-4}{x-2}$ is $(-\infty, 2) \cup (2, \infty)$ while the domain of $\tilde{f}(x) = x + 2$ is $\mathbb{R}$, so that $\tilde{f}$ is an extension of f. This seems like nitpicking and this example may appear silly, but the function f, not $\tilde{f}$, arises naturally in computing the derivative of $g(x) = x^2$ at $x = 2$. Indeed, using the definition of derivative we have

$$g'(2) = \lim_{x \to 2} \frac{g(x) - g(2)}{x - 2} = \lim_{x \to 2} \frac{x^2 - 4}{x - 2},$$

so our modest computation above shows $g'(2) = 4$. Of course, this is obvious from the formula $g'(x) = 2x$, but we are preparing the foundations of limits and derivatives, so we are beginning with simple examples. □

Example 3
Consider $\lim_{x \to 1} \frac{\sqrt{x}-1}{x-1}$. We employ a trick that should be familiar by now; we multiply the numerator and denominator by $\sqrt{x} + 1$ and obtain

$$\frac{\sqrt{x} - 1}{x - 1} = \frac{(\sqrt{x} - 1)(\sqrt{x} + 1)}{(x - 1)(\sqrt{x} + 1)} = \frac{x - 1}{(x - 1)(\sqrt{x} + 1)} = \frac{1}{\sqrt{x} + 1}$$

for $x \neq 1$. Hence $\lim_{x \to 1} \frac{\sqrt{x}-1}{x-1} = \lim_{x \to 1} \frac{1}{\sqrt{x}+1} = \frac{1}{2}$. We have just laboriously verified that if $h(x) = \sqrt{x}$, then $h'(1) = \frac{1}{2}$. □

Example 4
Let $f(x) = \frac{1}{(x-2)^3}$ for $x \neq 2$. Then $\lim_{x \to \infty} f(x) = \lim_{x \to -\infty} f(x) = 0$, $\lim_{x \to 2+} f(x) = +\infty$ and $\lim_{x \to 2-} f(x) = -\infty$.

To verify $\lim_{x \to \infty} f(x) = 0$, we consider a sequence (x_n) such that $\lim_{n \to \infty} x_n = +\infty$ and show $\lim_{n \to \infty} f(x_n) = 0$. This will show $\lim_{x \to \infty^S} f(x) = 0$ for $S = (2, \infty)$, for example. Exercise 9.11 and Theorem 9.9 show $\lim_{n \to \infty} (x_n - 2)^3 = +\infty$, and then Theorem 9.10

shows

$$\lim_{n \to \infty} f(x_n) = \lim_{n \to \infty} (x_n - 2)^{-3} = 0. \tag{1}$$

Here is a direct proof of (1). Let $\epsilon > 0$. For large n, we need $|x_n - 2|^{-3} < \epsilon$ or $\epsilon^{-1} < |x_n - 2|^3$ or $\epsilon^{-1/3} < |x_n - 2|$. The last inequality holds if $x_n > \epsilon^{-1/3} + 2$. Since $\lim_{n \to \infty} x_n = +\infty$, there exists N so that

$$n > N \quad \text{implies} \quad x_n > \epsilon^{-1/3} + 2.$$

Reversing the algebraic steps above, we find

$$n > N \quad \text{implies} \quad |x_n - 2|^{-3} < \epsilon.$$

This establishes (1).

Similar arguments prove $\lim_{x \to -\infty} f(x) = 0$ and $\lim_{x \to 2+} f(x) = +\infty$. To prove $\lim_{x \to 2-} f(x) = -\infty$, consider a sequence (x_n) such that $x_n < 2$ for all n and $\lim_{n \to \infty} x_n = 2$. Then $2 - x_n > 0$ for all n and $\lim_{n \to \infty}(2 - x_n) = 0$. Hence $\lim_{n \to \infty}(2 - x_n)^3 = 0$ by Theorem 9.4, and 9.10 implies $\lim_{n \to \infty}(2 - x_n)^{-3} = +\infty$. It follows [Exercise 9.10(b)] that

$$\lim_{n \to \infty} f(x_n) = \lim_{n \to \infty} (x_n - 2)^{-3} = -\infty. \tag{2}$$

This proves that $\lim_{x \to 2^S} f(x) = -\infty$ for $S = (-\infty, 2)$, so that $\lim_{x \to 2-} f(x) = -\infty$. Of course, a direct proof of (2) also can be given.

The limits discussed above are confirmed in Fig. 20.1. $\qquad \square$

We will discuss the various limits defined in Definition 20.3 further at the end of this section. First we prove some limit theorems in considerable generality.

20.4 Theorem.
Let f_1 and f_2 be functions for which the limits $L_1 = \lim_{x \to a^S} f_1(x)$ and $L_2 = \lim_{x \to a^S} f_2(x)$ exist and are finite. Then
   (i) $\lim_{x \to a^S}(f_1 + f_2)(x)$ *exists and equals* $L_1 + L_2$;
   (ii) $\lim_{x \to a^S}(f_1 f_2)(x)$ *exists and equals* $L_1 L_2$;
   (iii) $\lim_{x \to a^S}(f_1/f_2)(x)$ *exists and equals* L_1/L_2 *provided* $L_2 \neq 0$
      and $f_2(x) \neq 0$ *for* $x \in S$.

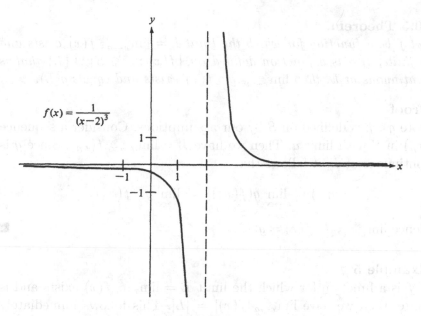

$$f(x) = \frac{1}{(x-2)^3}$$

FIGURE 20.1

Proof

The hypotheses imply both f_1 and f_2 are defined on S and a is the limit of some sequence in S. Clearly the functions $f_1 + f_2$ and $f_1 f_2$ are defined on S and so is f_1/f_2 if $f_2(x) \neq 0$ for $x \in S$.

Consider a sequence (x_n) in S with limit a. By hypotheses we have $L_1 = \lim_{n \to \infty} f_1(x_n)$ and $L_2 = \lim_{n \to \infty} f_2(x_n)$. Theorems 9.3 and 9.4 now show

$$\lim_{n \to \infty} (f_1 + f_2)(x_n) = \lim_{n \to \infty} f_1(x_n) + \lim_{n \to \infty} f_2(x_n) = L_1 + L_2$$

and

$$\lim_{n \to \infty} (f_1 f_2)(x_n) = \left[\lim_{n \to \infty} f_1(x_n) \right] \cdot \left[\lim_{n \to \infty} f_2(x_n) \right] = L_1 L_2.$$

Thus (2) in Definition 20.1 holds for $f_1 + f_2$ and $f_1 f_2$, so that (i) and (ii) hold. Likewise (iii) follows by an application of Theorem 9.6. ∎

Some of the infinite variations of Theorem 20.4 appear in Exercise 20.20. The next theorem is less general than might have been expected; Example 7 shows why.

20.5 Theorem.
Let f be a function for which the limit $L = \lim_{x \to a^S} f(x)$ exists and is finite. If g is a function defined on $\{f(x) : x \in S\} \cup \{L\}$ that is continuous at L, then $\lim_{x \to a^S} g \circ f(x)$ exists and equals $g(L)$.

Proof
Note $g \circ f$ is defined on S by our assumptions. Consider a sequence (x_n) in S with limit a. Then we have $L = \lim_{n \to \infty} f(x_n)$. Since g is continuous at L, it follows that

$$g(L) = \lim_{n \to \infty} g(f(x_n)) = \lim_{n \to \infty} g \circ f(x_n).$$

Hence $\lim_{x \to a^S} g \circ f(x) = g(L)$.   ∎

Example 5
If f is a function for which the limit $L = \lim_{x \to a} f(x)$ exists and is finite, then we have $\lim_{x \to a} |f(x)| = |L|$. This follows immediately from Theorem 20.5 with $g(x) = |x|$. Similarly, we have $\lim_{x \to a} e^{f(x)} = e^L$ if we accept the fact that $g(x) = e^x$ is continuous on $\mathbb{R}$.   □

Example 6
If f is a function for which $\lim_{x \to 0^+} f(x) = 0$ and $\lim_{x \to \infty} f(x) = \frac{\pi}{2}$, then we have $\lim_{x \to 0^+} e^{f(x)} = e^0 = 1$, $\lim_{x \to \infty} e^{f(x)} = e^{\frac{\pi}{2}}$, $\lim_{x \to 0^+} \sin(f(x)) = \sin(0) = 0$ and $\lim_{x \to \infty} \sin(f(x)) = \sin \frac{\pi}{2} = 1$.   □

Example 7
We give an example to show continuity of g is needed in Theorem 20.5. Explicitly, we give examples of functions f and g such that $\lim_{x \to 0} f(x) = 1$, $\lim_{x \to 1} g(x) = 4$ and yet $\lim_{x \to 0} g \circ f(x)$ does not exist. One would expect this limit to exist and to equal 4, but in the example $f(x)$ will *equal* 1 for arbitrarily small x while $g(1) \neq 4$. The functions f and g are defined by $f(x) = 1 + x \sin \frac{\pi}{x}$ for $x \neq 0$, $g(x) = 4$ for $x \neq 1$, and $g(1) = -4$. Clearly $\lim_{x \to 0} f(x) = 1$ and $\lim_{x \to 1} g(x) = 4$. Let $x_n = \frac{2}{n}$ for $n \in \mathbb{N}$. Then $f(x_n) = 1 + \frac{2}{n} \sin(\frac{n\pi}{2})$; hence $f(x_n) = 1$ for even n and $f(x_n) \neq 1$ for odd n. Therefore $g \circ f(x_n) = -4$ for even n and $g \circ f(x_n) = 4$ for odd n. Since $\lim_{n \to \infty} x_n = 0$, we conclude $\lim_{x \to 0} g \circ f(x)$ does not exist.   □

As in Theorem 17.2, the limits defined in Definitions 20.1 and 20.3 can be recast to avoid sequences. First we state and prove a typical result of this sort. Then, after Corollary 20.8, we give a general scheme without proof.

20.6 Theorem.
Let f be a function defined on a subset S of $\mathbb{R}$, let a be a real number that is the limit of some sequence in S, and let L be a real number. then $\lim_{x \to a} {}^S f(x) = L$ if and only if

$$\text{for each } \epsilon > 0 \text{ there exists } \delta > 0 \text{ such that} \tag{1}$$
$$x \in S \text{ and } |x - a| < \delta \text{ imply } |f(x) - L| < \epsilon.$$

Proof
We imitate our proof of Theorem 17.2. Suppose (1) holds, and consider a sequence (x_n) in S such that $\lim_{n \to \infty} x_n = a$. To show $\lim_{n \to \infty} f(x_n) = L$, consider $\epsilon > 0$. By (1) there exists $\delta > 0$ such that

$$x \in S \quad \text{and} \quad |x - a| < \delta \quad \text{imply} \quad |f(x) - L| < \epsilon.$$

Since $\lim_{n \to \infty} x_n = a$, there exists a number N such that $n > N$ implies $|x_n - a| < \delta$. Since $x_n \in S$ for all n, we conclude

$$n > N \quad \text{implies} \quad |f(x_n) - L| < \epsilon.$$

Thus $\lim_{n \to \infty} f(x_n) = L$.

Now assume $\lim_{x \to a} {}^S f(x) = L$, but (1) fails. Then for some $\epsilon > 0$ the implication

$$\text{``} x \in S \quad \text{and} \quad |x - a| < \delta \quad \text{imply} \quad |f(x) - L| < \epsilon \text{''}$$

fails for each $\delta > 0$. Then for each $n \in \mathbb{N}$ there exists x_n in S where $|x_n - a| < \frac{1}{n}$ while $|f(x_n) - L| \geq \epsilon$. Hence (x_n) is a sequence in S with limit a for which $\lim_{n \to \infty} f(x_n) = L$ fails. Consequently $\lim_{x \to a} {}^S f(x) = L$ fails to hold. ■

20.7 Corollary.
Let f be a function defined on $J \setminus \{a\}$ for some open interval J containing a, and let L be a real number. Then $\lim_{x \to a} f(x) = L$ if

and only if

$$\text{for each } \epsilon > 0 \text{ there exists } \delta > 0 \text{ such that} \\ 0 < |x - a| < \delta \text{ implies } |f(x) - L| < \epsilon. \tag{1}$$

20.8 Corollary.
Let f be a function defined on some interval (a, b), and let L be a real number. Then $\lim_{x \to a^+} f(x) = L$ if and only if

$$\text{for each } \epsilon > 0 \text{ there exists } \delta > 0 \text{ such that} \\ 0 < x < a + \delta \text{ implies } |f(x) - L| < \epsilon. \tag{1}$$

20.9 Discussion.
We now consider $\lim_{x \to s} f(x) = L$ where L can be finite, $+\infty$ or $-\infty$, and s is a symbol a, a^+, a^-, ∞ or $-\infty$ [here $a \in \mathbb{R}$]. Note we have 15 $[= 3 \cdot 5]$ different sorts of limits here. It turns out that $\lim_{x \to s} f(x) = L$ if and only if

$$\text{for each } \underline{\hspace{1cm}} \text{ there exists } \underline{\hspace{1cm}} \text{ such that} \\ \underline{\hspace{1cm}} \text{ implies } \underline{\hspace{1cm}}. \tag{1}$$

For finite limits L, the first and last blanks are filled in by "$\epsilon > 0$" and "$|f(x) - L| < \epsilon$." For $L = +\infty$, the first and last blanks are filled in by "$M > 0$" and "$f(x) > M$," while for $L = -\infty$ they are filled in by "$M < 0$" and "$f(x) < M$." When we consider $\lim_{x \to a} f(x)$, then f is defined on $J \setminus \{a\}$ for some open interval J containing a, and the second and third blanks are filled in by "$\delta > 0$" and "$0 < |x - a| < \delta$." For $\lim_{x \to a^+} f(x)$ we require f to be defined on an interval (a, b) and the second and third blanks are filled in by "$\delta > 0$" and "$a < x < a + \delta$." For $\lim_{x \to a^-} f(x)$ we require f to be defined on an interval (c, a) and the second and third blanks are filled in by "$\delta > 0$" and "$a - \delta < x < a$." For $\lim_{x \to \infty} f(x)$ we require f to be defined on an interval (c, ∞) and the second and third blanks are filled in by "$\alpha < \infty$ and "$\alpha < x$." A similar remark applies to $\lim_{x \to -\infty} f(x)$.

The assertions above with L finite and s equal to a or a^+ are contained in Corollaries 20.7 and 20.8. $\qquad \square$

20.10 Theorem.
Let f be a function defined on $J \setminus \{a\}$ for some open interval J containing a. Then $\lim_{x \to a} f(x)$ exists if and only if the limits

$\lim_{x \to a^+} f(x)$ and $\lim_{x \to a^-} f(x)$ both exist and are equal, in which case all three limits are equal.

Proof

Suppose $\lim_{x \to a} f(x) = L$ and L is finite. Then (1) in Corollary 20.7 holds, so (1) in Corollary 20.8 obviously holds. Thus we have $\lim_{x \to a^+} f(x) = L$; similarly $\lim_{x \to a^-} f(x) = L$.

Now suppose $\lim_{x \to a^+} f(x) = \lim_{x \to a^-} f(x) = L$ where L is finite. Consider $\epsilon > 0$; we apply Corollary 20.8 and its analogue for a^- to obtain $\delta_1 > 0$ and $\delta_2 > 0$ such that

$$a < x < a + \delta_1 \quad \text{implies} \quad |f(x) - L| < \epsilon$$

and

$$a - \delta_2 < x < a \quad \text{implies} \quad |f(x) - L| < \epsilon.$$

If $\delta = \min\{\delta_1, \delta_2\}$, then

$$0 < |x - a| < \delta \quad \text{implies} \quad |f(x) - L| < \epsilon,$$

so $\lim_{x \to a} f(x) = L$ by Corollary 20.7.

Similar arguments apply if the limits L are infinite. For example, suppose $\lim_{x \to a} f(x) = +\infty$ and consider $M > 0$. There exists $\delta > 0$ such that

$$0 < |x - a| < \delta \quad \text{implies} \quad f(x) > M. \tag{1}$$

Then clearly

$$a < x < a + \delta \quad \text{implies} \quad f(x) > M \tag{2}$$

and

$$a - \delta < x < a \quad \text{implies} \quad f(x) > M, \tag{3}$$

so that $\lim_{x \to a^+} f(x) = \lim_{x \to a^-} f(x) = +\infty$.

As a last example, suppose $\lim_{x \to a^+} f(x) = \lim_{x \to a^-} f(x) = +\infty$. For each $M > 0$ there exists $\delta_1 > 0$ so that (2) holds, and there exists $\delta_2 > 0$ so that (3) holds. Then (1) holds with $\delta = \min\{\delta_1, \delta_2\}$. We conclude $\lim_{x \to a} f(x) = +\infty$. ∎

20.11 Remark.

Note $\lim_{x\to-\infty} f(x)$ is very similar to the right-hand limits $\lim_{x\to a+} f(x)$. For example, if L is finite, then $\lim_{x\to a+} f(x) = L$ if and only if

$$\text{for each } \epsilon > 0 \text{ there exists } \alpha > a \text{ such that} \tag{1}$$
$$a < x < \alpha \text{ implies } |f(x) - L| < \epsilon,$$

since $\alpha > a$ if and only if $\alpha = a + \delta$ for some $\delta > 0$; see Corollary 20.8. If we set $a = -\infty$ in (1), we obtain the condition (1) in Discussion 20.9 equivalent to $\lim_{x\to-\infty} f(x) = L$.

In the same way, the limits $\lim_{x\to\infty} f(x)$ and $\lim_{x\to a-} f(x)$ will equal L [L finite] if and only if

$$\text{for each } \epsilon > 0 \text{ there exists } \alpha < a \text{ such that} \tag{2}$$
$$\alpha < x < a \text{ implies } |f(x) - L| < \epsilon.$$

Obvious changes are needed if L is infinite. □

Exercises

20.1 Sketch the function $f(x) = \frac{x}{|x|}$. Determine, by inspection, the limits $\lim_{x\to\infty} f(x)$, $\lim_{x\to 0+} f(x)$, $\lim_{x\to 0-} f(x)$, $\lim_{x\to-\infty} f(x)$ and $\lim_{x\to 0} f(x)$ *when they exist*. Also indicate when they do not exist.

20.2 Repeat Exercise 20.1 for $f(x) = \frac{x^3}{|x|}$.

20.3 Repeat Exercise 20.1 for $f(x) = \frac{\sin x}{x}$. See Example 9 of §19.

20.4 Repeat Exercise 20.1 for $f(x) = x \sin \frac{1}{x}$.

20.5 Prove the limit assertions in Exercise 20.1.

20.6 Prove the limit assertions in Exercise 20.2.

20.7 Prove the limit assertions in Exercise 20.3.

20.8 Prove the limit assertions in Exercise 20.4.

20.9 Repeat Exercise 20.1 for $f(x) = \frac{1-x^2}{x}$.

20.10 Prove the limit assertions in Exercise 20.9.

20.11 Find the following limits.
   (a) $\lim_{x\to a} \frac{x^2-a^2}{x-a}$ (b) $\lim_{x\to b} \frac{\sqrt{x}-\sqrt{b}}{x-b}$, $b > 0$
   (c) $\lim_{x\to a} \frac{x^3-a^3}{x-a}$ *Hint*: $x^3-a^3 = (x-a)(x^2+ax+a^2)$.

20.12 (a) Sketch the function $f(x) = (x-1)^{-1}(x-2)^{-2}$.

(b) Determine $\lim_{x\to 2+} f(x)$, $\lim_{x\to 2-} f(x)$, $\lim_{x\to 1+} f(x)$ and $\lim_{x\to 1-} f(x)$.

(c) Determine $\lim_{x\to 2} f(x)$ and $\lim_{x\to 1} f(x)$ if they exist.

20.13 Prove that if $\lim_{x\to a} f(x) = 3$ and $\lim_{x\to a} g(x) = 2$, then

(a) $\lim_{x\to a}[3f(x) + g(x)^2] = 13$,

(b) $\lim_{x\to a} \frac{1}{g(x)} = \frac{1}{2}$,

(c) $\lim_{x\to a} \sqrt{3f(x) + 8g(x)} = 5$.

20.14 Prove $\lim_{x\to 0+} \frac{1}{x} = +\infty$ and $\lim_{x\to 0-} \frac{1}{x} = -\infty$.

20.15 Prove $\lim_{x\to -\infty} f(x) = 0$ and $\lim_{x\to 2+} f(x) = +\infty$ for the function f in Example 4.

20.16 Suppose the limits $L_1 = \lim_{x\to a+} f_1(x)$ and $L_2 = \lim_{x\to a+} f_2(x)$ exist.

(a) Show if $f_1(x) \le f_2(x)$ for all x in some interval (a, b), then $L_1 \le L_2$.

(b) Suppose that, in fact, $f_1(x) < f_2(x)$ for all x in some interval (a, b). Can you conclude $L_1 < L_2$?

20.17 Show that if $\lim_{x\to a+} f_1(x) = \lim_{x\to a+} f_3(x) = L$ and if $f_1(x) \le f_2(x) \le f_3(x)$ for all x in some interval (a, b), then $\lim_{x\to a+} f_2(x) = L$. This is called the squeeze lemma. *Warning*: This is not immediate from Exercise 20.16(a), because we are not assuming $\lim_{x\to a+} f_2(x)$ exists; this must be proved.

20.18 Let $f(x) = \frac{\sqrt{1+3x^2}-1}{x^2}$ for $x \ne 0$. Show $\lim_{x\to 0} f(x)$ exists and determine its value. Justify all claims.

20.19 The limits defined in Definition 20.3 do not depend on the choice of the set S. As an example, consider $a < b_1 < b_2$ and suppose f is defined on (a, b_2). Show that if the limit $\lim_{x\to a} f(x)$ exists for either $S = (a, b_1)$ or $S = (a, b_2)$, then the limit exists for the other choice of S and these limits are identical. Their common value is what we write as $\lim_{x\to a+} f(x)$.

20.20 Let f_1 and f_2 be functions such that $\lim_{x\to a^S} f_1(x) = +\infty$ and such that the limit $L_2 = \lim_{x\to a^S} f_2(x)$ exists.

(a) Prove $\lim_{x\to a^S}(f_1 + f_2)(x) = +\infty$ if $L_2 \ne -\infty$. *Hint*: Use Exercise 9.11.

(b) Prove $\lim_{x \to a^s}(f_1 f_2)(x) = +\infty$ if $0 < L_2 \leq +\infty$. *Hint*: Use Theorem 9.9.

(c) Prove $\lim_{x \to a^s}(f_1 f_2)(x) = -\infty$ if $-\infty \leq L_2 < 0$.

(d) What can you say about $\lim_{x \to a^s}(f_1 f_2)(x)$ if $L_2 = 0$?

§21 * More on Metric Spaces: Continuity

In this section and the next section we continue the introduction to metric space ideas initiated in §13. More thorough treatments appear in [33,53] and [62]. In particular, for this brief introduction we avoid the technical and somewhat confusing matter of relative topologies that is not, and should not be, avoided in the more thorough treatments.

We are interested in functions between metric spaces (S, d) and (S^*, d^*). We will write "$f: S \to S^*$" to signify $\text{dom}(f) = S$ and f takes values in S^*, i.e., $f(x) \in S^*$ for all $s \in S$.

21.1 Definition.
Consider metric spaces (S, d) and (S^*, d^*). A function $f: S \to S^*$ is *continuous at s_0* in S if

$$\begin{array}{c} \text{for each } \epsilon > 0 \text{ there exists } \delta > 0 \text{ such that} \\ d(s, s_0) < \delta \text{ implies } d^*(f(s), f(s_0)) < \epsilon. \end{array} \tag{1}$$

We say f is *continuous on a subset E* of S if f is continuous at each point of E. The function f is *uniformly continuous on a subset E* of S if

$$\begin{array}{c} \text{for each } \epsilon > 0 \text{ there exists } \delta > 0 \text{ such that} \\ s, t \in E \text{ and } d(s, t) < \delta \text{ imply } d^*(f(s), f(t)) < \epsilon. \end{array} \tag{2}$$

Example 1
Let $S = S^* = \mathbb{R}$ and $d = d^* = \text{dist}$ where, as usual, $\text{dist}(a, b) = |a - b|$. The definition of continuity given above is equivalent to that in §17 in view of Theorem 17.2. The definition of uniform continuity is equivalent to that in Definition 19.1. $\square$

Example 2

In several variable calculus, real-valued functions with domain $\mathbb{R}^2$ or $\mathbb{R}^3$, or even $\mathbb{R}^k$, are extensively studied. This corresponds to the case $S = \mathbb{R}^k$,

$$d(\boldsymbol{x}, \boldsymbol{y}) = \left[\sum_{j=1}^{k} (x_j - y_j)^2 \right]^{1/2},$$

$S^* = \mathbb{R}$ and $d^* = \mathrm{dist}$. We will not develop the theory, but generally speaking, functions that look continuous will be. Some examples on $\mathbb{R}^2$ are $f(x_1, x_2) = x_1^2 + x_2^2$, $f(x_1, x_2) = x_1 x_2 \sqrt{x_1^2 + x_2^2 + 1}$, $f(x_1, x_2) = \cos(x_1 - x_2^5)$. Some examples on $\mathbb{R}^3$ are $g(x_1, x_2, x_3) = x_1^2 + x_2^2 + x_3^2$, $g(x_1, x_2, x_3) = x_1 x_2 + x_1 x_3 + x_2 x_3$, $g(x_1, x_2, x_3) = e^{x_1 + x_2} \log(x_3^2 + 2)$. □

Example 3

Functions with domain $\mathbb{R}$ and values in $\mathbb{R}^2$ or $\mathbb{R}^3$, or generally $\mathbb{R}^k$, are also studied in several variable calculus. This corresponds to the case $S = \mathbb{R}$, $d = \mathrm{dist}$, $S^* = \mathbb{R}^k$ and

$$d^*(\boldsymbol{x}, \boldsymbol{y}) = \left[\sum_{j=1}^{k} (x_j - y_j)^2 \right]^{1/2}.$$

The images of such functions are what nonmathematicians often call a "curve" or "path." In order to distinguish a function from its image, we will adhere to the following terminology. Suppose $\gamma : \mathbb{R} \to \mathbb{R}^k$ is continuous. Then we will call γ a *path*; its image $\gamma(\mathbb{R})$ in $\mathbb{R}^k$ will be called a *curve*. We will also use this terminology if γ is defined and continuous on some subinterval of $\mathbb{R}$, such as $[a, b]$; see Exercise 21.7.

As an example, consider γ where $\gamma(t) = (\cos t, \sin t)$. This function maps $\mathbb{R}$ onto the circle in $\mathbb{R}^2$ about $(0,0)$ with radius 1. More generally $\gamma_0(t) = (a \cos t, b \sin t)$ maps $\mathbb{R}$ onto the ellipse with equation $\frac{x^2}{a^2} + \frac{y^2}{b^2} = 1$; see Fig. 21.1.

The graph of an ordinary continuous function $f : \mathbb{R} \to \mathbb{R}$ looks like a curve, and it is! It is the curve for the path $\gamma(t) = (t, f(t))$.

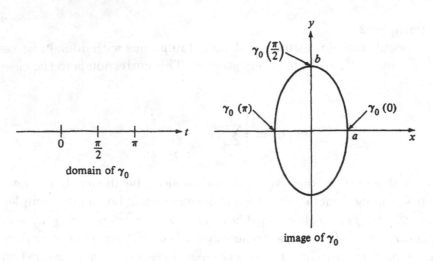

FIGURE 21.1

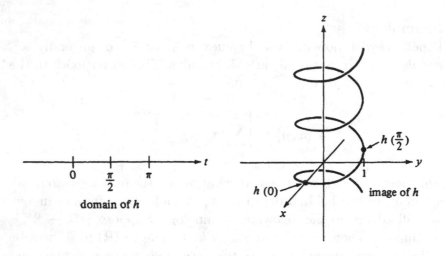

FIGURE 21.2

Curves in $\mathbb{R}^3$ can be quite exotic. For example, the curve for the path $h(t) = (\cos t, \sin t, \frac{t}{4})$ is a *helix*. See Fig. 21.2. $\square$

We did not prove that any of the paths above are continuous, because we can easily prove the following general fact.

21.2 Proposition.

Consider a function $\gamma : [a, b] \to \mathbb{R}^k$ and write

$$\gamma(t) = (f_1(t), f_2(t), \ldots, f_k(t)) \quad for \quad t \in [a, b];$$

thus $f_1, f_2, \ldots, f_k$ are real-valued functions on $[a, b]$. Then γ is continuous, i.e., a path, if and only if each f_j is continuous.

Proof

Both implications follow from formula (1) in the proof of Lemma 13.3 and Exercise 13.2:

$$|x_j - y_j| \le d^*(\boldsymbol{x}, \boldsymbol{y}) \le \sqrt{k}\max\{|x_j - y_j| : j = 1, 2, \ldots, k\}. \quad (1)$$

For t, t_0 in $[a, b]$, (1) implies

$$|f_j(t) - f_j(t_0)| \le d^*(\gamma(t), \gamma(t_0)) \quad for \quad j = 1, \ldots, k, \quad (2)$$

and $d^*(\gamma(t), \gamma(t_0)) \le \sqrt{k}\max\{|f_j(t) - f_j(t_0)| : j = 1, 2, \ldots, k\}.$ (3)

Suppose γ is continuous on $[a, b]$, and consider some j in $\{1, 2, \ldots, k\}$. To show f_j is continuous at t_0 in $[a, b]$, given $\epsilon > 0$, select $\delta > 0$ so that $d^*(\gamma(t), \gamma(t_0)) < \epsilon$ for $|t - t_0| < \delta$. Then by (2), $|t - t_0| < \delta$ implies $|f_j(t) - f_j(t_0)| < \epsilon$. So f_j is continuous on $[a, b]$.

Now suppose that each f_j is continuous on $[a, b]$, and consider $\epsilon > 0$. For each j, there is $\delta_j > 0$ such that

$$|t - t_0| < \delta_j \quad implies \quad |f_j(t) - f_j(t_0)| < \frac{\epsilon}{\sqrt{k}}.$$

For $\delta = \min\{\delta_1, \delta_2, \ldots, \delta_k\}$ and $|t - t_0| < \delta$, we have

$$\max\{|f_j(t) - f_j(t_0)| : j = 1, 2, \ldots, k\} < \frac{\epsilon}{\sqrt{k}},$$

so by (3) we have $d^*(\gamma(t), \gamma(t_0)) < \epsilon$. Thus γ is continuous at t_0. ∎

The next theorem shows continuity is a topological property; see Discussion 13.7.

21.3 Theorem.

Consider metric spaces (S, d) and (S^*, d^*). A function $f : S \to S^*$ is continuous on S if and only if

$$\begin{array}{c} f^{-1}(U) \text{ is an open subset of } S \\ \text{for every open subset } U \text{ of } S^*. \end{array} \quad (1)$$

Recall $f^{-1}(U) = \{s \in S : f(s) \in U\}$.

Proof

Suppose f is continuous on S. Let U be an open subset of S^*, and consider $s_0 \in f^{-1}(U)$. We need to show s_0 is interior to $f^{-1}(U)$. Since $f(s_0) \in U$ and U is open, we have

$$\{s^* \in S^* : d^*(s^*, f(s_0)) < \epsilon\} \subseteq U \tag{2}$$

for some $\epsilon > 0$. Since f is continuous at s_0, there exists $\delta > 0$ such that

$$d(s, s_0) < \delta \quad \text{implies} \quad d^*(f(s), f(s_0)) < \epsilon. \tag{3}$$

From (2) and (3) we conclude $d(s, s_0) < \delta$ implies $f(s)$ is in U; hence $s \in f^{-1}(U)$. That is,

$$\{s \in S : d(s, s_0) < \delta\} \subseteq f^{-1}(U),$$

so that s_0 is interior to $f^{-1}(U)$.

Conversely, suppose (1) holds, and consider $s_0 \in S$ and $\epsilon > 0$. Then $U = \{s^* \in S^* : d^*(s^*, f(s_0)) < \epsilon\}$ is open in S^*, so $f^{-1}(U)$ is open in S. Since $s_0 \in f^{-1}(U)$, for some $\delta > 0$ we have

$$\{s \in S : d(s, s_0) < \delta\} \subseteq f^{-1}(U).$$

It follows that

$$d(s, s_0) < \delta \quad \text{implies} \quad d^*(f(s), f(s_0)) < \epsilon.$$

Thus f is continuous at s_0. ∎

Continuity at a point is also a topological property; see Exercise 21.2. Uniform continuity is a topological property, too, but if we made this precise we would be led to a special class of topologies given by so-called "uniformities."

We will show continuous functions preserve two important topological properties: compactness, and connectedness which will be defined in the next section. The next theorem and corollary illustrate the power of compactness.

21.4 Theorem.
Consider metric spaces (S, d), (S^*, d^*) *and a continuous function* $f: S \to S^*$. *Let* E *be a compact subset of* S. *Then*
  (i) $f(E)$ *is a compact subset of* S^*, *and*
  (ii) f *is uniformly continuous on* E.

Proof
To prove (i), let $\mathcal{U}$ be an open cover of $f(E)$. For each $U \in \mathcal{U}$, $f^{-1}(U)$ is open in S. Moreover, $\{f^{-1}(U) : U \in \mathcal{U}\}$ is a cover of E. Hence there exist $U_1, U_2, \ldots, U_m$ in $\mathcal{U}$ such that

$$E \subseteq f^{-1}(U_1) \cup f^{-1}(U_2) \cup \cdots \cup f^{-1}(U_m).$$

Then

$$f(E) \subseteq U_1 \cup U_2 \cup \cdots \cup U_m,$$

so $\{U_1, U_2, \ldots, U_m\}$ is the desired finite subcover of $\mathcal{U}$ for $f(E)$. This proves (i).

To establish (ii), let $\epsilon > 0$. For each $s \in E$ there exists $\delta_s > 0$ [this δ depends on s] such that

$$d(s, t) < \delta_s \quad \text{implies} \quad d^*(f(s), f(t)) < \frac{\epsilon}{2}. \tag{1}$$

For each $s \in E$, let $V_s = \{t \in S : d(s, t) < \frac{1}{2}\delta_s\}$. Then the family $\mathcal{V} = \{V_s : s \in E\}$ is an open cover of E. By compactness, there exist finitely many points $s_1, s_2, \ldots, s_n$ in E such that

$$E \subseteq V_{s_1} \cup V_{s_2} \cup \cdots \cup V_{s_n}.$$

Let $\delta = \frac{1}{2} \min\{\delta_{s_1}, \delta_{s_2}, \ldots, \delta_{s_n}\}$. We complete the proof by showing

$$s, t \in E \quad \text{and} \quad d(s, t) < \delta \quad \text{imply} \quad d^*(f(s), f(t)) < \epsilon. \tag{2}$$

For some k in $\{1, 2, \ldots, n\}$ we have $s \in V_{s_k}$, i.e., $d(s, s_k) < \frac{1}{2}\delta_{s_k}$. Also we have

$$d(t, s_k) \leq d(t, s) + d(s, s_k) < \delta + \frac{1}{2}\delta_{s_k} \leq \delta_{s_k}.$$

Therefore applying (1) twice we have

$$d^*(f(s), f(s_k)) < \frac{\epsilon}{2} \quad \text{and} \quad d^*(f(t), f(s_k)) < \frac{\epsilon}{2}.$$

Hence $d^*(f(s), f(t)) < \epsilon$ as desired. ∎

Assertion (ii) in Theorem 21.4 generalizes Theorem 19.2. The next corollary should be compared with Theorem 18.1.

21.5 Corollary.
Let f be a continuous real-valued function on a metric space (S, d). If E is a compact subset of S, then
  (i) *f is bounded on E,*
  (ii) *f assumes its maximum and minimum on E.*

Proof
Since $f(E)$ is compact in $\mathbb{R}$, the set $f(E)$ is bounded by Theorem 13.12. This implies (i).

Since $f(E)$ is compact, it contains $\sup f(E)$ by Exercise 13.13. Thus there exists $s_0 \in E$ so that $f(s_0) = \sup f(E)$. This tells us f assumes its maximum value on E at the point s_0. Similarly, f assumes its minimum on E. ∎

Example 4
All the functions f in Example 2 are bounded on any compact subset of $\mathbb{R}^2$, i.e., on any closed and bounded set in $\mathbb{R}^2$. Likewise, all the functions g in Example 2 are bounded on each closed and bounded set in $\mathbb{R}^3$. □

Example 5
Let γ be any path in $\mathbb{R}^k$; see Example 3. For $-\infty < a < b < \infty$, the image $\gamma([a, b])$ is closed and bounded in $\mathbb{R}^k$ by Theorem 21.4. Note Corollary 21.5 does not apply in this case, since the set S^* in Theorem 21.4 is $\mathbb{R}^k$, not $\mathbb{R}$. Theorem 21.4 also tells us γ is uniformly continuous on $[a, b]$. Thus if $\epsilon > 0$, there exists $\delta > 0$ such that

$$s, t \in [a, b] \quad \text{and} \quad |s - t| < \delta \quad \text{imply} \quad d(\gamma(s), \gamma(t)) < \epsilon.$$

This fact is useful in several variable calculus, where one integrates along paths γ; compare Discussion 19.3. □

Example 6

Consider a function $f : E \to \mathbb{R}$, where E be a compact subset of $\mathbb{R}$. We will show f is continuous if and only if its graph $G = \{(t, f(t)) : t \in E\}$ is a compact subset of $\mathbb{R}^2$. Suppose first that f is continuous. The proof of Proposition 21.2 shows $\gamma(s) = (s, f(s))$ is a continuous function on E, so its image $\gamma(E) = G$ is compact by Theorem 21.4(i).

Now assume G is compact, but f is not continuous at t_0 in E. Then there exists an $\epsilon > 0$ and a sequence (t_n) in E such that

$$\lim_n t_n = t_0 \quad \text{and} \quad |f(t_n) - f(t_0)| \geq \epsilon \quad \text{for all} \quad n. \tag{1}$$

Since G is a bounded set, the Bolzano-Weierstrass Theorem 13.5 shows $(t_n, f(t_n))$ has a convergent subsequence $(t_{n_k}, f(t_{n_k}))$. Its limit is in G, because G is closed in $\mathbb{R}^2$ by the Heine-Borel Theorem 13.12. Thus there exists $\bar{t}$ in E so that $\lim_k (t_{n_k}, f(t_{n_k})) = (\bar{t}, f(\bar{t}))$. Since $\lim_n t_n = t_0$, we have $\bar{t} = t_0$, and therefore $\lim_k (t_{n_k}, f(t_{n_k})) = (t_0, f(t_0))$. In particular, $\lim_k f(t_{n_k}) = f(t_0)$, which contradicts the inequality in (1).

Mark Lynch [44] uses the fact above to give a very interesting construction of a continuous nowhere-differentiable function on a closed interval $[0, 1]$; see his construction on page 361. $\square$

21.6 Remark.

The remainder of this section will be devoted to the Baire Category Theorem and some interesting consequences. We will need some metric-space terminology, some of it from §13. Let (S, d) be a metric space. For a subset E of S, its complement is $S \backslash E = \{s \in S : s \notin E\}$. A point s is in the interior of a set E, or interior to E, if $s \in V \subseteq E$ for some open subset V of S. A subset E of S is closed if $S \setminus E$ is open. A point s is in the closure of a set E if every open set containing s also contains an element of E. We write E^- for the set of points in the closure of E.

A subset D of S is dense in S if every nonempty open set U intersects D, i.e., $D \cap U \neq \varnothing$. For example, $\mathbb{Q}$ is dense in $\mathbb{R}$, since every nonempty open interval in $\mathbb{R}$ contains rationals; see 4.7 on page 25. Finally, a totally new definition: a subset E of S is *nowhere dense* in S if its closure E^- has empty interior. Obviously E is nowhere dense if and only if E^- is nowhere dense. Also, E is nowhere dense in S if and only if $S \setminus E^-$ is dense in S.

Each part of the next theorem is a variant of what is called the Baire Category Theorem. After that theorem, we will give the ultimate version of this theorem and explain the terminology "category." Just as in Example 4 on page 88, open balls $B_r(x_0) = \{x \in S : d(x, x_0) < r\}$ in S are open sets, and closed balls $C_r(x_0) = \{x \in S : d(x, x_0) \leq r\}$ are closed sets. Note $C_s(x_0) \subseteq B_r(x_0)$ for $s < r$.

21.7 Theorem.
Let (S, d) be a complete metric space. Then
(a) *If (U_n) is a sequence of dense open subsets of S, then the intersection $X = \cap_{n=1}^{\infty} U_n$ is dense in S.*
(b) *If (F_n) is a sequence of closed subsets of S and if the union $F = \cup_{n=1}^{\infty} F_n$ contains a nonempty open set, then so does at least one of the sets F_n.*
(c) *The union of a sequence of nowhere dense subsets of S has dense complement.*
(d) *The space S is not a union of a sequence of nowhere dense subsets of S.*

Proof
All the hard work will be in proving: Part (a). Consider a nonempty open subset V of S; we need to show $X \cap V \neq \varnothing$. Since U_1 is open and dense and V is open, $U_1 \cap V$ is open and nonempty. So there exists $x_1 \in U_1 \cap V$ and $r_1 \leq 1$ so that $C_{r_1}(x_1) \subseteq U_1 \cap V$. Since U_2 is open and dense and $B_{r_1}(x_1)$ is open, there exists $x_2 \in B_{r_1}(x_1) \cap U_2$ and $r_2 \leq \frac{1}{2}$ so that $C_{r_2}(x_2) \subseteq B_{r_1}(x_1) \cap U_2$. Continuing, we obtain sequences (x_k) in S and positive numbers (r_k) so that

$$x_{k+1} \in B_{r_k}(x_k), \quad r_k \leq \frac{1}{2^{k-1}} \quad \text{and} \quad C_{r_{k+1}}(x_{k+1}) \subseteq B_{r_k}(x_k) \cap U_{k+1}.$$

Note $d(x_k, x_{k+1}) < r_k$ for all k, so for $m < n$ we have

$$d(x_m, x_n) \leq \sum_{k=m}^{n-1} d(x_k, x_{k+1}) < \sum_{k=m}^{n-1} \frac{1}{2^{k-1}} < \sum_{k=m}^{\infty} \frac{1}{2^{k-1}} = \frac{1}{2^{m-2}}.$$

Thus the sequence (x_k) is a Cauchy sequence in S. Since S is complete, we have $\lim_k x_k = x$ for some $x \in S$. Since we have

$$x_k \in C_{r_k}(x_k) \subseteq C_{r_n}(x_n) \quad \text{for} \quad k \geq n,$$

and since $C_{r_n}(x_n)$ is closed, x also belongs to C_{r_n}. This is true for all n, so

$$x \in \bigcap_{n=1}^{\infty} C_{r_n}(x_n) \subseteq \bigcap_{n=1}^{\infty} U_n = X.$$

Since we also have $x \in C_{r_1}(x_1) \subset V$, we conclude $x \in X \cap V$, as desired.

Part (b). Suppose $\cup_{n=1}^{\infty} F_n$ has nonempty interior. Then $S \setminus \cup_{n=1}^{\infty} F_n = \cap_{n=1}^{\infty}(S \setminus F_n)$ is not dense in S. Since the sets $S \setminus F_n$ are open, part (a) shows that at least one of them, say $S \setminus F_m$, is also not dense in S. Thus its complement F_m contains a nonempty open set.

Part (c). Let $A = \cup_{n=1}^{\infty} A_n$ be the union of a sequence of nowhere dense sets A_n. We need to show the complement of A, $S \setminus A$, is dense in S. If we replace each A_n by its closure, they will still be nowhere dense and $S \setminus A$ only gets smaller. In other words, we may assume each A_n is closed and nowhere dense. Part (b) implies that A contains no nonempty open set. Thus every nonempty open set intersects $S \setminus A$, and $S \setminus A$ is dense in S.

Part (d). This is obvious from part (c), since the complement of S is the empty set $\varnothing$. ∎

Theorem 21.7 tells us in different ways that the union of a sequence of nowhere dense subsets of a complete metric space is small relative to the size of S. For this reason, it is common to divide the subsets of S into two categories. Category 1 consists of sets that are unions of sequences of nowhere dense subsets of S, and Category 2 consists of the other subsets of S. Here is part (d) of Theorem 21.7 restated using this language.

21.8 Baire Category Theorem.
A complete metric space (S,d) is of the second category in itself.

The expression "in itself" here deserves a comment. Many topological properties about sets $E \subseteq S$ depend on the set S. For example, the set $(0,1]$ is not closed in $S = \mathbb{R}$, but it is closed in

$S = (0, \infty)$ viewed as a metric space. Category is another property that depends on S, so "in itself" in Theorem 21.8 stresses the reference to the complete metric space S.[1]

21.9 Corollary.
Each of the spaces $\mathbb{R}^k$ is of second category in itself.

Example 7
Any set that can be listed as a sequence is said to be "countable." Other sets are "uncountable."

(a) Countable unions of sets of first category are sets of first category. For example, countable subsets of $\mathbb{R}$, and more generally of $\mathbb{R}^k$, are of first category, since single points $\{x\}$ in these spaces are nowhere dense. This and Corollary 21.9 give another (rather heavy-handed) proof that $\mathbb{R}$ is uncountable; compare Exercise 16.8.

(b) Example 3 on page 70 shows the set $\mathbb{Q}$ of rationals is countable. Thus $\mathbb{Q}$ is of first category in $\mathbb{R}$, and the set of irrationals is of second category in $\mathbb{R}$.

(c) Each straight line in the plane $\mathbb{R}^2$ is of first category in $\mathbb{R}^2$, so a countable union of straight lines is of first category in $\mathbb{R}^2$. □

21.10 Discussion.
A closed set E in a metric space is said to be *perfect* if every point x in E is the limit of a sequence of points in $E \setminus \{x\}$; compare Proposition 13.9(b). This proposition also shows closed subsets of complete metric spaces are themselves complete metric spaces. In a perfect subset E of a complete metric space, single points are nowhere dense in E, so countable subsets of E are of first category. Therefore, nonempty perfect subsets of complete metric spaces are uncountable. The Cantor set (in Example 5 on page 89) is an interesting compact perfect set.   □

Recall Exercise 17.14, where a function on $\mathbb{R}$ is described that is continuous at each irrational and discontinuous at each rational. In

---

[1]One of the fine properties of compactness is: If a set is compact in some set, then it is compact in all sets.

the next example, we will show there are no functions on $\mathbb{R}$ that are continuous at *only* the rationals.

Example 8

(a) Consider any function f on $\mathbb{R}$. For each x in $\mathbb{R}$, let

$$\omega_f(x) = \inf_{\delta>0} \sup\{|f(y) - f(z)| : y, z \in (x - \delta, x + \delta)\}.$$

The function ω_f is the "oscillation function" of f. The function f is continuous at x if and only if $\omega_f(x) = 0$; see Exercise 21.13.

We show each set $\{x \in \mathbb{R} : \omega_f(x) < \epsilon\}$ is open in $\mathbb{R}$; $\epsilon > 0$. Suppose $\omega_f(x_0) < \epsilon$; we need to show this inequality holds in an interval containing x_0. By the definition, there exists $\delta > 0$ such that

$$\sup\{|f(y) - f(z)| : y, z \in (x_0 - \delta, x_0 + \delta)\} < \epsilon.$$

For $x \in (x_0 - \frac{\delta}{2}, x_0 + \frac{\delta}{2})$ we have $(x - \frac{\delta}{2}, x + \frac{\delta}{2}) \subseteq (x_0 - \delta, x_0 + \delta)$, and therefore

$$\omega_f(x) \leq \sup\left\{|f(y) - f(z)| : y, z \in \left(x - \frac{\delta}{2}, x + \frac{\delta}{2}\right)\right\}$$
$$\leq \sup\{|f(y) - f(z)| : y, z \in (x_0 - \delta, x_0 + \delta)\} < \epsilon.$$

(b) There is no function f on $\mathbb{R}$ that is continuous only on the set $\mathbb{Q}$ of rational numbers. Otherwise, by part (a) the equality

$$\mathbb{Q} = \{x \in \mathbb{R} : \omega_f(x) = 0\} = \bigcap_{n=1}^{\infty} \left\{x \in \mathbb{R} : \omega_f(x) < \frac{1}{n}\right\}$$

would express $\mathbb{Q}$ as the intersection of a sequence of open subsets of $\mathbb{R}$, contrary to Exercise 21.11. □

21.11 Theorem.

No nondegenerate interval I in $\mathbb{R}$ can be written as the disjoint union of two or more nondegenerate closed intervals. (An interval is "nondegenerate" if it has more than one point.)

Proof

We prove the theorem for $I = [a, b]$, $a < b$, and leave the other cases to Exercise 21.14. Assume I is the disjoint union of two or more nondegenerate closed intervals. It is clear there are infinitely many such intervals, because otherwise they could be listed in order, and there would be points in I between the first and second interval. It is also clear that the family can be listed as a sequence (i.e., is countable), since there is a one-to-one function from the collection of disjoint intervals into $\mathbb{Q}$.

So we assume $I = \cup_{k=1}^{\infty}[a_k, b_k]$. By Discussion 13.7(iii) on page 87, the union $\cup_{k=1}^{\infty}(a_k, b_k)$ is open, so that $K = I \setminus \cup_{k=1}^{\infty}(a_k, b_k)$ is closed. Then K is the set of endpoints of the removed intervals, so K is nonempty and countable. We will show K is perfect, and this will be a contradiction by Discussion 21.10. Consider x in K; we may assume $x \neq b$. Consider an interval $(x, x + h)$ where $h > 0$. Some closed interval $[a_m, b_m]$ must intersect $(x, x+h)$, and $[a_m, b_m]$ cannot contain x in K. Thus $x < a_m < x + h$. Since a_m is in K, the interval $(x, x + h)$ intersects K, and since $h > 0$ is arbitrary, x is a limit of points in $K \setminus \{x\}$. This shows K is perfect, completing the proof. ∎

A close examination of Example 8(b) and Theorem 21.11 shows they both depend on the Baire Category Theorem 21.8. Other applications of this theorem appear in Theorem 38.3 and twice in Theorem 38.5.

Exercises

21.1 Show that if the functions $f_1, f_2, \ldots, f_k$ in Proposition 21.2 are uniformly continuous, then so is γ.

21.2 Consider $f : S \to S^*$ where (S, d) and (S^*, d^*) are metric spaces. Show that f is continuous at $s_0 \in S$ if and only if

> for every open set U in S^* containing $f(s_0)$, there is an open set V in S containing s_0 such that $f(V) \subseteq U$.

21.3 Let (S, d) be a metric space and choose $s_0 \in S$. Show $f(s) = d(s, s_0)$ defines a uniformly continuous real-valued function f on S.

21.4 Consider $f : S \to \mathbb{R}$ where (S, d) is a metric space. Show the following are equivalent:
  (i) f is continuous;
  (ii) $f^{-1}((a, b))$ is open in S for all $a < b$;
  (iii) $f^{-1}((a, b))$ is open in S for all rational $a < b$.

21.5 Let E be a noncompact subset of $\mathbb{R}^k$.

(a) Show there is an unbounded continuous real-valued function on E. *Hint*: Either E is unbounded or else its closure E^- contains $x_0 \notin E$. In the latter case, use $\frac{1}{g}$ where $g(x) = d(x, x_0)$.

(b) Show there is a bounded continuous real-valued function on E that does not assume its maximum on E.

21.6 For metric spaces (S_1, d_1), (S_2, d_2), (S_3, d_3), prove that if $f\colon S_1 \to S_2$ and $g\colon S_2 \to S_3$ are continuous, then $g \circ f$ is continuous from S_1 into S_3. *Hint*: It is somewhat easier to use Theorem 21.3 than to use the definition.

21.7 (a) Observe that if $E \subseteq S$ where (S, d) is a metric space, then (E, d) is also a metric space. In particular, if $E \subseteq \mathbb{R}$, then $d(a, b) = |a - b|$ for $a, b \in E$ defines a metric on E.

(b) For $\gamma\colon [a, b] \to \mathbb{R}^k$, give the definition of continuity of γ.

21.8 Let (S, d) and (S^*, d^*) be metric spaces. Show that if $f\colon S \to S^*$ is uniformly continuous, and if (s_n) is a Cauchy sequence in S, then $(f(s_n))$ is a Cauchy sequence in S^*.

21.9 We say a function f *maps* a set E onto a set F provided $f(E) = F$.

(a) Show there is a continuous function mapping the unit square

$$\{(x_1, x_2) \in \mathbb{R}^2 : 0 \le x_1 \le 1, 0 \le x_2 \le 1\}$$

onto $[0, 1]$.

(b) Do you think there is a continuous function mapping $[0, 1]$ onto the unit square?

21.10 Show there exist continuous functions

(a) Mapping $(0, 1)$ onto $[0, 1]$,

(b) Mapping $(0, 1)$ onto $\mathbb{R}$,

(c) Mapping $[0, 1] \cup [2, 3]$ onto $[0, 1]$.

(d) Explain why there are no continuous functions mapping $[0, 1]$ onto $(0, 1)$ or $\mathbb{R}$.

21.11 Show the set $\mathbb{Q}$ of rational numbers is not the intersection of a sequence of open sets, so that the set $\mathbb{R} \setminus \mathbb{Q}$ of irrational numbers is not the union of a sequence of closed subsets of $\mathbb{R}$.

21.12 Give an example of an infinite disjoint sequence of subsets of $\mathbb{R}$, each of which is of second category in $\mathbb{R}$.

21.13 Let ω_f be the oscillation function in Example 8(a). Show that f is continuous at x in $\mathbb{R}$ if and only if $\omega_f(x) = 0$.

21.14 Complete the proof of Theorem 21.11. *Hint*: Assume a nondegenerate interval I in $\mathbb{R}$ is the union of two or more disjoint nondegenerate closed intervals. Select points $x < y$ in I, so that they are interior to two distinct such subintervals. Show $[x, y]$ is also the union of two or more disjoint nondegenerate closed intervals, contradicting the case covered in the proof of Theorem 21.11.

§22 * More on Metric Spaces: Connectedness

Consider a subset E of $\mathbb{R}$ that is not an interval. As noted in the proof of Corollary 18.3, the property

$$y_1, y_2 \in E \quad \text{and} \quad y_1 < y < y_2 \quad \text{imply} \quad y \in E$$

must fail. So there exist y_1, y_2, y in $\mathbb{R}$ such that

$$y_1 < y < y_2, \qquad y_1, y_2 \in E, \qquad y \notin E. \tag{*}$$

The set E is not "connected" because y separates E into two pieces. Put another way, if we set $U_1 = (-\infty, y)$ and $U_2 = (y, \infty)$, then we obtain open sets so that $E \cap U_1$ and $E \cap U_2$ are disjoint nonempty sets who union is E. The last observation can be promoted to a useful general definition.

22.1 Definition.
Let E be a subset of a metric space (S, d). The set E is *disconnected* if one of the following two equivalent conditions holds.
(a)[2] There are open subsets U_1 and U_2 of S such that

$$(E \cap U_1) \cap (E \cap U_2) = \varnothing \quad \text{and} \quad E = (E \cap U_1) \cup (E \cap U_2), \tag{1}$$
$$E \cap U_1 \neq \varnothing \quad \text{and} \quad E \cap U_2 \neq \varnothing. \tag{2}$$

---

[2]Readers familiar with relative topologies will recognize this as stating that E is a disjoint union of nonempty relatively-open subsets of E.

(b) There are disjoint nonempty subsets A and B of E such that $E = A \cup B$ and neither set intersects the closure of the other set, i.e.,

$$A^- \cap B = \varnothing \quad \text{and} \quad A \cap B^- = \varnothing. \tag{3}$$

A set E is *connected* if it is not disconnected.

Remark. We show that conditions (a) and (b) above are equivalent.

If (a) holds, set $A = E \cap U_1$ and $B = E \cap U_2$. To check (b), we need only verify (3). If, for example, $A^- \cap B \neq \varnothing$, then there exists $b \in A^- \cap B \subseteq U_2$, so there is $r > 0$ so that $\{s \in S : d(s,b) < r\} \subseteq U_2$. Since b is in A^-, Proposition 13.9(c) shows there is a in A such that $d(a,b) < r$. But then $a \in U_2 \cap E = B$, a contradiction, since A and B are disjoint. Thus (3) holds, and condition (b) holds.

Now suppose that (b) holds. Let $U_1 = S \setminus B^-$ and $U_2 = S \setminus A^-$. To verify (a), it suffices to show $E \cap U_1 = A$ and $E \cap U_2 = B$. For example, $x \in E \cap U_1$ implies $x \in S \setminus B$ implies $x \notin B$. Since $x \in A \cup B$, we conclude $x \in A$. Finally, if $x \in A$, then $x \notin B^-$ by (3), so $x \in S \setminus B^- = U_1$. Therefore x is in $E \cap U_1$. We've shown $E \cap U_1 = A$.

Example 1

As noted before the definition, sets in $\mathbb{R}$ that are not intervals are disconnected. Conversely, intervals in $\mathbb{R}$ are connected. To prove this from the definition, we will assume the contrary and obtain a contradiction. So we assume there is an interval I and open sets U_1 and U_2 disconnecting I as in Definition 22.1(a). Select $a \in I \cap U_1$ and $b \in I \cap U_2$. We may suppose $a < b$. Now the interval $[a,b]$ satisfies the same conditions in Definition 22.1(a) as I. So we may assume $I = [a,b]$, $a \in U_1$ and $b \in U_2$.

Let $t = \sup[a,b) \cap U_1$, so that $t > a$. Since $(b - \epsilon, b] \subseteq U_2$ for some $\epsilon > 0$, we also have $t < b$. If t is in U_1, then $[t, t+\epsilon) \subseteq U_1 \cap [a,b]$ for some $\epsilon > 0$, but then $t = \sup[a,b) \cap U_1 \geq t + \epsilon$. If $t \in U_2$, then $(t - \epsilon, t] \subseteq U_2 \cap [a,b]$, which would imply $t = \sup[a,b) \cap U_1 \leq t - \epsilon$. Each possibility leads to a contradiction, so $[a,b]$ and the original interval I are connected. $\qquad\square$

22.2 Theorem.
Consider metric spaces (S, d), (S^*, d^*), and let $f: S \to S^*$ be continuous. If E is a connected subset of S, then $f(E)$ is a connected subset of S^*.

Proof
Assume $f(E)$ is not connected in S^*. Then there exist open sets V_1 and V_2 in S^* such that

$$(f(E) \cap V_1) \cap (f(E) \cap V_2) = \varnothing, \quad f(E) = (f(E) \cap V_1) \cup (f(E) \cap V_2),$$
$$f(E) \cap V_1 \neq \varnothing \quad \text{and} \quad f(E) \cap V_2 \neq \varnothing.$$

Let $U_1 = f^{-1}(V_1)$ and $U_2 = f^{-1}(V_2)$. Then U_1 and U_2 are open sets in S that separate E as in Definition 22.1(a). Thus E is not connected, a contradiction. ∎

The next corollary generalizes Theorem 18.2 and its corollary.

22.3 Corollary.
Let f be a continuous real-valued function on a metric space (S, d). If E is a connected subset of S, then $f(E)$ is an interval in $\mathbb{R}$. In particular, f has the intermediate value property.

Example 2
Curves are connected. That is, if γ is a path in $\mathbb{R}^k$ as described in Example 3 of §21, page 165 and I is a subinterval of $\mathbb{R}$, then the image $\gamma(I)$ is connected in $\mathbb{R}^k$. □

22.4 Definition.
A subset E of a metric space (S, d) is said to be *path-connected* if, for each pair s, t of points in E, there exists a continuous function $\gamma: [a, b] \to E$ such that $\gamma(a) = s$ and $\gamma(b) = t$. We call γ a *path*.

22.5 Theorem.
If E in (S, d) is path-connected, then E is connected. [The failure of the converse is illustrated in Exercise 22.4.]

Proof
Assume E is disconnected by open sets U_1 and U_2:

$$(E \cap U_1) \cap (E \cap U_2) = \varnothing \quad \text{and} \quad E = (E \cap U_1) \cup (E \cap U_2), \quad (1)$$
$$E \cap U_1 \neq \varnothing \quad \text{and} \quad E \cap U_2 \neq \varnothing. \quad (2)$$

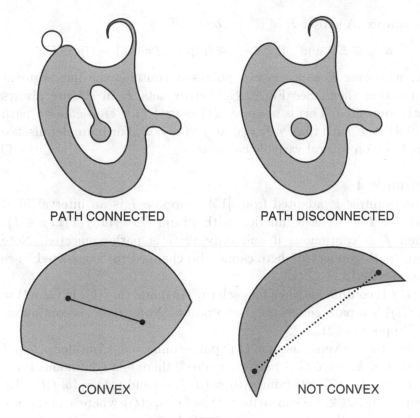

PATH CONNECTED   PATH DISCONNECTED

CONVEX   NOT CONVEX

FIGURE 22.1

Select $s \in E \cap U_1$ and $t \in E \cap U_2$. Let $\gamma \colon [a, b] \to E$ be a path where $\gamma(a) = s$ and $\gamma(b) = t$. Let $F = \gamma([a, b])$. Then (1) and (2) hold with F in place of E. Thus F is disconnected, but F is connected by Theorem 22.2. ∎

Figure 22.1 gives a path-connected set and a disconnected set in $\mathbb{R}^2$.

Example 3

Many familiar sets in $\mathbb{R}^k$ such as the open ball $\{x : d(x, 0) < r\}$, the closed ball $\{x : d(x, 0) \le r\}$ and the k-dimensional cube

$$\{x : \max\{|x_j| : j = 1, 2, \ldots, k\} \le 1\}$$

are convex. A subset E of $\mathbb{R}^k$ is *convex* if

$$x, y \in E \quad \text{and} \quad 0 < t < 1 \quad \text{imply} \quad tx + (1-t)y \in E,$$

i.e., whenever E contains two points it contains the line segment connecting them. See Fig. 22.1. Convex sets E in $\mathbb{R}^k$ are always path-connected. This is because $\gamma(t) = tx + (1-t)y$ defines a path $\gamma \colon [0, 1] \to E$ such that $\gamma(0) = y$ and $\gamma(1) = x$. For more details, see any book on several variable calculus. □

Example 4
This example is adapted from [15]. Suppose I is an interval in $\mathbb{R}$ and $f : I \to \mathbb{R}$ is a function with graph $G = \{(x, f(x)) : x \in I\}$. Then f is continuous if and only if G is path-connected. Note that "path-connected" here cannot be changed to "connected;" see Exercise 22.13.

If f is continuous, then for each $(a, f(a))$ and $(b, f(b))$ in G, $\gamma(t) = (t, f(t))$ is a path connecting these points. Note that γ is continuous by Proposition 21.2.

For the converse, assume G is path-connected. Consider a, b in I where $a < b$. Since G is path-connected, there is a continuous function $\gamma : [0, 1] \to G$ satisfying $\gamma(0) = (a, f(a))$ and $\gamma(1) = (b, f(b))$. By Proposition 21.2, we can write $\gamma(t) = (x(t), y(t))$ where x and y are continuous real-valued functions on $[0, 1]$. Of course, $y(t) = f(x(t))$ for all t. First, we claim

$$\{(z, f(z)) : z \in [a, b]\} \subseteq \gamma([0, 1]). \tag{1}$$

Since $x(t)$ is continuous on $[0, 1]$, $x(0) = a$ and $x(1) = b$, the Intermediate Value Theorem 18.2 for continuous functions shows that if z is in (a, b), then $x(t) = z$ for some $t \in (0, 1)$. Thus $(z, f(z)) = (x(t), f(x(t))) = \gamma(t)$. Since the cases $z = a$ and $z = b$ are trivial, this implies (1).

Next we show f satisfies IVP, the intermediate value property:

(IVP) If $a, b \in I$, $a < b$ and c is between $f(a)$ and $f(b)$, then $f(x) = c$ for some $x \in [a, b]$.

Let $\alpha = \sup\{t \in [0, 1] : x(t) \leq a\}$ and $\beta = \inf\{t \in [0, 1] : x(t) \geq b\}$. Thus $0 \leq \alpha < \beta \leq 1$. Since $x(t)$ is continuous, we see that $x([\alpha, \beta]) = [a, b]$. Now we apply the Intermediate Value Theorem again, this time to $f(x(t))$ on $[\alpha, \beta]$. Since $f(x(\alpha)) = f(a)$ and

$f(x(\beta)) = f(b)$, we have $f(x(t)) = c$ for some t in (α, β). For this t we have $x(t) \in [a, b]$, so we have verified IVP.

Now assume f is not continuous on I. Then there exists $x_0 \in I$, $\epsilon > 0$, and a sequence (x_n) in I that converges to x_0 satisfying $|f(x_n) - f(x_0)| > \epsilon$ for all n. Then $f(x_n) > f(x_0) + \epsilon > f(x_0)$ for infinitely many n, or else $f(x_n) < f(x_0) - \epsilon < f(x_0)$ for infinitely many n. Passing to a subsequence, we may assume $f(x_n) > f(x_0) + \epsilon > f(x_0)$ for all n, say. By the IVP, for each n there is y_n between x_n and x_0 so that $f(y_n) = f(x_0) + \epsilon$. Then $(y_n, f(y_n))$ is a sequence in G satisfying

$$\lim_n y_n = x_0 \quad \text{and} \quad \lim(y_n, f(y_n)) = (x_0, f(x_0) + \epsilon). \tag{2}$$

Moreover, there exist $a < b$ in I such that x_0 and all y_n belong to $[a, b]$. We apply (1) to this a and b and γ described there. So, for each $n \in \mathbb{N}$ there exists $t_n \in [0, 1]$ so that $\gamma(t_n) = (y_n, f(y_n))$. Passing to a subsequence $(t_{n_k})_{k \in \mathbb{N}}$ of (t_n), we may assume that $\lim_k t_{n_k} = t_0$ for some t_0 in $[0, 1]$. Since γ is continuous, we conclude from (2) that

$$\gamma(t_0) = \lim_k \gamma(t_{n_k}) = \lim_k (y_{n_k}, f(y_{n_k})) = (x_0, f(x_0) + \epsilon).$$

Since this limit is not in G, and $\gamma([0, 1]) \subset G$, we have a contradiction and f is continuous on I. $\qquad\qquad\qquad\qquad\qquad\qquad\qquad\qquad \square$

We end this section with a discussion of some very different metric spaces. The points in these spaces are actually functions themselves.

22.6 Definition.
Let S be a subset of $\mathbb{R}$. Let $C(S)$ be the set of all bounded continuous real-valued functions on S and, for $f, g \in C(S)$, let

$$d(f, g) = \sup\{|f(x) - g(x)| : x \in S\}.$$

With this definition, $C(S)$ becomes a metric space [Exercise 22.6]. In this metric space, a sequence (f_n) converges to a point [function!] f provided $\lim_{n \to \infty} d(f_n, f) = 0$, that is

$$\lim_{n \to \infty} \sup\{|f_n(x) - f(x)| : x \in S\} = 0. \tag{*}$$

Put another way, for each $\epsilon > 0$ there exists a number N such that

$$|f_n(x) - f(x)| < \epsilon \quad \text{for all} \quad x \in S \quad \text{and} \quad n > N.$$

We will study this important concept in the next chapter, but without using metric space terminology. See Definition 24.2 and Remark 24.4 where (*) is called *uniform convergence*.

A sequence (f_n) in $C(S)$ is a Cauchy sequence with respect to the metric d exactly when it is *uniformly Cauchy* as defined in Definition 25.3. In our metric space terminology, Theorems 25.4 and 24.3 tell us that $C(S)$ is a *complete* metric space.

Exercises

22.1 Show there do not exist continuous functions

   (a) Mapping $[0,1]$ onto $[0,1] \cup [2,3]$,

   (b) Mapping $(0,1)$ onto $\mathbb{Q}$.

22.2 Show $\{(x_1, x_2) \in \mathbb{R}^2 : x_1^2 + x_2^2 = 1\}$ is a connected subset of $\mathbb{R}^2$.

22.3 Prove that if E is a connected subset of a metric space (S, d), then its closure E^- is also connected.

22.4 Consider the following subset of $\mathbb{R}^2$:

$$E = \left\{ \left(x, \sin \frac{1}{x} \right) : x \in (0, 1] \right\};$$

E is simply the graph of $g(x) = \sin \frac{1}{x}$ along the interval $(0,1]$.

   (a) Determine its closure E^-. See Fig. 19.4.

   (b) Show E^- is connected.

   (c) Show E^- is not path-connected.

22.5 Let E and F be connected sets in some metric space.

   (a) Prove that if $E \cap F \neq \varnothing$, then $E \cup F$ is connected.

   (b) Give an example to show $E \cap F$ need not be connected. Incidentally, the empty set *is* connected.

22.6 **(a)** Show $C(S)$ given in Definition 22.6 is a metric space.

   (b) Why did we require the functions in $C(S)$ to be bounded when no such requirement appears in Definition 24.2?

22.7 Explain why the metric space B in Exercise 13.3 can be regarded as $C(\mathbb{N})$.

22.8 Consider $C(S)$ for a subset S of $\mathbb{R}$. For a fixed s_0 in S, define $F(f) = f(s_0)$. Show F is a uniformly continuous real-valued function on the metric space $C(S)$.

22.9 Consider $f, g \in C(S)$ where $S \subseteq \mathbb{R}$. Let $F(t) = tf + (1-t)g$. Show F is a uniformly continuous function from $\mathbb{R}$ into $C(S)$.

22.10 Let f be a (fixed) uniformly continuous function in $C(\mathbb{R})$. For each $x \in \mathbb{R}$, let f_x be the function defined by $f_x(y) = f(x+y)$ for $y \in \mathbb{R}$. Let $F(x) = f_x$; show F is uniformly continuous from $\mathbb{R}$ into $C(\mathbb{R})$.

22.11 Consider $C(S)$ where $S \subseteq \mathbb{R}$, and let $\mathcal{E}$ consist of all f in $C(S)$ such that $\sup\{|f(x)| : x \in S\} \le 1$.

    (a) Show $\mathcal{E}$ is closed in $C(S)$.

    (b) Show $C(S)$ is connected.

    (c) Show $\mathcal{E}$ is connected.

22.12 Consider a subset $\mathcal{E}$ of $C(S)$, $S \subseteq \mathbb{R}$. For this exercise, we say a function f_0 in $\mathcal{E}$ is *interior* to $\mathcal{E}$ if there exists a finite subset F of S and an $\epsilon > 0$ such that

$$\{f \in C(S) : |f(x) - f_0(x)| < \epsilon \text{ for } x \in F\} \subseteq \mathcal{E}.$$

The set $\mathcal{E}$ is *open* if every function in $\mathcal{E}$ is interior to $\mathcal{E}$.

    (a) Reread Discussion 13.7.

    (b) Show the family of open sets defined above forms a topology for $C(S)$. *Remarks.* This topology is different from the one given by the metric in Definition 22.6. In fact, this topology does not come from any metric at all! It is called the *topology of pointwise convergence* and can be used to study the convergence in Definition 24.1 just as the metric in Definition 22.6 can be used to study the convergence in Definition 24.2.

22.13 Let f be the function defined on $[0, \infty)$ by $f(x) = (x, \sin(1/x))$ for $x > 0$ and $f(0) = 0$. Show that the graph of f is connected but not path-connected; compare Exercise 22.4. Is f continuous on $[0, \infty)$?

22.14 Consider a bounded sequence (s_n) in $\mathbb{R}$ and its set S of subsequential limits as in Theorem 11.8. Show that if $\lim_n (s_{n+1} - s_n) = 0$, then S is a connected interval in $\mathbb{R}$. This is from [11]. *Hint:* Consider $a < c < b$ where a and b are in S, and show c is in S. Use Theorem 11.2.

4

CHAPTER

Sequences and Series of Functions

In this chapter we develop some of the basic properties of power series. In doing so, we will introduce uniform convergence and illustrate its importance. In §26 we prove power series can be differentiated and integrated term-by-term.

§23 Power Series

Given a sequence $(a_n)_{n=0}^{\infty}$ of real numbers, the series $\sum_{n=0}^{\infty} a_n x^n$ is called a *power series*. Observe the variable x. Thus the power series is a function of x provided it converges for some or all x. Of course, it converges for $x = 0$; note the convention $0^0 = 1$. Whether it converges for other values of x depends on the choice of *coefficients* (a_n). It turns out that, given any sequence (a_n), one of the following holds for its power series:

  (a) The power series converges for all $x \in \mathbb{R}$;
  (b) The power series converges only for $x = 0$;
  (c) The power series converges for all x in some bounded interval centered at 0; the interval may be open, half-open or closed.

K.A. Ross, *Elementary Analysis: The Theory of Calculus*,
Undergraduate Texts in Mathematics, DOI 10.1007/978-1-4614-6271-2_4,
© Springer Science+Business Media New York 2013

These remarks are consequences of the following important theorem.

23.1 Theorem.

For the power series $\sum a_n x^n$, let

$$\beta = \limsup |a_n|^{1/n} \quad and \quad R = \frac{1}{\beta}.$$

[If $\beta = 0$ we set $R = +\infty$, and if $\beta = +\infty$ we set $R = 0$.] Then
  (i) *The power series converges for $|x| < R$;*
  (ii) *The power series diverges for $|x| > R$.*

R is called the *radius of convergence* for the power series. Note that (i) is a vacuous statement if $R = 0$ and that (ii) is a vacuous statement if $R = +\infty$. Note also that (a) above corresponds to the case $R = +\infty$, (b) above corresponds to the case $R = 0$, and (c) above corresponds to the case $0 < R < +\infty$.

Proof of Theorem 23.1

The proof follows quite easily from the Root Test 14.9. Here are the details. We want to apply the Root Test to the series $\sum a_n x^n$. So for each $x \in \mathbb{R}$, let α_x be the number or symbol defined in 14.9 for the series $\sum a_n x^n$. Since the nth term of the series is $a_n x^n$, we have

$$\alpha_x = \limsup |a_n x^n|^{1/n} = \limsup |x||a_n|^{1/n} = |x| \cdot \limsup |a_n|^{1/n} = \beta|x|.$$

The third equality is justified by Exercise 12.6(a). Now we consider cases.

Case 1. Suppose $0 < R < +\infty$. In this case $\alpha_x = \beta|x| = \frac{|x|}{R}$. If $|x| < R$ then $\alpha_x < 1$, so the series converges by the Root Test. Likewise, if $|x| > R$, then $\alpha_x > 1$ and the series diverges.

Case 2. Suppose $R = +\infty$. Then $\beta = 0$ and $\alpha_x = 0$ no matter what x is. Hence the power series converges for all x by the Root Test.

Case 3. Suppose $R = 0$. Then $\beta = +\infty$ and $\alpha_x = +\infty$ for $x \neq 0$. Thus by the Root Test the series diverges for $x \neq 0$. ∎

Recall that if $\lim |\frac{a_{n+1}}{a_n}|$ exists, then this limit equals β of the last theorem by Corollary 12.3. This limit is often easier to calculate than $\limsup |a_n|^{1/n}$; see the examples below.

Example 1
Consider $\sum_{n=0}^{\infty} \frac{1}{n!} x^n$. If $a_n = \frac{1}{n!}$, then $\frac{a_{n+1}}{a_n} = \frac{1}{n+1}$, so $\lim |\frac{a_{n+1}}{a_n}| = 0$. Therefore $\beta = 0$, $R = +\infty$ and this series has radius of convergence $+\infty$. That is, it converges for all x in $\mathbb{R}$. In fact, it converges to e^x for all x, but that is another story; see Example 1 in §31, page 252, and also §37. $\square$

Example 2
Consider $\sum_{n=0}^{\infty} x^n$. Then $\beta = 1$ and $R = 1$. Note this series does not converge for $x = 1$ or $x = -1$, so the interval of convergence is exactly $(-1, 1)$. [By *interval of convergence* we mean the set of x for which the power series converges.] The series converges to $\frac{1}{1-x}$ by formula (2) of Example 1 in §14, page 96. $\square$

Example 3
Consider $\sum_{n=0}^{\infty} \frac{1}{n} x^n$. Since $\lim \frac{\frac{1}{n+1}}{\frac{1}{n}} = 1$, we again have $\beta = 1$ and $R = 1$. This series diverges for $x = 1$ [see Example 1 of §15], but it converges for $x = -1$ by the Alternating Series theorem 15.3 on page 108. Hence the interval of convergence is exactly $[-1, 1)$. $\square$

Example 4
Consider $\sum_{n=0}^{\infty} \frac{1}{n^2} x^n$. Once again $\beta = 1$ and $R = 1$. This series converges at both $x = 1$ and $x = -1$, so its interval of convergence is exactly $[-1, 1]$. $\square$

Example 5
The series $\sum_{n=0}^{\infty} n! x^n$ has radius of convergence $R = 0$ because we have $\lim |\frac{(n+1)!}{n!}| = +\infty$. It diverges for every $x \neq 0$. $\square$

Examples 1–5 illustrate all the possibilities discussed in (a)–(c) prior to Theorem 23.1.

Example 6

Consider $\sum_{n=0}^{\infty} 2^{-n} x^{3n}$. This is deceptive, and it is tempting to calculate $\beta = \lim \sup (2^{-n})^{1/n} = \frac{1}{2}$ and conclude $R = 2$. *This is wrong* because 2^{-n} is the coefficient of x^{3n} *not* x^n, and the calculation of β must involve the coefficients a_n of x^n. We need to handle this series more carefully. The series can be written $\sum_{n=0}^{\infty} a_n x^n$ where $a_{3k} = 2^{-k}$ and $a_n = 0$ if n is not a multiple of 3. We calculate β by using the subsequence of all nonzero terms, i.e., the subsequence given by $\sigma(k) = 3k$. This yields

$$\beta = \lim \sup |a_n|^{1/n} = \lim_{k \to \infty} |a_{3k}|^{1/3k} = \lim_{k \to \infty} (2^{-k})^{1/3k} = 2^{-1/3}.$$

Therefore the radius of convergence is $R = \frac{1}{\beta} = 2^{1/3}$. $\qquad\square$

One may consider more general power series of the form

$$\sum_{n=0}^{\infty} a_n (x - x_0)^n, \tag{*}$$

where x_0 is a fixed real number, but they reduce to series of the form $\sum_{n=0}^{\infty} a_n y^n$ by the change of variable $y = x - x_0$. The interval of convergence for the series (*) will be an interval centered at x_0.

Example 7

Consider the series

$$\sum_{n=1}^{\infty} \frac{(-1)^{n+1}}{n} (x - 1)^n. \tag{1}$$

The radius of convergence for the series $\sum_{n=1}^{\infty} \frac{(-1)^{n+1}}{n} y^n$ is $R = 1$, so the interval of convergence for the series (1) is the interval $(0, 2)$ plus perhaps an endpoint or two. Direct substitution shows the series (1) converges at $x = 2$ [it's an alternating series] and diverges to $-\infty$ at $x = 0$. So the exact interval of convergence is $(0, 2]$. It turns out that the series (1) represents the function $\log_e x$ on $(0, 2]$. See Examples 1 and 2 in §26. $\qquad\square$

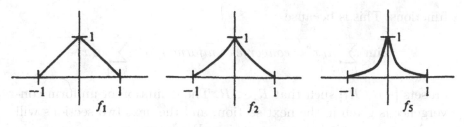

FIGURE 23.1

On of our major goals is to understand the function given by a power series

$$f(x) = \sum_{k=0}^{\infty} a_k x^k \quad \text{for} \quad |x| < R.$$

We are interested in questions like: Is f continuous? Is f differentiable? If so, can one differentiate f term-by-term:

$$f'(x) = \sum_{k=1}^{\infty} k a_k x^{k-1}?$$

Can one integrate f term-by-term?

Returning to the question of continuity, what reason is there to believe f is continuous? Its partial sums $f_n = \sum_{k=0}^{n} a_k x^k$ are continuous, since they are polynomials. Moreover, we have $\lim_{n \to \infty} f_n(x) = f(x)$ for $|x| < R$. Therefore f would be continuous *if* a result like the following were true: If (f_n) is a sequence of continuous functions on (a, b) and if $\lim_{n \to \infty} f_n(x) = f(x)$ for all $x \in (a, b)$, then f is continuous on (a, b). However, this fine sounding result is false!

Example 8
Let $f_n(x) = (1 - |x|)^n$ for $x \in (-1, 1)$; see Fig. 23.1. Let $f(x) = 0$ for $x \neq 0$ and let $f(0) = 1$. Then we have $\lim_{n \to \infty} f_n(x) = f(x)$ for all $x \in (-1, 1)$, since $\lim_{n \to \infty} a^n = 0$ if $|a| < 1$. Each f_n is a continuous function, but the limit function f is clearly discontinuous at $x = 0$. □

This example, as well as Exercises 23.7–23.9, may be discouraging, but it turns out that power series do converge to continuous functions. This is because

$$\lim_{n\to\infty} \sum_{k=0}^{n} a_k x^k \quad converges\ uniformly\ to \quad \sum_{k=0}^{\infty} a_k x^k$$

on sets $[-R_1, R_1]$ such that $R_1 < R$. The definition of uniform convergence is given in the next section, and the next two sections will be devoted to this important notion. We return to power series in §26, and again in §31.

Exercises

23.1 For each of the following power series, find the radius of convergence and determine the exact interval of convergence.

(a) $\sum n^2 x^n$　　　　　　　　　　(b) $\sum (\frac{x}{n})^n$

(c) $\sum (\frac{2^n}{n^2}) x^n$　　　　　　　　(d) $\sum (\frac{n^3}{3^n}) x^n$

(e) $\sum (\frac{2^n}{n!}) x^n$　　　　　　　　(f) $\sum (\frac{1}{(n+1)^2 2^n}) x^n$

(g) $\sum (\frac{3^n}{n \cdot 4^n}) x^n$　　　　　　(h) $\sum (\frac{(-1)^n}{n^2 \cdot 4^n}) x^n$

23.2 Repeat Exercise 23.1 for the following:

(a) $\sum \sqrt{n} x^n$　　　　　　　　(b) $\sum \frac{1}{n\sqrt{n}} x^n$

(c) $\sum x^{n!}$　　　　　　　　　　(d) $\sum \frac{3^n}{\sqrt{n}} x^{2n+1}$

23.3 Find the exact interval of convergence for the series in Example 6.

23.4 For $n = 0, 1, 2, 3, \ldots$, let $a_n = [\frac{4+2(-1)^n}{5}]^n$.

　(a) Find $\limsup (a_n)^{1/n}$, $\liminf (a_n)^{1/n}$, $\limsup |\frac{a_{n+1}}{a_n}|$ and $\liminf |\frac{a_{n+1}}{a_n}|$.

　(b) Do the series $\sum a_n$ and $\sum (-1)^n a_n$ converge? Explain briefly.

　(c) Now consider the power series $\sum a_n x^n$ with the coefficients a_n as above. Find the radius of convergence and determine the exact interval of convergence for the series.

23.5 Consider a power series $\sum a_n x^n$ with radius of convergence R.

　(a) Prove that if all the coefficients a_n are integers and if infinitely many of them are nonzero, then $R \leq 1$.

　(b) Prove that if $\limsup |a_n| > 0$, then $R \leq 1$.

23.6 **(a)** Suppose $\sum a_n x^n$ has finite radius of convergence R and $a_n \geq 0$ for all n. Show that if the series converges at R, then it also converges at $-R$.

(b) Give an example of a power series whose interval of convergence is exactly $(-1, 1]$.

The next three exercises are designed to show that the notion of convergence of functions discussed prior to Example 8 has many defects.

23.7 For each $n \in \mathbb{N}$, let $f_n(x) = (\cos x)^n$. Each f_n is a continuous function. Nevertheless, show

(a) $\lim f_n(x) = 0$ unless x is a multiple of π,

(b) $\lim f_n(x) = 1$ if x is an even multiple of π,

(c) $\lim f_n(x)$ does not exist if x is an odd multiple of π.

23.8 For each $n \in \mathbb{N}$, let $f_n(x) = \frac{1}{n} \sin nx$. Each f_n is a differentiable function. Show

(a) $\lim f_n(x) = 0$ for all $x \in \mathbb{R}$,

(b) But $\lim f_n'(x)$ need not exist [at $x = \pi$ for instance].

23.9 Let $f_n(x) = nx^n$ for $x \in [0, 1]$ and $n \in \mathbb{N}$. Show

(a) $\lim f_n(x) = 0$ for $x \in [0, 1)$. *Hint:* Use Exercise 9.12.

(b) However, $\lim_{n \to \infty} \int_0^1 f_n(x)\, dx = 1$.

§24 Uniform Convergence

We first formalize the notion of convergence discussed prior to Example 8 in the preceding section.

24.1 Definition.

Let (f_n) be a sequence of real-valued functions defined on a set $S \subseteq \mathbb{R}$. The sequence (f_n) *converges pointwise* [i.e., at each point] to a function f defined on S if

$$\lim_{n \to \infty} f_n(x) = f(x) \quad \text{for all} \quad x \in S.$$

We often write $\lim f_n = f$ *pointwise* [*on S*] or $f_n \to f$ *pointwise* [*on S*].

Example 1
All the functions f obtained in the last section as a limit of a sequence of functions were pointwise limits. See Example 8 of §23 and Exercises 23.7–23.9. In Exercise 23.8 we have $f_n \to 0$ pointwise on $\mathbb{R}$, and in Exercise 23.9 we have $f_n \to 0$ pointwise on $[0, 1)$. □

Example 2
Let $f_n(x) = x^n$ for $x \in [0, 1]$. Then $f_n \to f$ pointwise on $[0, 1]$ where $f(x) = 0$ for $x \in [0, 1)$ and $f(1) = 1$. □

Now observe $f_n \to f$ pointwise on S means exactly the following:

$$\text{for each } \epsilon > 0 \text{ and } x \text{ in } S \text{ there exists } N \text{ such that} \tag{1}$$
$$|f_n(x) - f(x)| < \epsilon \text{ for } n > N.$$

Note the value of N depends on both $\epsilon > 0$ and x in S. If for each $\epsilon > 0$ we could find N so that

$$|f_n(x) - f(x)| < \epsilon \quad \text{for } all \quad x \in S \quad \text{and} \quad n > N,$$

then the values $f_n(x)$ would be "uniformly" close to the values $f(x)$. Here N would depend on ϵ but not on x. This concept is extremely useful.

24.2 Definition.
Let (f_n) be a sequence of real-valued functions defined on a set $S \subseteq \mathbb{R}$. The sequence (f_n) *converges uniformly on S* to a function f defined on S if

$$\text{for each } \epsilon > 0 \text{ there exists a number } N \text{ such that} \tag{1}$$
$$|f_n(x) - f(x)| < \epsilon \text{ for all } x \in S \text{ and all } n > N.$$

We write $\lim f_n = f$ *uniformly on S* or $f_n \to f$ *uniformly on S*.

Note that if $f_n \to f$ uniformly on S and if $\epsilon > 0$, then there exists N such that $f(x) - \epsilon < f_n(x) < f(x) + \epsilon$ for *all* $x \in S$ and $n > N$. In other words, for $n > N$ the graph of f_n lies in the strip between the graphs of $f - \epsilon$ and $f + \epsilon$. In Fig. 24.1 the graphs of f_n for $n > N$ would all lie between the dotted curves.

We return to our earlier examples.

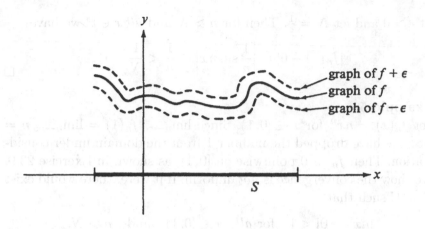

graph of $f + \epsilon$
graph of f
graph of $f - \epsilon$

S

FIGURE 24.1

Example 3

Let $f_n(x) = (1 - |x|)^n$ for $x \in (-1, 1)$. Also, let $f(x) = 0$ for $x \neq 0$ and $f(0) = 1$. As noted in Example 8 of §23, $f_n \to f$ pointwise on $(-1, 1)$. It turns out that the sequence (f_n) does not converge uniformly to f on $(-1, 1)$ in view of the next theorem. This can also be shown directly, as follows. Assume $f_n \to f$ uniformly on $(-1, 1)$. Then [with $\epsilon = \frac{1}{2}$ in mind] we see there exists N in $\mathbb{N}$ so that $|f(x) - f_n(x)| < \frac{1}{2}$ for *all* $x \in (-1, 1)$ and $n > N$. Hence

$$x \in (0, 1) \quad \text{and} \quad n > N \quad \text{imply} \quad |(1 - x)^n| < \frac{1}{2}.$$

In particular,

$$x \in (0, 1) \quad \text{implies} \quad (1 - x)^{N+1} < \frac{1}{2}.$$

However, this fails for sufficiently small x; for example, if we set $x = 1 - 2^{-1/(N+1)}$, then $1 - x = 2^{-1/(N+1)}$ and $(1-x)^{N+1} = 2^{-1} = \frac{1}{2}$. This contradiction shows (f_n) does not converge uniformly to f on $(-1, 1)$ as had been assumed. □

Example 4

Let $f_n(x) = \frac{1}{n} \sin nx$ for $x \in \mathbb{R}$. Then $f_n \to 0$ pointwise on $\mathbb{R}$ as shown in Exercise 23.8. In fact, $f_n \to 0$ uniformly on $\mathbb{R}$. To see this, let $\epsilon > 0$ and let $N = \frac{1}{\epsilon}$. Then for $n > N$ and *all* $x \in \mathbb{R}$ we have

$$|f_n(x) - 0| = \left|\frac{1}{n}\sin nx\right| \le \frac{1}{n} < \frac{1}{N} = \epsilon.$$

$\square$

Example 5
Let $f_n(x) = nx^n$ for $x \in [0,1)$. Since $\lim_{n\to\infty} f_n(1) = \lim_{n\to\infty} n = +\infty$, we have dropped the number 1 from the domain under consideration. Then $f_n \to 0$ pointwise on $[0,1)$, as shown in Exercise 23.9. We show the convergence is *not* uniform. If it were, there would exist N in $\mathbb{N}$ such that

$$|nx^n - 0| < 1 \quad \text{for } all \quad x \in [0,1) \quad \text{and} \quad n > N.$$

In particular, we would have $(N + 1)x^{N+1} < 1$ for all $x \in [0,1)$. But this fails for x sufficiently close to 1. Consider, for example, the reciprocal x of $(N + 1)^{1/(N+1)}$. $\square$

Example 6
As in Example 2, let $f_n(x) = x^n$ for $x \in [0,1]$, $f(x) = 0$ for $x \in [0,1)$ and $f(1) = 1$. Then $f_n \to f$ pointwise on $[0,1]$, but (f_n) does not converge uniformly to f on $[0,1]$, as can be seen directly or by applying the next theorem. $\square$

24.3 Theorem.
The uniform limit of continuous functions is continuous. More precisely, let (f_n) be a sequence of functions on a set $S \subseteq \mathbb{R}$, suppose $f_n \to f$ uniformly on S, and suppose $S = \mathrm{dom}(f)$. If each f_n is continuous at x_0 in S, then f is continuous at x_0. [So if each f_n is continuous on S, then f is continuous on S.]

Proof
This involves the famous "$\frac{\epsilon}{3}$ argument." The critical inequality is

$$|f(x) - f(x_0)| \le |f(x) - f_n(x)| + |f_n(x) - f_n(x_0)| + |f_n(x_0) - f(x_0)|.$$

$$(1)$$

If n is large enough, the first and third terms on the right side of (1) will be small, since $f_n \to f$ uniformly. *Once such n is selected*, the continuity of f_n implies that the middle term will be small provided x is close to x_0.

For the formal proof, let $\epsilon > 0$. There exists N in $\mathbb{N}$ such that

$$n > N \quad \text{implies} \quad |f_n(x) - f(x)| < \frac{\epsilon}{3} \quad \text{for all} \quad x \in S.$$

In particular,

$$|f_{N+1}(x) - f(x)| < \frac{\epsilon}{3} \quad \text{for all} \quad x \in S. \tag{2}$$

Since f_{N+1} is continuous at x_0 there is a $\delta > 0$ such that

$$x \in S \quad \text{and} \quad |x - x_0| < \delta \quad \text{imply} \quad |f_{N+1}(x) - f_{N+1}(x_0)| < \frac{\epsilon}{3}; \tag{3}$$

see Theorem 17.2. Now we apply (1) with $n = N+1$, (2) twice [once for x and once for x_0] and (3) to conclude

$$x \in S \quad \text{and} \quad |x - x_0| < \delta \quad \text{imply} \quad |f(x) - f(x_0)| < 3 \cdot \frac{\epsilon}{3} = \epsilon.$$

This proves that f is continuous at x_0. ∎

One might think this theorem would be useless in practice, since it should be easier to show a single function is continuous than to show a sequence (f_n) consists of continuous functions and the sequence converges to f uniformly. This would no doubt be true if f were given by a simple formula. But consider, for example,

$$f(x) = \sum_{n=1}^{\infty} \frac{1}{n^2} x^n \quad \text{for} \quad x \in [-1, 1]$$

or

$$J_0(x) = \sum_{n=0}^{\infty} \frac{(-1)^n (\frac{1}{2}x)^{2n}}{(n!)^2} \quad \text{for} \quad x \in \mathbb{R}.$$

The partial sums are clearly continuous, but neither f nor J_0 is given by a simple formula. Moreover, many functions that arise in mathematics and elsewhere, such as the Bessel function J_0, are defined by power series. It would be very useful to know when and where power series converge uniformly; an answer is given in §26.

24.4 Remark.
Uniform convergence can be reformulated as follows. *A sequence (f_n) of functions on a set $S \subseteq \mathbb{R}$ converges uniformly to a function f on*

S if and only if

$$\lim_{n \to \infty} \sup\{|f(x) - f_n(x)| : x \in S\} = 0. \tag{1}$$

We leave the straightforward proof to Exercise 24.12.

According to (1) we can decide whether a sequence (f_n) converges uniformly to f by calculating $\sup\{|f(x) - f_n(x)| : x \in S\}$ for each n. If $f - f_n$ is differentiable, we may use calculus to find these suprema. $\qquad \square$

Example 7

Let $f_n(x) = \frac{x}{1+nx^2}$ for $x \in \mathbb{R}$. Clearly we have $\lim_{n \to \infty} f_n(0) = 0$. If $x \neq 0$, then $\lim_{n \to \infty}(1 + nx^2) = +\infty$, so $\lim_{n \to \infty} f_n(x) = 0$. Thus $f_n \to 0$ pointwise on $\mathbb{R}$. To find the maximum and minimum of f_n, we calculate $f_n'(x)$ and set it equal to 0. This leads to $(1 + nx^2) \cdot 1 - x(2nx) = 0$ or $1 - nx^2 = 0$. Thus $f_n'(x) = 0$ if and only if $x = \frac{\pm 1}{\sqrt{n}}$. Further analysis or a sketch of f_n leads one to conclude f_n takes its maximum at $\frac{1}{\sqrt{n}}$ and its minimum at $-\frac{1}{\sqrt{n}}$. Since $f_n\left(\pm\frac{1}{\sqrt{n}}\right) = \pm\frac{1}{2\sqrt{n}}$, we conclude

$$\lim_{n \to \infty} \sup\{|f_n(x)| : x \in S\} = \lim_{n \to \infty} \frac{1}{2\sqrt{n}} = 0.$$

Therefore $f_n \to 0$ uniformly on $\mathbb{R}$ by Remark 24.4. $\qquad \square$

Example 8

Let $f_n(x) = n^2 x^n(1-x)$ for $x \in [0,1]$. Then we have $\lim_{n \to \infty} f_n(1) = 0$. For $x \in [0,1)$ we have $\lim_{n \to \infty} n^2 x^n = 0$ by applying Exercise 9.12, since for $x \neq 0$,

$$\frac{(n+1)^2 x^{n+1}}{n^2 x^n} = \left(\frac{n+1}{n}\right)^2 x \to x.$$

Hence $\lim_{n \to \infty} f_n(x) = 0$. Thus $f_n \to 0$ pointwise on $[0,1]$. Again, to find the maximum and minimum of f_n we set its derivative equal to 0. We obtain $x^n(-1) + (1-x)nx^{n-1} = 0$ or $x^{n-1}[n - (n+1)x] = 0$. Since f_n takes the value 0 at both endpoints of the interval $[0,1]$, it follows that f_n takes it maximum at $\frac{n}{n+1}$. We have

$$f_n\left(\frac{n}{n+1}\right) = n^2\left(\frac{n}{n+1}\right)^n\left(1 - \frac{n}{n+1}\right) = \frac{n^2}{n+1}\left(\frac{n}{n+1}\right)^n. \tag{1}$$

The reciprocal of $(\frac{n}{n+1})^n$ is $(1+\frac{1}{n})^n$, the nth term of a sequence which has limit e. This was mentioned, but not proved, in Example 3 of §7; a proof is given in Theorem 37.11. Therefore we have $\lim(\frac{n}{n+1})^n = \frac{1}{e}$. Since $\lim[\frac{n^2}{n+1}] = +\infty$, we conclude from (1) that $\lim f_n(\frac{n}{n+1}) = +\infty$; see Exercise 12.9(a). In particular, $\lim \sup\{|f_n(x)| : x \in [0,1]\} = +\infty$, so (f_n) does *not* converge uniformly to 0 on $[0,1]$. $\qquad\square$

Exercises

24.1 Let $f_n(x) = \frac{1+2\cos^2 nx}{\sqrt{n}}$. Prove carefully that (f_n) converges uniformly to 0 on $\mathbb{R}$.

24.2 For $x \in [0, \infty)$, let $f_n(x) = \frac{x}{n}$.

(a) Find $f(x) = \lim f_n(x)$.

(b) Determine whether $f_n \to f$ uniformly on $[0,1]$.

(c) Determine whether $f_n \to f$ uniformly on $[0, \infty)$.

24.3 Repeat Exercise 24.2 for $f_n(x) = \frac{1}{1+x^n}$.

24.4 Repeat Exercise 24.2 for $f_n(x) = \frac{x^n}{1+x^n}$.

24.5 Repeat Exercise 24.2 for $f_n(x) = \frac{x^n}{n+x^n}$.

24.6 Let $f_n(x) = (x - \frac{1}{n})^2$ for $x \in [0,1]$.

(a) Does the sequence (f_n) converge pointwise on the set $[0,1]$? If so, give the limit function.

(b) Does (f_n) converge uniformly on $[0,1]$? Prove your assertion.

24.7 Repeat Exercise 24.6 for $f_n(x) = x - x^n$.

24.8 Repeat Exercise 24.6 for $f_n(x) = \sum_{k=0}^{n} x^k$.

24.9 Consider $f_n(x) = nx^n(1-x)$ for $x \in [0,1]$.

(a) Find $f(x) = \lim f_n(x)$.

(b) Does $f_n \to f$ uniformly on $[0,1]$? Justify.

(c) Does $\int_0^1 f_n(x)\, dx$ converge to $\int_0^1 f(x)\, dx$? Justify.

24.10 (a) Prove that if $f_n \to f$ uniformly on a set S, and if $g_n \to g$ uniformly on S, then $f_n + g_n \to f + g$ uniformly on S.

(b) Do you believe the analogue of (a) holds for products? If so, see the next exercise.

24.11 Let $f_n(x) = x$ and $g_n(x) = \frac{1}{n}$ for all $x \in \mathbb{R}$. Let $f(x) = x$ and $g(x) = 0$ for $x \in \mathbb{R}$.

  (a) Observe $f_n \to f$ uniformly on $\mathbb{R}$ [obvious!] and $g_n \to g$ uniformly on $\mathbb{R}$ [almost obvious].

  (b) Observe the sequence $(f_n g_n)$ does not converge uniformly to fg on $\mathbb{R}$. Compare Exercise 24.2.

24.12 Prove the assertion in Remark 24.4.

24.13 Prove that if (f_n) is a sequence of uniformly continuous functions on an interval (a, b), and if $f_n \to f$ uniformly on (a, b), then f is also uniformly continuous on (a, b). *Hint*: Try an $\frac{\varepsilon}{3}$ argument as in the proof of Theorem 24.3.

24.14 Let $f_n(x) = \frac{nx}{1+n^2 x^2}$ and $f(x) = 0$ for $x \in \mathbb{R}$.

  (a) Show $f_n \to f$ pointwise on $\mathbb{R}$.

  (b) Does $f_n \to f$ uniformly on $[0, 1]$? Justify.

  (c) Does $f_n \to f$ uniformly on $[1, \infty)$? Justify.

24.15 Let $f_n(x) = \frac{nx}{1+nx}$ for $x \in [0, \infty)$.

  (a) Find $f(x) = \lim f_n(x)$.

  (b) Does $f_n \to f$ uniformly on $[0, 1]$? Justify.

  (c) Does $f_n \to f$ uniformly on $[1, \infty)$? Justify.

24.16 Repeat Exercise 24.15 for $f_n(x) = \frac{nx}{1+nx^2}$.

24.17 Let (f_n) be a sequence of continuous functions on $[a, b]$ that converges uniformly to f on $[a, b]$. Show that if (x_n) is a sequence in $[a, b]$ and if $x_n \to x$, then $\lim_{n \to \infty} f_n(x_n) = f(x)$.

§25   More on Uniform Convergence

Our next theorem shows one can interchange integrals and *uniform* limits. The adjective "uniform" here is important; compare Exercise 23.9.

25.1 Discussion.

To prove Theorem 25.2 below we merely use some basic facts about integration which should be familiar [or believable] even if your calculus is rusty. Specifically, we use:

(a) *If g and h are integrable on $[a,b]$ and if $g(x) \leq h(x)$ for all $x \in [a,b]$, then $\int_a^b g(x)\,dx \leq \int_a^b h(x)\,dx$.* See Theorem 33.4(i).

We also use the following corollary:

(b) *If g is integrable on $[a,b]$, then*

$$\left| \int_a^b g(x)\,dx \right| \leq \int_a^b |g(x)|\,dx.$$

Continuous functions on closed intervals are integrable, as noted in Discussion 19.3 and proved in Theorem 33.2. $\qquad\square$

25.2 Theorem.

Let (f_n) be a sequence of continuous functions on $[a,b]$, and suppose $f_n \to f$ uniformly on $[a,b]$. Then

$$\lim_{n\to\infty} \int_a^b f_n(x)\,dx = \int_a^b f(x)\,dx. \tag{1}$$

Proof

By Theorem 24.3 f is continuous, so the functions $f_n - f$ are all integrable on $[a,b]$. Let $\epsilon > 0$. Since $f_n \to f$ uniformly on $[a,b]$, there exists a number N such that $|f_n(x) - f(x)| < \frac{\epsilon}{b-a}$ for all $x \in [a,b]$ and all $n > N$. Consequently $n > N$ implies

$$\left| \int_a^b f_n(x)\,dx - \int_a^b f(x)\,dx \right| = \left| \int_a^b [f_n(x) - f(x)]\,dx \right|$$

$$\leq \int_a^b |f_n(x) - f(x)|\,dx \leq \int_a^b \frac{\epsilon}{b-a}\,dx = \epsilon.$$

The first $\leq$ follows from Discussion 25.1(b) applied to $g = f_n - f$ and the second $\leq$ follows from Discussion 25.1(a) applied to $g = |f_n - f|$ and $h = \frac{\epsilon}{b-a}$; h happens to be a constant function, but this does no harm.

The last paragraph shows that given $\epsilon > 0$, there exists N such that $|\int_a^b f_n(x)\,dx - \int_a^b f(x)\,dx| \leq \epsilon$ for $n > N$. Therefore (1) holds. $\qquad\blacksquare$

Recall one of the advantages of the notion of Cauchy sequence: A sequence (s_n) of real numbers can be shown to converge *without knowing its limit* by simply verifying that it is a Cauchy sequence. Clearly a similar result for sequences of functions would be valuable, since it is likely that we will not know the limit function in advance. What we need is the idea of "uniformly Cauchy."

25.3 Definition.

A sequence (f_n) of functions defined on a set $S \subseteq \mathbb{R}$ is *uniformly Cauchy on S* if

$$\text{for each } \epsilon > 0 \text{ there exists a number } N \text{ such that} \quad |f_n(x) - f_m(x)| < \epsilon \text{ for all } x \in S \text{ and all } m, n > N. \tag{1}$$

Compare this definition with that of a Cauchy sequence of real numbers [Definition 10.8] and that of uniform convergence [Definition 24.2]. It is an easy exercise to show uniformly convergent sequences of functions are uniformly Cauchy; see Exercise 25.4. The interesting and useful result is the converse, just as in the case of sequences of real numbers.

25.4 Theorem.

Let (f_n) be a sequence of functions defined and uniformly Cauchy on a set $S \subseteq \mathbb{R}$. Then there exists a function f on S such that $f_n \to f$ uniformly on S.

Proof

First we have to "find" f. We begin by showing

$$\text{for each } x_0 \in S \text{ the sequence } (f_n(x_0)) \text{ is a Cauchy} \quad \text{sequence of real numbers.} \tag{1}$$

For each $\epsilon > 0$, there exists N such that

$$|f_n(x) - f_m(x)| < \epsilon \quad \text{for} \quad x \in S \quad \text{and} \quad m, n > N.$$

In particular, we have

$$|f_n(x_0) - f_m(x_0)| < \epsilon \quad \text{for} \quad m, n > N.$$

This shows $(f_n(x_0))$ is a Cauchy sequence, so (1) holds.

Now for each x in S, assertion (1) implies $\lim_{n \to \infty} f_n(x)$ exists; this is proved in Theorem 10.11 which in the end depends on the Completeness Axiom 4.4. Hence we *define* $f(x) = \lim_{n \to \infty} f_n(x)$. This defines a function f on S such that $f_n \to f$ *pointwise* on S.

Now that we have "found" f, we need to prove $f_n \to f$ uniformly on S. Let $\epsilon > 0$. There is a number N such that

$$|f_n(x) - f_m(x)| < \frac{\epsilon}{2} \quad \text{for all} \quad x \in S \quad \text{and all} \quad m, n > N. \quad (2)$$

Consider $m > N$ and $x \in S$. Assertion (2) tells us that $f_n(x)$ lies in the open interval $(f_m(x) - \frac{\epsilon}{2}, f_m(x) + \frac{\epsilon}{2})$ for all $n > N$. Therefore, as noted in Exercise 8.9, the limit $f(x) = \lim_{n \to \infty} f_n(x)$ lies in the closed interval $[f_m(x) - \frac{\epsilon}{2}, f_m(x) + \frac{\epsilon}{2}]$. In other words,

$$|f(x) - f_m(x)| \leq \frac{\epsilon}{2} \quad \text{for all} \quad x \in S \quad \text{and} \quad m > N.$$

Then of course

$$|f(x) - f_m(x)| < \epsilon \quad \text{for all} \quad x \in S \quad \text{and} \quad m > N.$$

This shows $f_m \to f$ uniformly on S, as desired. ∎

Theorem 25.4 is especially useful for "series of functions." Let us recall what $\sum_{k=1}^{\infty} a_k$ signifies when the a_k's are real numbers. This signifies $\lim_{n \to \infty} \sum_{k=1}^{n} a_k$ provided this limit exists [as a real number, $+\infty$ or $-\infty$]. Otherwise the symbol $\sum_{k=1}^{\infty} a_k$ has no meaning. Thus the infinite series is the limit of the sequence of partial sums $\sum_{k=1}^{n} a_k$. Similar remarks apply to series of functions. A *series of functions* is an expression $\sum_{k=0}^{\infty} g_k$ or $\sum_{k=0}^{\infty} g_k(x)$ which makes sense provided the sequence of partial sums $\sum_{k=0}^{n} g_k$ converges, or diverges to $+\infty$ or $-\infty$ pointwise. If the sequence of partial sums converges uniformly on a set S to $\sum_{k=0}^{\infty} g_k$, then we say the *series is uniformly convergent on S*.

Example 1

Any power series is a series of functions, since $\sum_{k=0}^{\infty} a_k x^k$ has the form $\sum_{k=0}^{\infty} g_k$ where $g_k(x) = a_k x^k$ for all x. □

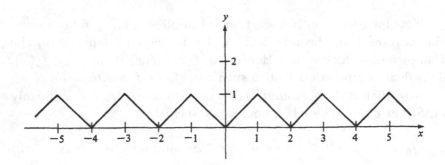

FIGURE 25.1

Example 2

$\sum_{k=0}^{\infty} \frac{x^k}{1+x^k}$ is a series of functions, but is not a power series, at least not in its present form. This is a series $\sum_{k=0}^{\infty} g_k$ where $g_0(x) = \frac{1}{2}$ for all x, $g_1(x) = \frac{x}{1+x}$ for all x, $g_2(x) = \frac{x^2}{1+x^2}$ for all x, etc. □

Example 3

Let g be the function drawn in Fig. 25.1, and let $g_n(x) = g(4^n x)$ for all $x \in \mathbb{R}$. Then $\sum_{n=0}^{\infty} (\frac{3}{4})^n g_n(x)$ is a series of functions. The limit function f is continuous on $\mathbb{R}$, but has the amazing property that it is not differentiable at any point! The proof of the nondifferentiability of f is somewhat delicate; see [62, 7.18]. A similar example is given in Example 38.1 on page 348. □

Theorems for sequences of functions translate easily into theorems for series of functions. Here is an example.

25.5 Theorem.

Consider a series $\sum_{k=0}^{\infty} g_k$ of functions on a set $S \subseteq \mathbb{R}$. Suppose each g_k is continuous on S and the series converges uniformly on S. Then the series $\sum_{k=0}^{\infty} g_k$ represents a continuous function on S.

Proof

Each partial sum $f_n = \sum_{k=1}^{n} g_k$ is continuous and the sequence (f_n) converges uniformly on S. Hence the limit function is continuous by Theorem 24.3. ∎

Recall the Cauchy criterion for series $\sum a_k$ given in Definition 14.3:

> for each $\epsilon > 0$ there exists a number N such that
> $n \geq m > N$ implies $|\sum_{k=m}^{n} a_k| < \epsilon$. $\qquad (*)$

The analogue for series of functions is also useful. The sequence of partial sums of a series $\sum_{k=0}^{\infty} g_k$ of functions is uniformly Cauchy on a set S if and only if the series satisfies the *Cauchy criterion* [*uniformly on S*]:

> for each $\epsilon > 0$ there exists a number N such that
> $n \geq m > N$ implies $|\sum_{k=m}^{n} g_k(x)| < \epsilon$ for all $x \in S$. $\qquad (**)$

25.6 Theorem.
If a series $\sum_{k=0}^{\infty} g_k$ of functions satisfies the Cauchy criterion uniformly on a set S, then the series converges uniformly on S.

Proof
Let $f_n = \sum_{k=0}^{n} g_k$. The sequence (f_n) of partial sums is uniformly Cauchy on S, so (f_n) converges uniformly on S by Theorem 25.4. ∎

Here is a useful corollary.

25.7 Weierstrass M-test.
Let (M_k) be a sequence of nonnegative real numbers where $\sum M_k < \infty$. If $|g_k(x)| \leq M_k$ for all x in a set S, then $\sum g_k$ converges uniformly on S.

Proof
To verify the Cauchy criterion on S, let $\epsilon > 0$. Since the series $\sum M_k$ converges, it satisfies the Cauchy criterion in Definition 14.3. So there exists a number N such that

$$n \geq m > N \quad \text{implies} \quad \sum_{k=m}^{n} M_k < \epsilon.$$

Hence if $n \geq m > N$ and x is in S, then

$$\left| \sum_{k=m}^{n} g_k(x) \right| \leq \sum_{k=m}^{n} |g_k(x)| \leq \sum_{k=m}^{n} M_k < \epsilon.$$

Thus the series $\sum g_k$ satisfies the Cauchy criterion uniformly on S, and Theorem 25.6 shows it converges uniformly on S. ∎

Example 4
Show $\sum_{n=1}^{\infty} 2^{-n} x^n$ represents a continuous function f on $(-2, 2)$, but the convergence is not uniform. □

Solution
This is a power series with radius of convergence 2. Clearly the series does not converge at $x = 2$ or at $x = -2$, so its interval of convergence is $(-2, 2)$.

Consider $0 < a < 2$ and note $\sum_{n=1}^{\infty} 2^{-n} a^n = \sum_{n=1}^{\infty} (\frac{a}{2})^n$ converges. Since $|2^{-n} x^n| \leq 2^{-n} a^n = (\frac{a}{2})^n$ for $x \in [-a, a]$, the Weierstrass M-test 25.7 shows the series $\sum_{n=1}^{\infty} 2^{-n} x^n$ converges uniformly to a function on $[-a, a]$. By Theorem 25.5 the limit function f is continuous at each point of the set $[-a, a]$. Since a can be any number less than 2, we conclude f represents a continuous function on $(-2, 2)$.

Since we have $\sup\{|2^{-n} x^n| : x \in (-2, 2)\} = 1$ for each n, the convergence of the series cannot be uniform on $(-2, 2)$ in view of the next example. □

Example 5
Show that if the series $\sum g_n$ converges uniformly on a set S, then

$$\lim_{n \to \infty} \sup\{|g_n(x)| : x \in S\} = 0. \tag{1}$$

□

Solution
Let $\epsilon > 0$. Since the series $\sum g_n$ satisfies the Cauchy criterion, there exists N such that

$$n \geq m > N \quad \text{implies} \quad \left| \sum_{k=m}^{n} g_k(x) \right| < \epsilon \quad \text{for all} \quad x \in S.$$

In particular,

$$n > N \quad \text{implies} \quad |g_n(x)| < \epsilon \quad \text{for all} \quad x \in S.$$

Therefore

$$n > N \quad \text{implies} \quad \sup\{|g_n(x)| : x \in S\} \leq \epsilon.$$

This establishes (1). □

Exercises

25.1 Derive Discussions 25.1(b) from 25.1(a). *Hint*: Apply (a) twice, once to g and $|g|$ and once to $-|g|$ and g.

25.2 Let $f_n(x) = \frac{x^n}{n}$. Show (f_n) is uniformly convergent on $[-1, 1]$ and specify the limit function.

25.3 Let $f_n(x) = \frac{n + \cos x}{2n + \sin^2 x}$ for all real numbers x.

    (a) Show (f_n) converges uniformly on $\mathbb{R}$. *Hint*: First decide what the limit function is; then show (f_n) converges uniformly to it.

    (b) Calculate $\lim_{n \to \infty} \int_2^7 f_n(x)\, dx$. *Hint*: Don't integrate f_n.

25.4 Let (f_n) be a sequence of functions on a set $S \subseteq \mathbb{R}$, and suppose $f_n \to f$ uniformly on S. Prove (f_n) is uniformly Cauchy on S. *Hint*: Use the proof of Lemma 10.9 on page 63 as a model, but be careful.

25.5 Let (f_n) be a sequence of bounded functions on a set S, and suppose $f_n \to f$ uniformly on S. Prove f is a bounded function on S.

25.6 (a) Show that if $\sum |a_k| < \infty$, then $\sum a_k x^k$ converges uniformly on $[-1, 1]$ to a continuous function.

    (b) Does $\sum_{n=1}^{\infty} \frac{1}{n^2} x^n$ represent a continuous function on $[-1, 1]$?

25.7 Show $\sum_{n=1}^{\infty} \frac{1}{n^2} \cos nx$ converges uniformly on $\mathbb{R}$ to a continuous function.

25.8 Show $\sum_{n=1}^{\infty} \frac{x^n}{n^2 2^n}$ has radius of convergence 2 and the series converges uniformly to a continuous function on $[-2, 2]$.

25.9 (a) Let $0 < a < 1$. Show the series $\sum_{n=0}^{\infty} x^n$ converges uniformly on $[-a, a]$ to $\frac{1}{1-x}$.

    (b) Does the series $\sum_{n=0}^{\infty} x^n$ converge uniformly on $(-1, 1)$ to $\frac{1}{1-x}$? Explain.

25.10 (a) Show $\sum \frac{x^n}{1+x^n}$ converges for $x \in [0, 1)$.

    (b) Show that the series converges uniformly on $[0, a]$ for each a, $0 < a < 1$.

    (c) Does the series converge uniformly on $[0, 1)$? Explain.

25.11 (a) Sketch the functions g_0, g_1, g_2 and g_3 in Example 3.

    (b) Prove the function f in Example 3 is continuous.

25.12 Suppose $\sum_{k=1}^{\infty} g_k$ is a series of continuous functions g_k on $[a, b]$ that converges uniformly to g on $[a, b]$. Prove

$$\int_a^b g(x)\, dx = \sum_{k=1}^{\infty} \int_a^b g_k(x)\, dx.$$

25.13 Suppose $\sum_{k=1}^{\infty} g_k$ and $\sum_{k=1}^{\infty} h_k$ converge uniformly on a set S. Show $\sum_{k=1}^{\infty}(g_k + h_k)$ converges uniformly on S.

25.14 Prove that if $\sum g_k$ converges uniformly on a set S and if h is a bounded function on S, then $\sum h g_k$ converges uniformly on S.

25.15 Let (f_n) be a sequence of continuous functions on $[a, b]$.

(a) Suppose that, for each x in $[a, b]$, $(f_n(x))$ is a decreasing sequence of real numbers. Prove that if $f_n \to 0$ pointwise on $[a, b]$, then $f_n \to 0$ uniformly on $[a, b]$. *Hint*: If not, there exists $\epsilon > 0$ and a sequence (x_n) in $[a, b]$ such that $f_n(x_n) \geq \epsilon$ for all n. Obtain a contradiction.

(b) Suppose that, for each x in $[a, b]$, $(f_n(x))$ is an increasing sequence of real numbers. Prove that if $f_n \to f$ pointwise on $[a, b]$ and if f is continuous on $[a, b]$, then $f_n \to f$ uniformly on $[a, b]$. This is *Dini's theorem*.

§26 Differentiation and Integration of Power Series

The following result was mentioned in §23 after Example 8.

26.1 Theorem.
Let $\sum_{n=0}^{\infty} a_n x^n$ be a power series with radius of convergence $R > 0$ [possibly $R = +\infty$]. If $0 < R_1 < R$, then the power series converges uniformly on $[-R_1, R_1]$ to a continuous function.

Proof
Consider $0 < R_1 < R$. A glance at Theorem 23.1 shows the series $\sum a_n x^n$ and $\sum |a_n| x^n$ have the same radius of convergence, since β and R are defined in terms of $|a_n|$. Since $|R_1| < R$, we have $\sum |a_n| R_1^n < \infty$. Clearly we have $|a_n x^n| \leq |a_n| R_1^n$ for all x in

$[-R_1, R_1]$, so the series $\sum a_n x^n$ converges uniformly on $[-R_1, R_1]$ by the Weierstrass M-test 25.7. The limit function is continuous at each point of $[-R_1, R_1]$ by Theorem 25.5. ∎

26.2 Corollary.
The power series $\sum a_n x^n$ converges to a continuous function on the open interval $(-R, R)$.

Proof
If $x_0 \in (-R, R)$, then $x_0 \in (-R_1, R_1)$ for some $R_1 < R$. The theorem shows the limit of the series is continuous at x_0. ∎

We emphasize that a power series need *not* converge uniformly on its interval of convergence though it might; see Example 4 of §25 and Exercise 25.8.

We are going to differentiate and integrate power series term-by-term, so clearly it would be useful to know where the new series converge. The next lemma tells us.

26.3 Lemma.
If the power series $\sum_{n=0}^{\infty} a_n x^n$ has radius of convergence R, then the power series

$$\sum_{n=1}^{\infty} n a_n x^{n-1} \quad \text{and} \quad \sum_{n=0}^{\infty} \frac{a_n}{n+1} x^{n+1}$$

also have radius of convergence R.

Proof
First observe the series $\sum n a_n x^{n-1}$ and $\sum n a_n x^n$ have the same radius of convergence: since the second series is x times the first series, they converge for exactly the same values of x. Likewise $\sum \frac{a_n}{n+1} x^{n+1}$ and $\sum \frac{a_n}{n+1} x^n$ have the same radius of convergence.

Next recall $R = \frac{1}{\beta}$ where $\beta = \limsup |a_n|^{1/n}$. For the series $\sum n a_n x^n$, we consider $\limsup (n|a_n|)^{1/n} = \limsup n^{1/n} |a_n|^{1/n}$. By Theorem 9.7(c) on page 48, we have $\lim n^{1/n} = 1$, so $\limsup (n|a_n|)^{1/n} = \beta$ by Theorem 12.1 on page 78. Hence the series $\sum n a_n x^n$ has radius of convergence R.

For the series $\sum \frac{a_n}{n+1}x^n$, we consider $\limsup(\frac{|a_n|}{n+1})^{1/n}$. It is easy to show $\lim(n+1)^{1/n} = 1$; therefore $\lim(\frac{1}{n+1})^{1/n} = 1$. Hence by Theorem 12.1 we have $\limsup(\frac{|a_n|}{n+1})^{1/n} = \beta$, so the series $\sum \frac{a_n}{n+1}x^n$ has radius of convergence R. ∎

26.4 Theorem.
Suppose $f(x) = \sum_{n=0}^{\infty} a_n x^n$ has radius of convergence $R > 0$. Then

$$\int_0^x f(t)\, dt = \sum_{n=0}^{\infty} \frac{a_n}{n+1} x^{n+1} \quad for \quad |x| < R. \tag{1}$$

Proof
We fix x and assume $x < 0$; the case $x > 0$ is similar [Exercise 26.1]. On the interval $[x, 0]$, the sequence of partial sums $\sum_{k=0}^{n} a_k t^k$ converges uniformly to $f(t)$ by Theorem 26.1. Consequently, by Theorem 25.2 we have

$$\int_x^0 f(t)\, dt = \lim_{n\to\infty} \int_x^0 \left(\sum_{k=0}^{n} a_k t^k \right) dt = \lim_{n\to\infty} \sum_{k=0}^{n} a_k \int_x^0 t^k\, dt$$

$$= \lim_{n\to\infty} \sum_{k=0}^{n} a_k \left[\frac{0^{k+1} - x^{k+1}}{k+1} \right] = -\sum_{k=0}^{\infty} \frac{a_k}{k+1} x^{k+1}. \tag{2}$$

The second equality is valid because we can interchange integrals and *finite* sums; this is a basic property of integrals [Theorem 33.3]. Since $\int_0^x f(t)\, dt = -\int_x^0 f(t)\, dt$, Eq. (2) implies Eq. (1). ∎

The theorem just proved shows that a power series can be integrated term-by-term inside its interval of convergence. Term-by-term differentiation is also legal.

26.5 Theorem.
Let $f(x) = \sum_{n=0}^{\infty} a_n x^n$ have radius of convergence $R > 0$. Then f is differentiable on $(-R, R)$ and

$$f'(x) = \sum_{n=1}^{\infty} n a_n x^{n-1} \quad for \quad |x| < R. \tag{1}$$

The proof of Theorem 26.4 was a straightforward application of Theorem 25.2, but the direct analogue of Theorem 25.2 for derivatives is not true [see Exercise 23.8 and Example 4 of §24]. So we give a devious indirect proof of the theorem.

Proof
We begin with the series $g(x) = \sum_{n=1}^{\infty} n a_n x^{n-1}$ and observe this series converges for $|x| < R$ by Lemma 26.3. Theorem 26.4 shows that we can *integrate* g term-by-term:

$$\int_0^x g(t)\, dt = \sum_{n=1}^{\infty} a_n x^n = f(x) - a_0 \quad \text{for} \quad |x| < R.$$

Thus if $0 < R_1 < R$, then

$$f(x) = \int_{-R_1}^x g(t)\, dt + k \quad \text{for} \quad |x| \le R_1,$$

where k is a constant; in fact, $k = a_0 - \int_{-R_1}^0 g(t)\, dt$. Since g is continuous, one of the versions of the Fundamental Theorem of Calculus [Theorem 34.3] shows f is differentiable and $f'(x) = g(x)$. Thus

$$f'(x) = g(x) = \sum_{n=1}^{\infty} n a_n x^{n-1} \quad \text{for} \quad |x| < R.$$

■

Example 1
Recall

$$\sum_{n=0}^{\infty} x^n = \frac{1}{1-x} \quad \text{for} \quad |x| < 1. \tag{1}$$

Differentiating term-by-term, we obtain

$$\sum_{n=1}^{\infty} n x^{n-1} = \frac{1}{(1-x)^2} \quad \text{for} \quad |x| < 1.$$

Integrating (1) term-by-term, we get

$$\sum_{n=0}^{\infty} \frac{1}{n+1} x^{n+1} = \int_0^x \frac{1}{1-t}\, dt = -\log_e(1-x)$$

or

$$\log_e(1-x) = -\sum_{n=1}^{\infty} \frac{1}{n} x^n \quad \text{for} \quad |x| < 1. \tag{2}$$

Replacing x by $-x$, we find

$$\log_e(1+x) = x - \frac{x^2}{2} + \frac{x^3}{3} - \frac{x^4}{4} + \cdots \quad \text{for} \quad |x| < 1. \tag{3}$$

It turns out that this equality is also valid for $x = 1$ [see Example 2], so we have the interesting identity

$$\log_e 2 = 1 - \frac{1}{2} + \frac{1}{3} - \frac{1}{4} + \frac{1}{5} - \frac{1}{6} + \cdots. \tag{4}$$

In Eq. (2) set $x = \frac{m-1}{m}$. Then

$$\sum_{n=1}^{\infty} \frac{1}{n} \left(\frac{m-1}{m} \right)^n = -\log_e \left(1 - \frac{m-1}{m} \right) = -\log_e \left(\frac{1}{m} \right) = \log_e m.$$

Hence we have

$$\sum_{n=1}^{\infty} \frac{1}{n} \geq \sum_{n=1}^{\infty} \frac{1}{n} \left(\frac{m-1}{m} \right)^n = \log_e m \quad \text{for all} \quad m.$$

Here is yet another proof that $\sum_{n=1}^{\infty} \frac{1}{n} = +\infty$. □

To establish (4) we need a relatively difficult theorem about convergence of a power series at the endpoints of its interval of convergence.

26.6 Abel's Theorem.
Let $f(x) = \sum_{n=0}^{\infty} a_n x^n$ be a power series with finite positive radius of convergence R. If the series converges at $x = R$, then f is continuous at $x = R$. If the series converges at $x = -R$, then f is continuous at $x = -R$.

Example 2
As promised, we return to (3) in Example 1:

$$\log_e(1+x) = x - \frac{x^2}{2} + \frac{x^3}{3} - \frac{x^4}{4} + \cdots \quad \text{for} \quad |x| < 1.$$

For $x = 1$ the series converges by the Alternating Series Theorem 15.3. Thus the series represents a function f on $(-1, 1]$ that is continuous at $x = 1$ by Abel's theorem. The function $\log_e(1 + x)$ is also continuous at $x = 1$, so the functions agree at $x = 1$. [In detail, if (x_n) is a sequence in $(-1, 1)$ converging to 1, then $f(1) = \lim_{n \to \infty} f(x_n) = \lim_{n \to \infty} \log_e(1 + x_n) = \log_e 2$.] Therefore we have

$$\log_e 2 = 1 - \frac{1}{2} + \frac{1}{3} - \frac{1}{4} + \frac{1}{5} - \frac{1}{6} + \cdots .$$

Another proof of this identity is given in Example 2 of §31. $\qquad \square$

Example 3
Recall $\sum_{n=0}^{\infty} x^n = \frac{1}{1-x}$ for $|x| < 1$. Note that at $x = -1$ the function $\frac{1}{1-x}$ is continuous and takes the value $\frac{1}{2}$. However, the series does *not* converge for $x = -1$, so Abel's theorem does not apply. $\qquad \square$

Proof of Abel's Theorem
The heart of the proof is in Case 1.

Case 1. Suppose $f(x) = \sum_{n=0}^{\infty} a_n x^n$ has radius of convergence 1 and the series converges at $x = 1$. We will prove f is continuous on $[0, 1]$. By subtracting a constant from f, we may assume $f(1) = \sum_{n=0}^{\infty} a_n = 0$. Let $f_n(x) = \sum_{k=0}^{n} a_k x^k$ and $s_n = \sum_{k=0}^{n} a_k = f_n(1)$ for $n = 0, 1, 2, \ldots$. Since $f_n(x) \to f(x)$ pointwise on $[0, 1]$ and each f_n is continuous, Theorem 24.3 on page 196 shows it suffices to show $f_n \to f$ uniformly on $[0, 1]$. Theorem 25.4 on page 202 shows it suffices to show the convergence is uniformly Cauchy.

For $m < n$, we have

$$f_n(x) - f_m(x) = \sum_{k=m+1}^{n} a_k x^k = \sum_{k=m+1}^{n} (s_k - s_{k-1}) x^k$$

$$= \sum_{k=m+1}^{n} s_k x^k - x \sum_{k=m+1}^{n} s_{k-1} x^{k-1}$$

$$= \sum_{k=m+1}^{n} s_k x^k - x \sum_{k=m}^{n-1} s_k x^k ,$$

and therefore

$$f_n(x) - f_m(x) = s_n x^n - s_m x^{m+1} + (1 - x) \sum_{k=m+1}^{n-1} s_k x^k. \qquad (1)$$

Since $\lim s_n = \sum_{k=0}^{\infty} a_k = f(1) = 0$, given $\epsilon > 0$, there is an integer N so that $|s_n| < \frac{\epsilon}{3}$ for all $n \geq N$. Then for $n > m \geq N$ and x in $[0, 1)$, we have

$$\left| (1 - x) \sum_{k=m+1}^{n-1} s_k x^k \right| \leq \frac{\epsilon}{3} (1 - x) \sum_{k=m+1}^{n-1} x^k$$

$$= \frac{\epsilon}{3} (1 - x) x^{m+1} \frac{1 - x^{n-m-1}}{1 - x} < \frac{\epsilon}{3}. \qquad (2)$$

The first term in inequality (2) is also less than $\frac{\epsilon}{3}$ for $x = 1$. Therefore, for $n > m \geq N$ and x in $[0, 1]$, (1) and (2) show

$$|f_n(x) - f_m(x)| \leq |s_n| x^n + |s_m| x^{m+1} + \frac{\epsilon}{3} < \frac{\epsilon}{3} + \frac{\epsilon}{3} + \frac{\epsilon}{3} = \epsilon.$$

Thus the sequence (f_n) is uniformly Cauchy on $[0, 1]$, and its limit f is continuous.

Case 2. Suppose $f(x) = \sum_{n=0}^{\infty} a_n x^n$ has radius of convergence R, $0 < R < \infty$, and the series converges at $x = R$. Let $g(x) = f(Rx)$ and note that

$$g(x) = \sum_{n=0}^{\infty} a_n R^n x^n \quad \text{for} \quad |x| < 1.$$

This series has radius of convergence 1, and it converges at $x = 1$. By Case 1, g is continuous at $x = 1$. Since $f(x) = g(\frac{x}{R})$, it follows that f is continuous at $x = R$.

Case 3. Suppose $f(x) = \sum_{n=0}^{\infty} a_n x^n$ has radius of convergence R, $0 < R < \infty$, and the series converges at $x = -R$. Let $h(x) = f(-x)$ and note that

$$h(x) = \sum_{n=0}^{\infty} (-1)^n a_n x^n \quad \text{for} \quad |x| < R.$$

The series for h converges at $x = R$, so h is continuous at $x = R$ by Case 2. It follows that $f(x) = h(-x)$ is continuous at $x = -R$. ∎

The point of view in our extremely brief introduction to power series has been: For a given power series $\sum a_n x^n$, what can one say about the function $f(x) = \sum a_n x^n$? This point of view was misleading. Often, in real life, one begins with a function f and seeks a power series that represents the function for some or all values of x. This is because power series, being limits of polynomials, are in some sense basic objects.

If we have $f(x) = \sum_{n=0}^{\infty} a_n x^n$ for $|x| < R$, then we can differentiate f term-by-term forever. At each step, we may calculate the kth derivative of f at 0, written $f^{(k)}(0)$. It is easy to show $f^{(k)}(0) = k! a_k$ for $k \geq 0$. This tells us that *if f can be represented by a power series*, then that power series must be $\sum_{k=0}^{\infty} \frac{f^{(k)}(0)}{k!} x^k$. This is the *Taylor series* for f about 0. Frequently, but not always, the Taylor series will agree with f on the interval of convergence. This turns out to be true for many familiar functions. Thus the following relations can be proved:

$$e^x = \sum_{k=0}^{\infty} \frac{1}{k!} x^k, \quad \cos x = \sum_{k=0}^{\infty} \frac{(-1)^k}{(2k)!} x^{2k}, \quad \sin x = \sum_{k=0}^{\infty} \frac{(-1)^k}{(2k+1)!} x^{2k+1}$$

for all x in $\mathbb{R}$. A detailed study of Taylor series is given in §31.

Exercises

26.1 Prove Theorem 26.4 for $x > 0$.

26.2 (a) Observe $\sum_{n=1}^{\infty} n x^n = \frac{x}{(1-x)^2}$ for $|x| < 1$; see Example 1.

(b) Evaluate $\sum_{n=1}^{\infty} \frac{n}{2^n}$. Compare with Exercise 14.13(d).

(c) Evaluate $\sum_{n=1}^{\infty} \frac{n}{3^n}$ and $\sum_{n=1}^{\infty} \frac{(-1)^n n}{3^n}$.

26.3 (a) Use Exercise 26.2 to derive an explicit formula for $\sum_{n=1}^{\infty} n^2 x^n$.

(b) Evaluate $\sum_{n=1}^{\infty} \frac{n^2}{2^n}$ and $\sum_{n=1}^{\infty} \frac{n^2}{3^n}$.

26.4 (a) Observe $e^{-x^2} = \sum_{n=0}^{\infty} \frac{(-1)^n}{n!} x^{2n}$ for $x \in \mathbb{R}$, since we have $e^x = \sum_{n=1}^{\infty} \frac{1}{n!} x^n$ for $x \in \mathbb{R}$.

(b) Express $F(x) = \int_0^x e^{-t^2} dt$ as a power series.

26.5 Let $f(x) = \sum_{n=0}^{\infty} \frac{1}{n!} x^n$ for $x \in \mathbb{R}$. Show $f' = f$. Do *not* use the fact that $f(x) = e^x$; this is true but has not been established at this point in the text.

26.6 Let $s(x) = x - \frac{x^3}{3!} + \frac{x^5}{5!} - \cdots$ and $c(x) = 1 - \frac{x^2}{2!} + \frac{x^4}{4!} - \cdots$ for $x \in \mathbb{R}$.

(a) Prove $s' = c$ and $c' = -s$.

(b) Prove $(s^2 + c^2)' = 0$.

(c) Prove $s^2 + c^2 = 1$.

Actually $s(x) = \sin x$ and $c(x) = \cos x$, but you do *not* need these facts.

26.7 Let $f(x) = |x|$ for $x \in \mathbb{R}$. Is there a power series $\sum a_n x^n$ such that $f(x) = \sum_{n=0}^{\infty} a_n x^n$ for all x? Discuss.

26.8 (a) Show $\sum_{n=0}^{\infty} (1)^n x^{2n} = \frac{1}{1+x^2}$ for $x \in (-1, 1)$. *Hint:* $\sum_{n=0}^{\infty} y^n = \frac{1}{1-y}$. Let $y = -x^2$.

(b) Show $\arctan x = \sum_{n=0}^{\infty} \frac{(-1)^n}{2n+1} x^{2n+1}$ for $x \in (-1, 1)$.

(c) Show the equality in (b) also holds for $x = 1$. Use this to find a nice formula for π.

(d) What happens at $x = -1$?

§27 * Weierstrass's Approximation Theorem

Suppose a power series has radius of convergence greater than 1, and let f denote the function given by the power series. Theorem 26.1 tells us that the partial sums of the power series get uniformly close to f on $[-1, 1]$. In other words, f can be approximated uniformly on $[-1, 1]$ by polynomials. Weierstrass's approximation theorem is a generalization of this last observation, for it tells us that *any* continuous function on $[-1, 1]$ can be uniformly approximated by polynomials on $[-1, 1]$. This result is quite different because such a function need not be given by a power series; see Exercise 26.7. The approximation theorem is valid for any closed interval $[a, b]$ and can be deduced easily from the case $[0, 1]$; see Exercise 27.1.

We give the beautiful proof due to S. N. Bernstein. Bernstein was motivated by probabilistic considerations, but we will not use any probability here. One of the attractive features of Bernstein's proof is that the approximating polynomials will be given explicitly. There are more abstract proofs in which this is not the case. On the other hand, the abstract proofs lead to far-reaching and important generalizations. See the treatment in [31] or [62].

We need some preliminary facts about polynomials involving binomial coefficients.

27.1 Lemma.

For every $x \in \mathbb{R}$ and $n \geq 0$, we have

$$\sum_{k=0}^{n} \binom{n}{k} x^k (1-x)^{n-k} = 1.$$

Proof

This is just the binomial theorem [Exercise 1.12] applied to $a = x$ and $b = 1 - x$, since in this case $(a+b)^n = 1^n = 1$. ∎

27.2 Lemma.

For $x \in \mathbb{R}$ and $n \geq 0$, we have

$$\sum_{k=0}^{n} (nx - k)^2 \binom{n}{k} x^k (1-x)^{n-k} = nx(1-x) \leq \frac{n}{4}. \qquad (1)$$

Proof

Since $k\binom{n}{k} = n\binom{n-1}{k-1}$ for $k \geq 1$, we have

$$\sum_{k=0}^{n} k \binom{n}{k} x^k (1-x)^{n-k} = n \sum_{k=1}^{n} \binom{n-1}{k-1} x^k (1-x)^{n-k}$$

$$= nx \sum_{j=0}^{n-1} \binom{n-1}{j} x^j (1-x)^{n-1-j}$$

$$= nx. \qquad (2)$$

Since $k(k-1)\binom{n}{k} = n(n-1)\binom{n-2}{k-2}$ for $k \geq 2$, we have

$$\sum_{k=0}^{n} k(k-1) \binom{n}{k} x^k (1-x)^{n-k} = n(n-1)x^2 \sum_{j=0}^{n-2} \binom{n-2}{j} x^j (1-x)^{n-2-j}$$

$$= n(n-1)x^2. \qquad (3)$$

Adding the results in (2) and (3), we find

$$\sum_{k=0}^{n} k^2 \binom{n}{k} x^k (1-x)^{n-k} = n(n-1)x^2 + nx = n^2 x^2 + nx(1-x). \quad (4)$$

Since $(nx-k)^2 = n^2 x^2 - 2nx \cdot k + k^2$, we use Lemma 27.1, (2) and (4) to obtain

$$\sum_{k=0}^{n} (nx-k)^2 \binom{n}{k} x^k (1-x)^{n-k} = n^2 x^2 - 2nx(nx) + [n^2 x^2 + nx(1-x)]$$

$$= nx(1-x).$$

This establishes the equality in (1). The inequality in (1) simply reflects the inequality $x(1-x) \leq \frac{1}{4}$, which is equivalent to $4x^2 - 4x + 1 \geq 0$ or $(2x-1)^2 \geq 0$. ∎

27.3 Definition.
Let f be a function defined on $[0,1]$. The polynomials $B_n f$ defined by

$$B_n f(x) = \sum_{k=0}^{n} f\left(\frac{k}{n}\right) \cdot \binom{n}{k} x^k (1-x)^{n-k}$$

are called *Bernstein polynomials* for the function f.

Here is Bernstein's version of the Weierstrass approximation theorem.

27.4 Theorem.
For every continuous function f on $[0,1]$, we have

$$B_n f \to f \quad \text{uniformly on} \quad [0,1].$$

Proof
We assume f is not identically zero, and we let

$$M = \sup\{|f(x)| : x \in [0,1]\}.$$

Consider $\epsilon > 0$. Since f is uniformly continuous by Theorem 19.2, there exists $\delta > 0$ such that

$$x, y \in [0,1] \quad \text{and} \quad |x-y| < \delta \quad \text{imply} \quad |f(x) - f(y)| < \frac{\epsilon}{2}. \quad (1)$$

Let $N = \frac{M}{\epsilon \delta^2}$. This choice of N is unmotivated at this point, but we make it here to emphasize that it does not depend on the choice of x. We will show

$$|B_n f(x) - f(x)| < \epsilon \quad \text{for all} \quad x \in [0,1] \quad \text{and all} \quad n > N, \quad (2)$$

completing the proof of the theorem.

To prove (2), consider a fixed $x \in [0,1]$ and $n > N$. In view of Lemma 27.1, we have

$$f(x) = \sum_{k=0}^{n} f(x) \binom{n}{k} x^k (1-x)^{n-k},$$

so

$$|B_n f(x) - f(x)| \leq \sum_{k=0}^{n} \left| f\left(\frac{k}{n}\right) - f(x) \right| \cdot \binom{n}{k} x^k (1-x)^{n-k}. \quad (3)$$

To estimate this sum, we divide the set $\{0, 1, 2, \ldots, n\}$ into two sets:

$$k \in A \quad \text{if} \quad \left| \frac{k}{n} - x \right| < \delta \quad \text{while} \quad k \in B \quad \text{if} \quad \left| \frac{k}{n} - x \right| \geq \delta.$$

For $k \in A$ we have $|f(\frac{k}{n}) - f(x)| < \frac{\epsilon}{2}$ by (1), so

$$\sum_{k \in A} \left| f\left(\frac{k}{n}\right) - f(x) \right| \cdot \binom{n}{k} x^k (1-x)^{n-k}$$

$$\leq \sum_{k \in A} \frac{\epsilon}{2} \binom{n}{k} x^k (1-x)^{n-k} \leq \frac{\epsilon}{2} \quad (4)$$

using Lemma 27.1. For $k \in B$, we have $|\frac{k-nx}{n}| \geq \delta$ or $(k - nx)^2 \geq n^2 \delta^2$, so

$$\sum_{k \in B} \left| f\left(\frac{k}{n}\right) - f(x) \right| \cdot \binom{n}{k} x^k (1-x)^{n-k} \leq 2M \sum_{k \in B} \binom{n}{k} x^k (1-x)^{n-k}$$

$$\leq \frac{2M}{n^2 \delta^2} \sum_{k \in B} (k - nx)^2 \binom{n}{k} x^k (1-x)^{n-k}.$$

By Lemma 27.2, this is bounded by

$$\frac{2M}{n^2 \delta^2} \cdot \frac{n}{4} = \frac{M}{2n\delta^2} < \frac{M}{2N\delta^2} = \frac{\epsilon}{2}.$$

This observation, (4) and (3) show

$$|B_n f(x) - f(x)| < \epsilon.$$

That is, (2) holds. ∎

27.5 Weierstrass's Approximation Theorem.
Every continuous function on a closed interval $[a, b]$ can be uniformly approximated by polynomials on $[a, b]$.

In other words, given a continuous function f on $[a, b]$, there exists a sequence (p_n) of polynomials such that $p_n \to f$ uniformly on $[a, b]$.

27.6 Corollary.
Given a continuous function f on $[a, b]$, there exists a sequence (p_n) of polynomials such that $p_n \to f$ uniformly on $[a, b]$, and for each n, $p_n(a) = f(a)$ and $p_n(b) = f(b)$.

Proof
See Exercise 27.4. ∎

Exercises

27.1 Prove Theorem 27.5 from Theorem 27.4. *Hint*: Let $\phi(x) = (b-a)x+a$ so that ϕ maps $[0, 1]$ onto $[a, b]$. If f is continuous on $[a, b]$, then $f \circ \phi$ is continuous on $[0, 1]$.

27.2 Show that if f is continuous on $\mathbb{R}$, then there exists a sequence (p_n) of polynomials such that $p_n \to f$ uniformly on each bounded subset of $\mathbb{R}$. *Hint*: Arrange for $|f(x) - p_n(x)| < \frac{1}{n}$ for $|x| \le n$.

27.3 Show there does not exist a sequence of polynomials converging uniformly on $\mathbb{R}$ to f if
    (a) $f(x) = \sin x$,   (b) $f(x) = e^x$.

27.4 Prove Corollary 27.6. *Hint*: Select a sequence (q_n) of polynomials such that $q_n \to f$ uniformly on $[a, b]$. For each n, let s_n be the function on $\mathbb{R}$ whose graph is the straight line passing through the points $(a, f(a) - q_n(a))$ and $(b, f(b) - q_n(b))$. Set $p_n = q_n + s_n$.

27.5 Find the sequence $(B_n f)$ of Bernstein polynomials in case
    (a) $f(x) = x$,   (b) $f(x) = x^2$.

27.6 The Bernstein polynomials were defined for any function f on $[0, 1]$. Show that if $B_n f \to f$ uniformly on $[0, 1]$, then f is continuous on $[0, 1]$.

27.7 Let f be a bounded function on $[0, 1]$, say $|f(x)| \leq M$ for all $x \in [0, 1]$. Show that all the Bernstein polynomials $B_n f$ are bounded by M.

5

CHAPTER

Differentiation

In this chapter we give a theoretical treatment of differentiation and related concepts, most or all of which will be familiar from the standard calculus course. Three of the most useful results are the Mean Value Theorem which is treated in §29, L'Hospital's Rule which is treated in §30, and Taylor's Theorem which is given in §31.

§28 Basic Properties of the Derivative

The reader may wish to review the theory of limits treated in §20.

28.1 Definition.
Let f be a real-valued function defined on an open interval containing a point a. We say f is *differentiable at* a, or f *has a derivative at* a, if the limit

$$\lim_{x \to a} \frac{f(x) - f(a)}{x - a}$$

K.A. Ross, *Elementary Analysis: The Theory of Calculus*,   223
Undergraduate Texts in Mathematics, DOI 10.1007/978-1-4614-6271-2_5,
© Springer Science+Business Media New York 2013

exists and is finite. We will write $f'(a)$ for the derivative of f at a:

$$f'(a) = \lim_{x \to a} \frac{f(x) - f(a)}{x - a} \tag{1}$$

whenever this limit exists and is finite.

Generally speaking, we will be interested in f' as a function in its own right. The domain of f' is the set of points at which f is differentiable; thus $\mathrm{dom}(f') \subseteq \mathrm{dom}(f)$.

Example 1
The derivative of the function $g(x) = x^2$ at $x = 2$ was calculated in Example 2 of §20:

$$g'(2) = \lim_{x \to 2} \frac{x^2 - 4}{x - 2} = \lim_{x \to 2} (x + 2) = 4.$$

We can calculate $g'(a)$ just as easily:

$$g'(a) = \lim_{x \to a} \frac{x^2 - a^2}{x - a} = \lim_{x \to a} (x + a) = 2a.$$

This computation is even valid for $a = 0$. We may write $g'(x) = 2x$ since the name of the variable a or x is immaterial. Thus the derivative of the function given by $g(x) = x^2$ is the function given by $g'(x) = 2x$, as every calculus student knows. $\square$

Example 2
The derivative of $h(x) = \sqrt{x}$ at $x = 1$ was calculated in Example 3 of §20: $h'(1) = \frac{1}{2}$. In fact, $h(x) = x^{1/2}$ for $x \geq 0$ and $h'(x) = \frac{1}{2}x^{-1/2}$ for $x > 0$; see Exercise 28.3. $\square$

Example 3
Let n be a positive integer, and let $f(x) = x^n$ for all $x \in \mathbb{R}$. We show $f'(x) = nx^{n-1}$ for all $x \in \mathbb{R}$. Fix a in $\mathbb{R}$ and observe

$$f(x) - f(a) = x^n - a^n = (x-a)(x^{n-1} + ax^{n-2} + a^2 x^{n-3} + \cdots + a^{n-2}x + a^{n-1}),$$

so

$$\frac{f(x) - f(a)}{x - a} = x^{n-1} + ax^{n-2} + a^2 x^{n-3} + \cdots + a^{n-2}x + a^{n-1}$$

for $x \neq a$. It follows that

$$f'(a) = \lim_{x \to a} \frac{f(x) - f(a)}{x - a}$$
$$= a^{n-1} + aa^{n-2} + a^2 a^{n-3} + \cdots + a^{n-2}a + a^{n-1} = na^{n-1};$$

we are using Theorem 20.4 and the fact that $\lim_{x \to a} x^k = a^k$ for k in $\mathbb{N}$. $\qquad\square$

We first prove differentiability at a point implies continuity at the point. This may seem obvious from all the pictures of familiar differentiable functions. However, Exercise 28.8 contains an example of a function that is differentiable at 0 and of course continuous at 0 [by the next theorem], but is discontinuous at all other points.

28.2 Theorem.
If f is differentiable at a point a, then f is continuous at a.

Proof
We are given $f'(a) = \lim_{x \to a} \frac{f(x) - f(a)}{x - a}$, and we need to prove $\lim_{x \to a} f(x) = f(a)$. We have

$$f(x) = (x - a)\frac{f(x) - f(a)}{x - a} + f(a)$$

for $x \in \operatorname{dom}(f)$, $x \neq a$. Since $\lim_{x \to a}(x - a) = 0$ and $\lim_{x \to a} \frac{f(x) - f(a)}{x - a}$ exists and is finite, Theorem 20.4(ii) shows $\lim_{x \to a}(x - a) \cdot \frac{f(x) - f(a)}{x - a} = 0$. Therefore $\lim_{x \to a} f(x) = f(a)$, as desired. $\qquad\blacksquare$

We next prove some results about sums, products, etc. of derivatives. Let us first recall why the product rule is *not* $(fg)' = f'g'$ [as many naive calculus students wish!] even though the product of limits does behave as expected:

$$\lim_{x \to a}(f_1 f_2)(x) = \left[\lim_{x \to a} f_1(x)\right] \cdot \left[\lim_{x \to a} f_2(x)\right]$$

provided the limits on the right side exist and are finite; see Theorem 20.4(ii). The difficulty is that the limit for the derivative of the product is not the product of the limits of the derivatives, i.e.,

$$\frac{f(x)g(x) - f(a)g(a)}{x - a} \neq \frac{f(x) - f(a)}{x - a} \cdot \frac{g(x) - g(a)}{x - a}.$$

The correct product rule is obtained by shrewdly writing the left hand side in terms of $\frac{f(x)-f(a)}{x-a}$ and $\frac{g(x)-g(a)}{x-a}$ as in the proof of Theorem 28.3(iii) below.

28.3 Theorem.

Let f and g be functions that are differentiable at the point a. Each of the functions cf [c a constant], f+g, fg and f/g is also differentiable at a, except f/g if g(a) = 0 since f/g is not defined at a in this case. The formulas are

   (i) $(cf)'(a) = c \cdot f'(a)$;

   (ii) $(f + g)'(a) = f'(a) + g'(a)$;

   (iii) *[product rule]* $(fg)'(a) = f(a)g'(a) + f'(a)g(a)$;

   (iv) *[quotient rule]* $(f/g)'(a) = [g(a)f'(a) - f(a)g'(a)]/g^2(a)$ *if* $g(a) \neq 0$.

Proof

(i) By definition of cf we have $(cf)(x) = c \cdot f(x)$ for all $x \in \text{dom}(f)$; hence

$$(cf)'(a) = \lim_{x \to a} \frac{(cf)(x) - (cf)(a)}{x - a} = \lim_{x \to a} c \cdot \frac{f(x) - f(a)}{x - a} = c \cdot f'(a).$$

(ii) This follows from the identity

$$\frac{(f + g)(x) - (f + g)(a)}{x - a} = \frac{f(x) - f(a)}{x - a} + \frac{g(x) - g(a)}{x - a}$$

upon taking the limit as $x \to a$ and applying Theorem 20.4(i).

(iii) Observe

$$\frac{(fg)(x) - (fg)(a)}{x - a} = f(x)\frac{g(x) - g(a)}{x - a} + g(a)\frac{f(x) - f(a)}{x - a}$$

for $x \in \text{dom}(fg)$, $x \neq a$. We take the limit as $x \to a$ and note that $\lim_{x \to a} f(x) = f(a)$ by Theorem 28.2. We obtain [again using Theorem 20.4]

$$(fg)'(a) = f(a)g'(a) + g(a)f'(a).$$

(iv) Since $g(a) \neq 0$ and g is continuous at a, there exists an open interval I containing a such that $g(x) \neq 0$ for $x \in I$. For $x \in I$ we can write

$$(f/g)(x) - (f/g)(a) = \frac{f(x)}{g(x)} - \frac{f(a)}{g(a)} = \frac{g(a)f(x) - f(a)g(x)}{g(x)g(a)}$$
$$= \frac{g(a)f(x) - g(a)f(a) + g(a)f(a) - f(a)g(x)}{g(x)g(a)},$$

so

$$\frac{(f/g)(x) - (f/g)(a)}{x - a}$$
$$= \left\{ g(a)\frac{f(x) - f(a)}{x - a} - f(a)\frac{g(x) - g(a)}{x - a} \right\} \frac{1}{g(x)g(a)}$$

for $x \in I$, $x \neq a$. Now we take the limit as $x \to a$ to obtain (iv); note that $\lim_{x \to a} \frac{1}{g(x)g(a)} = \frac{1}{g^2(a)}$. ∎

Example 4
Let m be a positive integer, and let $h(x) = x^{-m}$ for $x \neq 0$. Then $h(x) = f(x)/g(x)$ where $f(x) = 1$ and $g(x) = x^m$ for all x. By the quotient rule,

$$h'(a) = \frac{g(a)f'(a) - f(a)g'(a)}{g^2(a)} = \frac{a^m \cdot 0 - 1 \cdot ma^{m-1}}{a^{2m}}$$
$$= \frac{-m}{a^{m+1}} = -ma^{-m-1}$$

for $a \neq 0$. If we write n for $-m$, then we see the derivative of x^n is nx^{n-1} for negative integers n as well as for positive integers. The result is also trivially valid for $n = 0$. For fractional exponents, see Exercise 29.15. ☐

28.4 Theorem [Chain Rule].
If f is differentiable at a and g is differentiable at $f(a)$, then the composite function $g \circ f$ is differentiable at a and we have $(g \circ f)'(a) = g'(f(a)) \cdot f'(a)$.

Discussion. Here is a faulty "proof" which nevertheless contains the essence of a valid proof. We write

$$\frac{g \circ f(x) - g \circ f(a)}{x - a} = \frac{g(f(x)) - g(f(a))}{f(x) - f(a)} \cdot \frac{f(x) - f(a)}{x - a} \qquad (1)$$

for $x \neq a$. Since $\lim_{x \to a} f(x) = f(a)$, we have

$$\lim_{x \to a} \frac{g(f(x)) - g(f(a))}{f(x) - f(a)} = \lim_{y \to f(a)} \frac{g(y) - g(f(a))}{y - f(a)} = g'(f(a)). \quad (2)$$

We also have $\lim_{x \to a} \frac{f(x) - f(a)}{x - a} = f'(a)$, so (1) shows $(g \circ f)'(a) = g'(f(a)) \cdot f'(a)$.

The main problem with this "proof" is that $f(x) - f(a)$ in Eq. (2) might be 0 for x arbitrarily close to a. If, however, $f(x) \neq f(a)$ for x near a, then this proof can be made rigorous. If $f(x) = f(a)$ for some x's near a, the "proof" cannot be repaired using (2). In fact, Exercise 28.5 gives an example of differentiable functions f and g for which $\lim_{x \to 0} \frac{g(f(x)) - g(f(0))}{f(x) - f(0)}$ is meaningless.

We now give a rigorous proof, but we will use the sequential definition of a limit. The proof reflects ideas in the article by Stephen Kenton [36].

Proof

The hypotheses include the assumptions that f is defined on an open interval J containing a, and g is defined on an open interval I containing $f(a)$. It is easy to check $g \circ f$ is defined on some open interval containing a (see Exercise 28.13), so by taking J smaller if necessary, we may assume $g \circ f$ is defined on J.

By Definitions 20.3(a) and 20.1, it suffices to consider a sequence (x_n) in $J \setminus \{a\}$ where $\lim_n x_n = a$ and show

$$\lim_{n \to \infty} \frac{g \circ f(x_n) - g \circ f(a)}{x_n - a} = g'(f(a)) \cdot f'(a). \quad (3)$$

For each n, let $y_n = f(x_n)$. Since f is continuous at $x = a$, we have $\lim_n y_n = f(a)$. For $f(x_n) \neq f(a)$,

$$\frac{(g \circ f)(x_n) - (g \circ f)(a)}{x_n - a} = \frac{g(y_n) - g(f(a))}{y_n - f(a)} \cdot \frac{f(x_n) - f(a)}{x_n - a}. \quad (4)$$

Case 1. Suppose $f(x) \neq f(a)$ for x near a. Then $y_n = f(x_n) \neq f(a)$ for large n, so taking the limit in (4), as $n \to \infty$, we obtain $(g \circ f)'(a) = g'(f(a)) \cdot f'(a)$.

Case 2. Suppose $f(x) = f(a)$ for x arbitrarily close to a. Then there is a sequence (z_n) in $J \setminus \{a\}$ such that $\lim_n z_n = a$ and $f(z_n) = f(a)$ for all n. Then

$$f'(a) = \lim_{n \to \infty} \frac{f(z_n) - f(a)}{z_n - a} = \lim_{n \to \infty} \frac{0}{z_n - a} = 0,$$

and it suffices to show $(g \circ f)'(a) = 0$. We do have

$$\lim_{n \to \infty} \frac{g(f(z_n)) - g(f(a))}{z_n - a} = \lim_{n \to \infty} \frac{0}{z_n - a} = 0,$$

but this only assures us that $(g \circ f)'(a) = 0$ *provided* we know this derivative exists. To help us prove this, we note $g'(f(a))$ exists and so the difference quotients $\frac{g(y) - g(f(a))}{y - f(a)}$ are bounded near $f(a)$. Thus, replacing the open interval I by a smaller one if necessary, there is a constant $C > 0$ so that

$$\left| \frac{g(y) - g(f(a))}{y - f(a)} \right| \le C \quad \text{for} \quad y \in I \setminus f(a).$$

Therefore

$$\left| \frac{(g \circ f)(x_n) - (g \circ f)(a)}{x_n - a} \right| \le C \left| \frac{f(x_n) - f(a)}{x_n - a} \right| \quad \text{for large} \quad n, \quad (5)$$

which is clear if $f(x_n) = f(a)$ and otherwise follows from Eq. (4). Since $f'(a) = 0$, the right side of (5) tends to 0 as $n \to \infty$; thus the left side also tends to 0. Since (x_n) is any sequence in $J \setminus \{a\}$ converging to a, we conclude $(g \circ f)'(a) = 0$, as desired. ∎

It is worth emphasizing that if f is differentiable on an interval I and if g is differentiable on $\{f(x) : x \in I\}$, then $(g \circ f)'$ is exactly the function $(g' \circ f) \cdot f'$ on I.

Example 5

Let $h(x) = \sin(x^3 + 7x)$ for $x \in \mathbb{R}$. The reader can undoubtedly verify $h'(x) = (3x^2 + 7) \cos(x^3 + 7x)$ for $x \in \mathbb{R}$ using some automatic technique learned in calculus. Whatever the automatic technique, it is justified by the chain rule. In this case, $h = g \circ f$ where $f(x) = x^3 + 7x$ and $g(y) = \sin y$. Then $f'(x) = 3x^2 + 7$ and $g'(y) = \cos y$ so that

$$h'(x) = g'(f(x)) \cdot f'(x) = [\cos f(x)] \cdot f'(x) = [\cos(x^3 + 7x)] \cdot (3x^2 + 7).$$

We do not want the reader to unlearn the automatic technique, but the reader should be aware that the chain rule stands behind it. $\quad\square$

Exercises

28.1 For each of the following functions defined on $\mathbb{R}$, give the set of points at which it is *not* differentiable. Sketches will be helpful.
   (a) $e^{|x|}$ (b) $\sin|x|$
   (c) $|\sin x|$ (d) $|x| + |x - 1|$
   (e) $|x^2 - 1|$ (f) $|x^3 - 8|$

28.2 Use the *definition* of derivative to calculate the derivatives of the following functions at the indicated points.

   (a) $f(x) = x^3$ at $x = 2$;

   (b) $g(x) = x + 2$ at $x = a$;

   (c) $f(x) = x^2 \cos x$ at $x = 0$;

   (d) $r(x) = \frac{3x+4}{2x-1}$ at $x = 1$.

28.3 (a) Let $h(x) = \sqrt{x} = x^{1/2}$ for $x \geq 0$. Use the *definition* of derivative to prove $h'(x) = \frac{1}{2}x^{-1/2}$ for $x > 0$.

   (b) Let $f(x) = x^{1/3}$ for $x \in \mathbb{R}$ and use the definition of derivative to prove $f'(x) = \frac{1}{3}x^{-2/3}$ for $x \neq 0$.

   (c) Is the function f in part (b) differentiable at $x = 0$? Explain.

28.4 Let $f(x) = x^2 \sin \frac{1}{x}$ for $x \neq 0$ and $f(0) = 0$.

   (a) Use Theorems 28.3 and 28.4 to show f is differentiable at each $a \neq 0$ and calculate $f'(a)$. Use, without proof, the fact that $\sin x$ is differentiable and that $\cos x$ is its derivative.

   (b) Use the definition to show f is differentiable at $x = 0$ and $f'(0) = 0$.

   (c) Show f' is not continuous at $x = 0$.

28.5 Let $f(x) = x^2 \sin \frac{1}{x}$ for $x \neq 0$, $f(0) = 0$, and $g(x) = x$ for $x \in \mathbb{R}$.

   (a) Observe f and g are differentiable on $\mathbb{R}$.

   (b) Calculate $f(x)$ for $x = \frac{1}{\pi n}$, $n = \pm 1, \pm 2, \ldots$.

   (c) Explain why $\lim_{x \to 0} \frac{g(f(x)) - g(f(0))}{f(x) - f(0)}$ is meaningless.

28.6 Let $f(x) = x \sin \frac{1}{x}$ for $x \neq 0$ and $f(0) = 0$. See Fig. 19.3.

   (a) Observe f is continuous at $x = 0$ by Exercise 17.9(c).

   (b) Is f differentiable at $x = 0$? Justify your answer.

28.7 Let $f(x) = x^2$ for $x \geq 0$ and $f(x) = 0$ for $x < 0$.

   (a) Sketch the graph of f.

   (b) Show f is differentiable at $x = 0$. *Hint*: You will have to use the definition of derivative.

   (c) Calculate f' on $\mathbb{R}$ and sketch its graph.

   (d) Is f' continuous on $\mathbb{R}$? differentiable on $\mathbb{R}$?

28.8 Let $f(x) = x^2$ for x rational and $f(x) = 0$ for x irrational.

   (a) Prove f is continuous at $x = 0$.

   (b) Prove f is discontinuous at all $x \neq 0$.

   (c) Prove f is differentiable at $x = 0$. *Warning*: You cannot simply claim $f'(x) = 2x$.

28.9 Let $h(x) = (x^4 + 13x)^7$.

   (a) Calculate $h'(x)$.

   (b) Show how the chain rule justifies your computation in part (a) by writing $h = g \circ f$ for suitable f and g.

28.10 Repeat Exercise 28.9 for the function $h(x) = (\cos x + e^x)^{12}$.

28.11 Suppose f is differentiable at a, g is differentiable at $f(a)$, and h is differentiable at $g \circ f(a)$. State and prove the chain rule for $(h \circ g \circ f)'(a)$. *Hint*: Apply Theorem 28.4 twice.

28.12 **(a)** Differentiate the function whose value at x is $\cos(e^{x^5 - 3x})$.

   (b) Use Exercise 28.11 or Theorem 28.4 to justify your computation in part (a).

28.13 Show that if f is defined on an open interval containing a, if g is defined on an open interval containing $f(a)$, and if f is continuous at a, then $g \circ f$ is defined on an open interval containing a.

28.14 Suppose f is differentiable at a. Prove
   (a) $\lim_{h \to 0} \frac{f(a+h) - f(a)}{h} = f'(a)$,   **(b)** $\lim_{h \to 0} \frac{f(a+h) - f(a-h)}{2h} = f'(a)$.

28.15 Prove Leibniz' rule

$$(fg)^{(n)}(a) = \sum_{k=0}^{n} \binom{n}{k} f^{(k)}(a)g^{(n-k)}(a)$$

provided both f and g have n derivatives at a. Here $h^{(j)}$ signifies the jth derivative of h so that $h^{(0)} = h$, $h^{(1)} = h'$, $h^{(2)} = h''$, etc. Also, $\binom{n}{k}$ is the binomial coefficient that appears in the binomial expansion; see Exercise 1.12. *Hint*: Use mathematical induction. For $n = 1$, apply Theorem 28.3(iii).

28.16 Let f be a function defined on an open interval I containing a. Show $f'(a)$ exists if and only if there is a function $\epsilon(x)$ defined on I such that

$$f(x) - f(a) = (x - a)[f'(a) - \epsilon(x)] \quad \text{and} \quad \lim_{x \to a} \epsilon(x) = 0.$$

§29 The Mean Value Theorem

Our first result justifies the following strategy in calculus: To find the maximum and minimum of a continuous function f on an interval $[a, b]$ it suffices to consider (a) the points x where $f'(x) = 0$; (b) the points where f is not differentiable; and (c) the endpoints a and b. These are the candidates for maxima and minima.

29.1 Theorem.
If f is defined on an open interval containing x_0, if f assumes its maximum or minimum at x_0, and if f is differentiable at x_0, then $f'(x_0) = 0$.

Proof
We suppose f is defined on (a, b) where $a < x_0 < b$. Since either f or $-f$ assumes its maximum at x_0, we may assume f assumes its maximum at x_0.

Assume first that $f'(x_0) > 0$. Since

$$f'(x_0) = \lim_{x \to x_0} \frac{f(x) - f(x_0)}{x - x_0},$$

there exists $\delta > 0$ such that $a < x_0 - \delta < x_0 + \delta < b$ and

$$0 < |x - x_0| < \delta \quad \text{implies} \quad \frac{f(x) - f(x_0)}{x - x_0} > 0; \qquad (1)$$

see Corollary 20.7. If we select x so that $x_0 < x < x_0 + \delta$, then (1) shows $f(x) > f(x_0)$, contrary to the assumption that f assumes its maximum at x_0. Likewise, if $f'(x_0) < 0$, there exists $\delta > 0$ such that

$$0 < |x - x_0| < \delta \quad \text{implies} \quad \frac{f(x) - f(x_0)}{x - x_0} < 0. \qquad (2)$$

If we select x so that $x_0 - \delta < x < x_0$, then (2) implies $f(x) > f(x_0)$, again a contradiction. Thus we have $f'(x_0) = 0$. ■

Our next result is fairly obvious except for one subtle point: one needs to know or believe that a continuous function on a closed interval assumes its maximum and minimum. We proved this in Theorem 18.1 using the Bolzano-Weierstrass theorem.

29.2 Rolle's Theorem.
Let f be a continuous function on $[a, b]$ that is differentiable on (a, b) and satisfies $f(a) = f(b)$. There exists [at least one] x in (a, b) such that $f'(x) = 0$.

Proof
By Theorem 18.1, there exist $x_0, y_0 \in [a, b]$ such that $f(x_0) \le f(x) \le f(y_0)$ for all $x \in [a, b]$. If x_0 and y_0 are both endpoints of $[a, b]$, then f is a constant function [since $f(a) = f(b)$] and $f'(x) = 0$ for *all* $x \in (a, b)$. Otherwise, f assumes either a maximum or a minimum at a point x in (a, b), in which case $f'(x) = 0$ by Theorem 29.1. ■

The Mean Value Theorem tells us that a differentiable function on $[a, b]$ must somewhere have its derivative equal to the slope of the line connecting $(a, f(a))$ to $(b, f(b))$, namely $\frac{f(b) - f(a)}{b - a}$. See Fig. 29.1.

29.3 Mean Value Theorem.
Let f be a continuous function on $[a, b]$ that is differentiable on (a, b). Then there exists [at least one] x in (a, b) such that

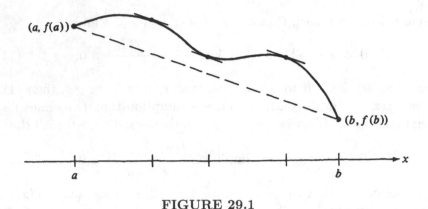

FIGURE 29.1

$$f'(x) = \frac{f(b) - f(a)}{b - a}. \tag{1}$$

Note that Rolle's Theorem is the special case of the Mean Value Theorem where $f(a) = f(b)$.

Proof
Let L be the function whose graph is the straight line connecting $(a, f(a))$ to $(b, f(b))$, i.e., the dotted line in Fig. 29.1. Observe $L(a) = f(a)$, $L(b) = f(b)$ and $L'(x) = \frac{f(b)-f(a)}{b-a}$ for all x. Let $g(x) = f(x) - L(x)$ for $x \in [a, b]$. Clearly g is continuous on $[a, b]$ and differentiable on (a, b). Also $g(a) = 0 = g(b)$, so $g'(x) = 0$ for some $x \in (a, b)$ by Rolle's Theorem 29.2. For this x, we have $f'(x) = L'(x) = \frac{f(b)-f(a)}{b-a}$. ■

29.4 Corollary.
Let f be a differentiable function on (a, b) such that $f'(x) = 0$ for all $x \in (a, b)$. Then f is a constant function on (a, b).

Proof
If f is not constant on (a, b), then there exist x_1, x_2 such that

$$a < x_1 < x_2 < b \quad \text{and} \quad f(x_1) \neq f(x_2).$$

By the Mean Value Theorem, for some $x \in (x_1, x_2)$ we have $f'(x) = \frac{f(x_2)-f(x_1)}{x_2-x_1} \neq 0$, a contradiction. ■

29.5 Corollary.
Let f and g be differentiable functions on (a, b) such that $f' = g'$ on (a, b). Then there exists a constant c such that $f(x) = g(x) + c$ for all $x \in (a, b)$.

Proof
Apply Corollary 29.4 to the function $f - g$. ∎

Corollary 29.5 is important for integral calculus because it guarantees all antiderivatives, alias indefinite integrals, for a function differ by a constant. Old integral tables and modern computational software programs provide formulas like

$$\int x^2 \cos x \, dx = 2x \cos x + (x^2 - 2) \sin x + C.$$

It is straightforward to show the derivative of each function $2x \cos x + (x^2 - 2) \sin x + C$ is in fact $x^2 \cos x$. Corollary 29.5 shows that *these are the only antiderivatives of $x^2 \cos x$.*

We need some terminology in order to give another useful corollary of the Mean Value Theorem.

29.6 Definition.
Let f be a real-valued function defined on an interval I. We say f is *strictly increasing on I* if

$$x_1, x_2 \in I \quad \text{and} \quad x_1 < x_2 \quad \text{imply} \quad f(x_1) < f(x_2),$$

strictly decreasing on I if

$$x_1, x_2 \in I \quad \text{and} \quad x_1 < x_2 \quad \text{imply} \quad f(x_1) > f(x_2),$$

increasing on I if

$$x_1, x_2 \in I \quad \text{and} \quad x_1 < x_2 \quad \text{imply} \quad f(x_1) \leq f(x_2),$$

and *decreasing on I* if

$$x_1, x_2 \in I \quad \text{and} \quad x_1 < x_2 \quad \text{imply} \quad f(x_1) \geq f(x_2).$$

Example 1
The functions e^x on $\mathbb{R}$ and $\sqrt{x}$ on $[0, \infty)$ are strictly increasing. The function $\cos x$ is strictly decreasing on $[0, \pi]$. The signum function and the postage-stamp function in Exercises 17.10 and 17.16 are increasing functions but not strictly increasing functions. □

29.7 Corollary.
Let f be a differentiable function on an interval (a,b). Then
  (i) *f is strictly increasing if $f'(x) > 0$ for all $x \in (a,b)$;*
  (ii) *f is strictly decreasing if $f'(x) < 0$ for all $x \in (a,b)$;*
  (iii) *f is increasing if $f'(x) \geq 0$ for all $x \in (a,b)$;*
  (iv) *f is decreasing if $f'(x) \leq 0$ for all $x \in (a,b)$.*

Proof
  (i) Consider x_1, x_2 where $a < x_1 < x_2 < b$. By the Mean Value
  Theorem, for some $x \in (x_1, x_2)$ we have

$$\frac{f(x_2) - f(x_1)}{x_2 - x_1} = f'(x) > 0.$$

  Since $x_2 - x_1 > 0$, we see $f(x_2) - f(x_1) > 0$ or $f(x_2) > f(x_1)$.
  The remaining cases are left to Exercise 29.8.   ∎

Exercise 28.4 shows the derivative f' of a differentiable function
f need not be continuous. Nevertheless, like a continuous function,
f' has the intermediate value property [see Theorem 18.2].

29.8 Intermediate Value Theorem for Derivatives.
*Let f be a differentiable function on (a,b). If $a < x_1 < x_2 < b$, and
if c lies between $f'(x_1)$ and $f'(x_2)$, there exists [at least one] x in
(x_1, x_2) such that $f'(x) = c$.*

Proof
We may assume $f'(x_1) < c < f'(x_2)$. Let $g(x) = f(x) - cx$ for
$x \in (a,b)$. Then we have $g'(x_1) < 0 < g'(x_2)$. Theorem 18.1 shows g
assumes its minimum on $[x_1, x_2]$ at some point $x_0 \in [x_1, x_2]$. Since

$$g'(x_1) = \lim_{y \to x_1} \frac{g(y) - g(x_1)}{y - x_1} < 0,$$

$g(y) - g(x_1)$ is negative for y close to and larger than x_1. In particular,
there exists y_1 in (x_1, x_2) such that $g(y_1) < g(x_1)$. Therefore g does
not take its minimum at x_1, so we have $x_0 \neq x_1$. Similarly, there
exists y_2 in (x_1, x_2) such that $g(y_2) < g(x_2)$, so $x_0 \neq x_2$. We have
shown x_0 is in (x_1, x_2), so $g'(x_0) = 0$ by Theorem 29.1. Therefore
$f'(x_0) = g'(x_0) + c = c$.   ∎

We next show how to differentiate the inverse of a differentiable function. Let f be a one-to-one differentiable function on an open interval I. By Theorem 18.6, f is strictly increasing or strictly decreasing on I, and by Corollary 18.3 the image $f(I)$ is an interval J. The set J is the domain of f^{-1} and

$$f^{-1} \circ f(x) = x \quad \text{for} \quad x \in I; \qquad f \circ f^{-1}(y) = y \quad \text{for} \quad y \in J.$$

The formula for the derivative of f^{-1} is easy to obtain [or remember] from the Chain Rule: $x = f^{-1} \circ f(x)$, so

$$1 = (f^{-1})'(f(x)) \cdot f'(x) \quad \text{for all} \quad x \in I.$$

If $x_0 \in I$ and $y_0 = f(x_0)$, then we can write $1 = (f^{-1})'(y_0) \cdot f'(x_0)$ or

$$(f^{-1})'(y_0) = \frac{1}{f'(x_0)} \quad \text{where} \quad y_0 = f(x_0).$$

This is *not* a proof because the Chain Rule requires the functions, f^{-1} and f in this case, be differentiable. We assumed f is differentiable, but we must *prove* f^{-1} is also differentiable. In addition, observe $f'(x_0)$ might be 0 [consider $f(x) = x^3$ at $x_0 = 0$], so our final result will have to avoid this possibility.

29.9 Theorem.
Let f be a one-to-one continuous function on an open interval I, and let $J = f(I)$. If f is differentiable at $x_0 \in I$ and if $f'(x_0) \neq 0$, then f^{-1} is differentiable at $y_0 = f(x_0)$ and

$$(f^{-1})'(y_0) = \frac{1}{f'(x_0)}.$$

Proof
Note that J is also an open interval. We have $\lim_{x \to x_0} \frac{f(x) - f(x_0)}{x - x_0} = f'(x_0)$. Since $f'(x_0) \neq 0$ and since $f(x) \neq f(x_0)$ for x near x_0, we can write

$$\lim_{x \to x_0} \frac{x - x_0}{f(x) - f(x_0)} = \frac{1}{f'(x_0)}; \tag{1}$$

see Theorem 20.4(iii). Let $\epsilon > 0$. By (1) and Corollary 20.7, there exists $\delta > 0$ such that

$$0 < |x - x_0| < \delta \quad \text{implies} \quad \left| \frac{x - x_0}{f(x) - f(x_0)} - \frac{1}{f'(x_0)} \right| < \epsilon. \tag{2}$$

Let $g = f^{-1}$ and observe that g is continuous at y_0 by Theorems 18.6 and 18.4 [or Exercise 18.11]. Hence there exists $\eta > 0$ [lower case Greek eta] such that

$$0 < |y - y_0| < \eta \quad \text{implies} \quad |g(y) - g(y_0)| < \delta, \quad \text{i.e.,} \quad |g(y) - x_0| < \delta. \tag{3}$$

Combining (3) and (2) we obtain

$$0 < |y - y_0| < \eta \quad \text{implies} \quad \left| \frac{g(y) - x_0}{f(g(y)) - f(x_0)} - \frac{1}{f'(x_0)} \right| < \epsilon.$$

Since $\frac{g(y)-x_0}{f(g(y))-f(x_0)} = \frac{g(y)-g(y_0)}{y-y_0}$, this shows

$$\lim_{y \to y_0} \frac{g(y) - g(y_0)}{y - y_0} = \frac{1}{f'(x_0)}.$$

Hence $g'(y_0)$ exists and equals $\frac{1}{f'(x_0)}$. ∎

Example 2
Let n be a positive integer, and let $g(y) = \sqrt[n]{y} = y^{1/n}$. If n is even, the domain of g is $[0, \infty)$ and, if n is odd, the domain is $\mathbb{R}$. In either case, g is strictly increasing and its inverse is $f(x) = x^n$; here $\mathrm{dom}(f) = [0, \infty)$ if n is even. Consider y_0 in $\mathrm{dom}(g)$ where $y_0 \neq 0$, and write $y_0 = x_0^n$ where $x_0 \in \mathrm{dom}(f)$. Since $f'(x_0) = nx_0^{n-1}$, Theorem 29.9 shows

$$g'(y_0) = \frac{1}{nx_0^{n-1}} = \frac{1}{ny_0^{(n-1)/n}} = \frac{1}{n}y_0^{1/n-1}.$$

This shows the function g is differentiable for $y \neq 0$ and the rule for differentiating x^n holds for exponents of the form $1/n$; see also Exercise 29.15. □

Theorem 29.9 applies to the various inverse functions encountered in calculus. We give one example.

Example 3
The function $f(x) = \sin x$ is one-to-one on $[-\frac{\pi}{2}, \frac{\pi}{2}]$, and it is traditional to use the inverse g of f restricted to this domain; g is usually denoted Sin^{-1} or arcsin. Note that $\mathrm{dom}(g) = [-1, 1]$. For $y_0 = \sin x_0$ in $(-1, 1)$ where $x_0 \in (-\frac{\pi}{2}, \frac{\pi}{2})$, Theorem 29.9 shows $g'(y_0) = \frac{1}{\cos x_0}$.

Since $1 = \sin^2 x_0 + \cos^2 x_0 = y_0^2 + \cos^2 x_0$ and $\cos x_0 > 0$, we may write

$$g'(y_0) = \frac{1}{\sqrt{1 - y_0^2}} \quad \text{for} \quad y_0 \in (-1, 1).$$

□

Exercises

29.1 Determine whether the conclusion of the Mean Value Theorem holds for the following functions on the specified intervals. If the conclusion holds, give an example of a point x satisfying (1) of Theorem 29.3. If the conclusion fails, state which *hypotheses* of the Mean Value Theorem fail.
(a) x^2 on $[-1, 2]$,   (b) $\sin x$ on $[0, \pi]$,
(c) $|x|$ on $[-1, 2]$,   (d) $\frac{1}{x}$ on $[-1, 1]$,
(e) $\frac{1}{x}$ on $[1, 3]$,   (f) $\text{sgn}(x)$ on $[-2, 2]$.
The function sgn is defined in Exercise 17.10.

29.2 Prove $|\cos x - \cos y| \leq |x - y|$ for all $x, y \in \mathbb{R}$.

29.3 Suppose f is differentiable on $\mathbb{R}$ and $f(0) = 0$, $f(1) = 1$ and $f(2) = 1$.

(a) Show $f'(x) = \frac{1}{2}$ for some $x \in (0, 2)$.

(b) Show $f'(x) = \frac{1}{7}$ for some $x \in (0, 2)$.

29.4 Let f and g be differentiable functions on an open interval I. Suppose a, b in I satisfy $a < b$ and $f(a) = f(b) = 0$. Show $f'(x) + f(x)g'(x) = 0$ for some $x \in (a, b)$. *Hint:* Consider $h(x) = f(x)e^{g(x)}$.

29.5 Let f be defined on $\mathbb{R}$, and suppose $|f(x) - f(y)| \leq (x - y)^2$ for all $x, y \in \mathbb{R}$. Prove f is a constant function.

29.6 Give the equation of the straight line used in the proof of the Mean Value Theorem 29.3.

29.7 (a) Suppose f is twice differentiable on an open interval I and $f''(x) = 0$ for all $x \in I$. Show f has the form $f(x) = ax + b$ for suitable constants a and b.

(b) Suppose f is three times differentiable on an open interval I and $f''' = 0$ on I. What form does f have? Prove your claim.

29.8 Prove (ii)–(iv) of Corollary 29.7.

29.9 Show $ex \leq e^x$ for all $x \in \mathbb{R}$.

29.10 Let $f(x) = x^2 \sin(\frac{1}{x}) + \frac{x}{2}$ for $x \neq 0$ and $f(0) = 0$.

    (a) Show $f'(0) > 0$; see Exercise 28.4.

    (b) Show f is not increasing on any open interval containing 0.

    (c) Compare this example with Corollary 29.7(i).

29.11 Show $\sin x \leq x$ for all $x \geq 0$. *Hint*: Show $f(x) = x - \sin x$ is increasing on $[0, \infty)$.

29.12 (a) Show $x < \tan x$ for all $x \in (0, \frac{\pi}{2})$.

    (b) Show $\frac{x}{\sin x}$ is a strictly increasing function on $(0, \frac{\pi}{2})$.

    (c) Show $x \leq \frac{\pi}{2} \sin x$ for $x \in [0, \frac{\pi}{2}]$.

29.13 Prove that if f and g are differentiable on $\mathbb{R}$, if $f(0) = g(0)$ and if $f'(x) \leq g'(x)$ for all $x \in \mathbb{R}$, then $f(x) \leq g(x)$ for $x \geq 0$.

29.14 Suppose f is differentiable on $\mathbb{R}$, $1 \leq f'(x) \leq 2$ for $x \in \mathbb{R}$, and $f(0) = 0$. Prove $x \leq f(x) \leq 2x$ for all $x \geq 0$.

29.15 Let r be a nonzero rational number $\frac{m}{n}$ where n is a positive integer, m is any nonzero integer, and m and n have no common factors. Let $h(x) = x^r$ where $\text{dom}(h) = [0, \infty)$ if n is even and $m > 0$, $\text{dom}(h) = (0, \infty)$ if n is even and $m < 0$, $\text{dom}(h) = \mathbb{R}$ if n is odd and $m > 0$, and $\text{dom}(h) = \mathbb{R} \setminus \{0\}$ if n is odd and $m < 0$. Show $h'(x) = rx^{r-1}$ for $x \in \text{dom}(h)$, $x \neq 0$. *Hint*: Use Example 2.

29.16 Use Theorem 29.9 to obtain the derivative of the inverse $g = \text{Tan}^{-1} = \arctan$ of f where $f(x) = \tan x$ for $x \in (-\frac{\pi}{2}, \frac{\pi}{2})$.

29.17 Let f and g be differentiable on an open interval I and consider $a \in I$. Define h on I by the rules: $h(x) = f(x)$ for $x < a$, and $h(x) = g(x)$ for $x \geq a$. Prove h is differentiable at a if and only if both $f(a) = g(a)$ and $f'(a) = g'(a)$ hold. *Suggestion*: Draw a picture to see what is going on.

29.18 Let f be differentiable on $\mathbb{R}$ with $a = \sup\{|f'(x)| : x \in \mathbb{R}\} < 1$.

    (a) Select $s_0 \in \mathbb{R}$ and define $s_n = f(s_{n-1})$ for $n \geq 1$. Thus $s_1 = f(s_0)$, $s_2 = f(s_1)$, etc. Prove (s_n) is a convergent sequence. *Hint*: To show that (s_n) is Cauchy, first show $|s_{n+1} - s_n| \leq a|s_n - s_{n-1}|$ for $n \geq 1$.

    (b) Show f has a *fixed point*, i.e., $f(s) = s$ for some s in $\mathbb{R}$.

§30 * L'Hospital's Rule

In analysis one frequently encounters limits of the form

$$\lim_{x \to s} \frac{f(x)}{g(x)}$$

where s signifies a, a^+, a^-, ∞ or $-\infty$. See Definition 20.3 concerning such limits. The limit exists and is simply $\frac{\lim_{x \to s} f(x)}{\lim_{x \to s} g(x)}$ provided the limits $\lim_{x \to s} f(x)$ and $\lim_{x \to s} g(x)$ exist and are finite and provided $\lim_{x \to s} g(x) \neq 0$; see Theorem 20.4. If these limits lead to an indeterminate form such as $\frac{0}{0}$ or $\frac{\infty}{\infty}$, then L'Hospital's rule can often be used. Moreover, other indeterminate forms, such as $\infty - \infty$, 1^∞, ∞^0, 0^0 or $0 \cdot \infty$, can usually be reformulated so as to take the form $\frac{0}{0}$ or $\frac{\infty}{\infty}$; see Examples 5–9. Before we state and prove L'Hospital's rule, we will prove a generalized mean value theorem.

30.1 Generalized Mean Value Theorem.
Let f and g be continuous functions on $[a, b]$ that are differentiable on (a, b). Then there exists [at least one] x in (a, b) such that

$$f'(x)[g(b) - g(a)] = g'(x)[f(b) - f(a)]. \tag{1}$$

This result reduces to the standard Mean Value Theorem 29.3 when g is the function given by $g(x) = x$ for all x.

Proof
The trick is to look at the difference of the two quantities in (1) and hope Rolle's Theorem will help. Thus we define

$$h(x) = f(x)[g(b) - g(a)] - g(x)[f(b) - f(a)];$$

it suffices to show $h'(x) = 0$ for some $x \in (a, b)$. Note

$$h(a) = f(a)[g(b) - g(a)] - g(a)[f(b) - f(a)] = f(a)g(b) - g(a)f(b)$$

and

$$h(b) = f(b)[g(b) - g(a)] - g(b)[f(b) - f(a)] = -f(b)g(a) + g(b)f(a) = h(a).$$

Clearly h is continuous on $[a, b]$ and differentiable on (a, b), so Rolle's Theorem 29.2 shows $h'(x) = 0$ for at least one x in (a, b). ∎

Our proof of L'Hospital's rule below is somewhat wordy but is really quite straightforward. It is based on the elegant presentation in Rudin [62]. Many texts give more complicated proofs.

30.2 L'Hospital's Rule.
Let s signify a, a^+, a^-, ∞ or $-\infty$ where $a \in \mathbb{R}$, and suppose f and g are differentiable functions for which the following limit exists:

$$\lim_{x \to s} \frac{f'(x)}{g'(x)} = L. \tag{1}$$

If

$$\lim_{x \to s} f(x) = \lim_{x \to s} g(x) = 0 \tag{2}$$

or if

$$\lim_{x \to s} |g(x)| = +\infty, \tag{3}$$

then

$$\lim_{x \to s} \frac{f(x)}{g(x)} = L. \tag{4}$$

Note that the hypothesis (1) includes some implicit assumptions: f and g must be defined and differentiable "near" s and $g'(x)$ must be *nonzero* "near" s. For example, if $\lim_{x \to a^+} \frac{f'(x)}{g'(x)}$ exists, then there is an interval (a, b) on which f and g are differentiable and g' is nonzero. The requirement that g' be nonzero is crucial; see Exercise 30.7.

Proof
We first make some reductions. The case of $\lim_{x \to a}$ follows from the cases $\lim_{x \to a^+}$ and $\lim_{x \to a^-}$, since $\lim_{x \to a} h(x)$ exists if and only if the limits $\lim_{x \to a^+} h(x)$ and $\lim_{x \to a^-} h(x)$ exist and are equal; see Theorem 20.10. In fact, we restrict our attention to $\lim_{x \to a^+}$ and $\lim_{x \to -\infty}$, since the other two cases are treated in an entirely analogous manner. Finally, we are able to handle these cases together in view of Remark 20.11.

We assume $a \in \mathbb{R}$ or $a = -\infty$. We will show that if $-\infty \leq L < \infty$ and $L_1 > L$, then there exists $\alpha_1 > a$ such that

$$a < x < \alpha_1 \quad \text{implies} \quad \frac{f(x)}{g(x)} < L_1. \tag{5}$$

A similar argument [which we omit] shows that if $-\infty < L \leq \infty$ and $L_2 < L$, then there exists $\alpha_2 > a$ such that

$$a < x < \alpha_2 \quad \text{implies} \quad \frac{f(x)}{g(x)} > L_2. \tag{6}$$

We now show how to complete the proof *using* (5) and (6); (5) will be proved in the next paragraph. If L is finite and $\epsilon > 0$, we can apply (5) to $L_1 = L + \epsilon$ and (6) to $L_2 = L - \epsilon$ to obtain $\alpha_1 > a$ and $\alpha_2 > a$ satisfying

$$a < x < \alpha_1 \quad \text{implies} \quad \tfrac{f(x)}{g(x)} < L + \epsilon,$$
$$a < x < \alpha_2 \quad \text{implies} \quad \tfrac{f(x)}{g(x)} > L - \epsilon.$$

Consequently if $\alpha = \min\{\alpha_1, \alpha_2\}$ then

$$a < x < \alpha \quad \text{implies} \quad \left| \frac{f(x)}{g(x)} - L \right| < \epsilon;$$

in view of Remark 20.11 this shows $\lim_{x \to a^+} \frac{f(x)}{g(x)} = L$ [if $a = -\infty$, then $a^+ = -\infty$]. If $L = -\infty$, then (5) and the fact that L_1 is arbitrary show $\lim_{x \to a^+} \frac{f(x)}{g(x)} = -\infty$. If $L = \infty$, then (6) and the fact that L_2 is arbitrary show $\lim_{x \to a^+} \frac{f(x)}{g(x)} = \infty$.

It remains for us to consider $L_1 > L \geq -\infty$ and show there exists $\alpha_1 > a$ satisfying (5). Let (a, b) be an interval on which f and g are differentiable and on which g' never vanishes. Theorem 29.8 shows either g' is positive on (a, b) or else g' is negative on (a, b). The former case can be reduced to the latter case by replacing g by $-g$. So we assume $g'(x) < 0$ for $x \in (a, b)$, so that g is strictly decreasing on (a, b) by Corollary 29.7. Since g is one-to-one on (a, b), $g(x)$ can equal 0 for at most one x in (a, b). By choosing b smaller if necessary, we may assume g never vanishes on (a, b). Now select K

so that $L < K < L_1$. By (1) there exists $\alpha > a$ such that

$$a < x < \alpha \quad \text{implies} \quad \frac{f'(x)}{g'(x)} < K.$$

If $a < x < y < \alpha$, then Theorem 30.1 shows

$$\frac{f(x) - f(y)}{g(x) - g(y)} = \frac{f'(z)}{g'(z)} \quad \text{for some} \quad z \in (x, y).$$

Therefore

$$a < x < y < \alpha \quad \text{implies} \quad \frac{f(x) - f(y)}{g(x) - g(y)} < K. \tag{7}$$

If hypothesis (2) holds, then

$$\lim_{x \to a^+} \frac{f(x) - f(y)}{g(x) - g(y)} = \frac{f(y)}{g(y)},$$

so (7) shows

$$\frac{f(y)}{g(y)} \leq K < L_1 \quad \text{for} \quad a < y < \alpha;$$

hence (5) holds in this case. If hypothesis (3) holds, then $\lim_{x \to a^+} g(x) = +\infty$ since g is strictly decreasing on (a, b). Also $g(x) > 0$ for $x \in (a, b)$ since g never vanishes on (a, b). We multiply both sides of (7) by $\frac{g(x) - g(y)}{g(x)}$, which is positive, to see

$$a < x < y < \alpha \quad \text{implies} \quad \frac{f(x) - f(y)}{g(x)} < K \cdot \frac{g(x) - g(y)}{g(x)}$$

and hence

$$\frac{f(x)}{g(x)} < K \cdot \frac{g(x) - g(y)}{g(x)} + \frac{f(y)}{g(x)} = K + \frac{f(y) - Kg(y)}{g(x)}.$$

We regard y as fixed and observe

$$\lim_{x \to a^+} \frac{f(y) - Kg(y)}{g(x)} = 0.$$

Hence there exists $\alpha_2 > a$ such that $\alpha_2 \leq y < \alpha$ and

$$a < x < \alpha_2 \quad \text{implies} \quad \frac{f(x)}{g(x)} \leq K < L_1.$$

Thus again (5) holds. ∎

Example 1

If we assume familiar properties of the trigonometric functions, then $\lim_{x\to 0} \frac{\sin x}{x}$ is easy to calculate by L'Hospital's rule:

$$\lim_{x\to 0} \frac{\sin x}{x} = \lim_{x\to 0} \frac{\cos x}{1} = \cos(0) = 1. \tag{1}$$

Note $f(x) = \sin x$ and $g(x) = x$ satisfy the hypotheses in Theorem 30.2. This particular computation is really dishonest, because the limit (1) is needed to *prove* the derivative of $\sin x$ is $\cos x$. This fact reduces to the assertion that the derivative of $\sin x$ at 0 is 1, i.e., to the assertion

$$\lim_{x\to 0} \frac{\sin x - \sin 0}{x - 0} = \lim_{x\to 0} \frac{\sin x}{x} = 1.$$

$\square$

Example 2

We calculate $\lim_{x\to 0} \frac{\cos x - 1}{x^2}$. L'Hospital's rule will apply provided the limit $\lim_{x\to 0} \frac{-\sin x}{2x}$ exists. but $\frac{-\sin x}{2x} = -\frac{1}{2} \frac{\sin x}{x}$ and this has limit $-\frac{1}{2}$ by Example 1. We conclude

$$\lim_{x\to 0} \frac{\cos x - 1}{x^2} = -\frac{1}{2}.$$

$\square$

Example 3

We show $\lim_{x\to\infty} \frac{x^2}{e^{3x}} = 0$. As written we have an indeterminate of the form $\frac{\infty}{\infty}$. By L'Hospital's rule, this limit will exist provided $\lim_{x\to\infty} \frac{2x}{3e^{3x}}$ exists and by L'Hospital's rule again, this limit will exist provided $\lim_{x\to\infty} \frac{2}{9e^{3x}}$ exists. The last limit is 0, so we conclude $\lim_{x\to\infty} \frac{x^2}{e^{3x}} = 0$. $\square$

Example 4

Consider $\lim_{x\to 0+} \frac{\log_e x}{x}$ if it exists. By L'Hospital's rule, this appears to be

$$\lim_{x\to 0+} \frac{1/x}{1} = +\infty$$

and yet this is *incorrect*. The difficulty is that we should have checked the hypotheses. Since $\lim_{x\to 0+} \log_e x = -\infty$ and $\lim_{x\to 0+} x = 0$, neither of the hypotheses (2) or (3) in Theorem 30.2 hold. To find the limit, we rewrite $\frac{\log_e x}{x}$ as $-\frac{\log_e(1/x)}{x}$. It is easy to show $\lim_{x \to 0+} \frac{\log_e(1/x)}{x}$ will agree with $\lim_{y \to \infty} y \log_e y$ provided the latter limit exists; see Exercise 30.4. It follows that $\lim_{x \to 0+} \frac{\log_e x}{x} = -\lim_{y \to \infty} y \log_e y = -\infty$. $\qquad \square$

The next five examples illustrate how indeterminate limits of various forms can be modified so that L'Hospital's rule applies.

Example 5

Consider $\lim_{x \to 0+} x \log_e x$. As written this limit is of the indeterminate form $0 \cdot (-\infty)$ since $\lim_{x \to 0+} x = 0$ and $\lim_{x \to 0+} \log_e x = -\infty$. By writing $x \log_e x$ as $\frac{\log_e x}{1/x}$ we obtain an indeterminate of the form $\frac{-\infty}{\infty}$, so we may apply L'Hospital's rule:

$$\lim_{x \to 0+} x \log_e x = \lim_{x \to 0+} \frac{\log_e x}{\frac{1}{x}} = \lim_{x \to 0+} \frac{\frac{1}{x}}{-\frac{1}{x^2}} = -\lim_{x \to 0+} x = 0.$$

We could also write $x \log_e x$ as $\frac{x}{1/\log_e x}$ to obtain an indeterminate of the form $\frac{0}{0}$. However, an attempt to apply L'Hospital's rule only makes the problem more complicated:

$$\lim_{x \to 0+} x \log_e x = \lim_{x \to 0+} \frac{x}{\frac{1}{\log_e x}} = \lim_{x \to 0+} \frac{1}{\frac{-1}{x(\log_e x)^2}} = -\lim_{x \to 0+} x(\log_e x)^2.$$
$\qquad \square$

Example 6

The limit $\lim_{x \to 0+} x^x$ is of the indeterminate form 0^0. We write x^x as $e^{x \log_e x}$ [remember $x = e^{\log_e x}$] and note that $\lim_{x \to 0+} x \log_e x = 0$ by Example 5. Since $g(x) = e^x$ is continuous at 0, Theorem 20.5 shows

$$\lim_{x \to 0+} x^x = \lim_{x \to 0+} e^{x \log_e x} = e^0 = 1.$$
$\qquad \square$

Example 7

The limit $\lim_{x \to \infty} x^{1/x}$ is of the indeterminate form ∞^0. We write $x^{1/x}$ as $e^{(\log_e x)/x}$. By L'Hospital's rule

$$\lim_{x \to \infty} \frac{\log_e x}{x} = \lim_{x \to \infty} \frac{\frac{1}{x}}{1} = 0.$$

Theorem 20.5 now shows $\lim_{x \to \infty} x^{1/x} = e^0 = 1.$ $\qquad \square$

Example 8

The limit $\lim_{x\to\infty}(1 - \frac{1}{x})^x$ is indeterminate of the form 1^∞. Since

$$\left(1 - \frac{1}{x}\right)^x = e^{x\log_e(1-1/x)}$$

we evaluate

$$\lim_{x\to\infty} x\log_e\left(1 - \frac{1}{x}\right) = \lim_{x\to\infty}\frac{\log_e\left(1 - \frac{1}{x}\right)}{\frac{1}{x}} = \lim_{x\to\infty}\frac{\left(1 - \frac{1}{x}\right)^{-1}x^{-2}}{-x^{-2}}$$

$$= \lim_{x\to\infty} -\left(1 - \frac{1}{x}\right)^{-1} = -1.$$

So by Theorem 20.5 we have

$$\lim_{x\to\infty}\left(1 - \frac{1}{x}\right)^x = e^{-1},$$

as should have been expected since $\lim_{n\to\infty}(1 - \frac{1}{n})^n = e^{-1}$. $\square$

Example 9

Consider $\lim_{x\to 0} h(x)$ where

$$h(x) = \frac{1}{e^x - 1} - \frac{1}{x} = (e^x - 1)^{-1} - x^{-1} \quad \text{for} \quad x \neq 0.$$

Neither of the limits $\lim_{x\to 0}(e^x - 1)^{-1}$ or $\lim_{x\to 0} x^{-1}$ exists, so $\lim_{x\to 0} h(x)$ is not an indeterminate form as written. However, $\lim_{x\to 0^+} h(x)$ is indeterminate of the form $\infty - \infty$ and $\lim_{x\to 0^-} h(x)$ is indeterminate of the form $(-\infty) - (-\infty)$. By writing

$$h(x) = \frac{x - e^x + 1}{x(e^x - 1)}$$

the limit $\lim_{x\to 0} h(x)$ becomes an indeterminate of the form $\frac{0}{0}$. By L'Hospital's rule this should be

$$\lim_{x\to 0}\frac{1 - e^x}{xe^x + e^x - 1},$$

which is still indeterminate $\frac{0}{0}$. Note $xe^x + e^x - 1 \neq 0$ for $x \neq 0$ so that the hypotheses of Theorem 30.2 hold. Applying L'Hospital's rule again, we obtain

$$\lim_{x\to 0}\frac{-e^x}{xe^x + 2e^x} = -\frac{1}{2}.$$

Note we have $xe^x + 2e^x \neq 0$ for x in $(-2, \infty)$. We conclude $\lim_{x \to 0} h(x) = -\frac{1}{2}$. $\qquad \square$

Exercises

30.1 Find the following limits if they exist.

(a) $\lim_{x \to 0} \frac{e^{2x} - \cos x}{x}$

(b) $\lim_{x \to 0} \frac{1 - \cos x}{x^2}$

(c) $\lim_{x \to \infty} \frac{x^3}{e^{2x}}$

(d) $\lim_{x \to 0} \frac{\sqrt{1+x} - \sqrt{1-x}}{x}$

30.2 Find the following limits if they exist.

(a) $\lim_{x \to 0} \frac{x^3}{\sin x - x}$

(b) $\lim_{x \to 0} \frac{\tan x - x}{x^3}$

(c) $\lim_{x \to 0} (\frac{1}{\sin x} - \frac{1}{x})$

(d) $\lim_{x \to 0} (\cos x)^{1/x^2}$

30.3 Find the following limits if they exist.

(a) $\lim_{x \to \infty} \frac{x - \sin x}{x}$

(b) $\lim_{x \to \infty} x^{\sin(1/x)}$

(c) $\lim_{x \to 0+} \frac{1 + \cos x}{e^x - 1}$

(d) $\lim_{x \to 0} \frac{1 - \cos 2x - 2x^2}{x^4}$

30.4 Let f be a function defined on some interval $(0, a)$, and define $g(y) = f(\frac{1}{y})$ for $y \in (a^{-1}, \infty)$; here we set $a^{-1} = 0$ if $a = \infty$. Show $\lim_{x \to 0+} f(x)$ exists if and only if $\lim_{y \to \infty} g(y)$ exists, in which case these limits are equal.

30.5 Find the limits

(a) $\lim_{x \to 0} (1 + 2x)^{1/x}$

(b) $\lim_{y \to \infty} (1 + \frac{2}{y})^y$

(c) $\lim_{x \to \infty} (e^x + x)^{1/x}$

30.6 Let f be differentiable on some interval (c, ∞) and suppose $\lim_{x \to \infty} [f(x) + f'(x)] = L$, where L is finite. Prove $\lim_{x \to \infty} f(x) = L$ and $\lim_{x \to \infty} f'(x) = 0$. Hint: $f(x) = \frac{f(x)e^x}{e^x}$.

30.7 This example is taken from [65, p. 188] and is due to Otto Stolz [64]. The requirement in Theorem 30.2 that $g'(x) \neq 0$ for x "near" s is important. In a careless application of L'Hospital's rule in which the zeros of g' "cancel" the zeros of f', erroneous results can be obtained. For $x \in \mathbb{R}$, let

$$f(x) = x + \cos x \sin x \quad \text{and} \quad g(x) = e^{\sin x}(x + \cos x \sin x).$$

(a) Show $\lim_{x \to \infty} f(x) = \lim_{x \to \infty} g(x) = +\infty$.

(b) Show $f'(x) = 2(\cos x)^2$ and $g'(x) = e^{\sin x} \cos x [2 \cos x + f(x)]$.

(c) Show $\frac{f'(x)}{g'(x)} = \frac{2e^{-\sin x}\cos x}{2\cos x + f(x)}$ if $\cos x \neq 0$ and $x > 3$.

(d) Show $\lim_{x \to \infty} \frac{2e^{-\sin x}\cos x}{2\cos x + f(x)} = 0$ and yet the limit $\lim_{x \to \infty} \frac{f(x)}{g(x)}$ does *not* exist.

§31 Taylor's Theorem

31.1 Discussion.
Consider a power series with radius of convergence $R > 0$ [R may be $+\infty$]:

$$f(x) = \sum_{k=0}^{\infty} a_k x^k. \qquad (1)$$

By Theorem 26.5, f is differentiable in the interval $|x| < R$ and

$$f'(x) = \sum_{k=1}^{\infty} k a_k x^{k-1}.$$

The same theorem shows f' is differentiable for $|x| < R$ and

$$f''(x) = \sum_{k=2}^{\infty} k(k-1) a_k x^{k-2}.$$

Continuing in this way, we find the nth derivative $f^{(n)}$ exists for $|x| < R$ and

$$f^{(n)}(x) = \sum_{k=n}^{\infty} k(k-1)\cdots(k-n+1) a_k x^{k-n}.$$

In particular,

$$f^{(n)}(0) = n(n-1)\cdots(n-n+1) a_n = n! a_n.$$

This relation even holds for $n = 0$ if we make the convention $f^{(0)} = f$ and recall the convention $0! = 1$. Since $f^{(k)}(0) = k! a_k$, the original power series (1) has the form

$$f(x) = \sum_{k=0}^{\infty} \frac{f^{(k)}(0)}{k!} x^k, \qquad |x| < R. \qquad (2)$$

As suggested at the end of §26, we now begin with a function f and seek a power series for f. The last paragraph shows f should possess derivatives of all orders at 0, i.e., $f'(0), f''(0), f'''(0), \dots$ should all exist. For such f, formula (2) might hold for some $R > 0$, in which case we have found a power series for f.   □

One can consider Taylor series that are not centered at 0, just as we did with the power series in Example 7 on 190.

31.2 Definition.
Let f be a function defined on some open interval containing c. If f possesses derivatives of all orders at c, then the series

$$\sum_{k=0}^{\infty} \frac{f^{(k)}(c)}{k!}(x-c)^k \qquad (1)$$

is called the *Taylor series for f about c*. For $n \geq 1$, *remainder $R_n(x)$* is defined by

$$R_n(x) = f(x) - \sum_{k=0}^{n-1} \frac{f^{(k)}(c)}{k!}(x-c)^k. \qquad (2)$$

Of course the remainder R_n depends on f and c, so a more accurate notation would be something like $R_n(f; c; x)$. The remainder is important because, for any x,

$$f(x) = \sum_{k=0}^{\infty} \frac{f^{(k)}(c)}{k!}(x-c)^k \quad \text{if and only if} \quad \lim_{n \to \infty} R_n(x) = 0.$$

We will show in Example 3 that f need not be given by its Taylor series, i.e., that $\lim_{n\to\infty} R_n(x) = 0$ can fail. Since we want to know when f is given by its Taylor series, our various versions of Taylor's Theorem all concern the nature of the remainder R_n.

31.3 Taylor's Theorem.
Let f be defined on (a,b) where $a < c < b$; here we allow $a = -\infty$ or $b = \infty$. Suppose the nth derivative $f^{(n)}$ exists on (a,b). Then for each $x \neq c$ in (a,b) there is some y between c and x such that

$$R_n(x) = \frac{f^{(n)}(y)}{n!}(x-c)^n.$$

The proof we give is due to James Wolfe[1] [71]; compare Exercise 31.6.

Proof

Fix $x \neq c$ and $n \geq 1$. Let M be the unique solution of

$$f(x) = \sum_{k=0}^{n-1} \frac{f^{(k)}(c)}{k!}(x - c)^k + \frac{M(x - c)^n}{n!} \qquad (1)$$

and observe we need only show

$$f^{(n)}(y) = M \quad \text{for some} \quad y \quad \text{between} \quad c \quad \text{and} \quad x. \qquad (2)$$

[To see this, replace M by $f^{(n)}(y)$ in Eq. (1) and recall the definition of $R_n(x)$.] To prove (2), consider the difference

$$g(t) = \sum_{k=0}^{n-1} \frac{f^{(k)}(c)}{k!}(t - c)^k + \frac{M(t - c)^n}{n!} - f(t). \qquad (3)$$

A direct calculation shows $g(c) = 0$ and $g^{(k)}(c) = 0$ for $k < n$. Also $g(x) = 0$ by the choice of M in (1). By Rolle's theorem 29.2 we have $g'(x_1) = 0$ for some x_1 between c and x. Since $g'(c) = 0$, a second application of Rolle's theorem shows $g''(x_2) = 0$ for some x_2 between c and x_1. Again, since $g''(c) = 0$ we have $g'''(x_3) = 0$ for some x_3 between c and x_2. This process continues until we obtain x_n between c and x_{n-1} such that $g^{(n)}(x_n) = 0$. From (3) it follows that $g^{(n)}(t) = M - f^{(n)}(t)$ for all $t \in (a, b)$, so (2) holds with $y = x_n$. ∎

31.4 Corollary.

Let f be defined on (a, b) where $a < c < b$. If all the derivatives $f^{(n)}$ exist on (a, b) and are bounded by a single constant C, then

$$\lim_{n \to \infty} R_n(x) = 0 \quad \text{for all} \quad x \in (a, b).$$

Proof

Consider x in (a, b). From Theorem 31.3 we see

$$|R_n(x)| \leq \frac{C}{n!}|x - c|^n \quad \text{for all} \quad n.$$

---

[1]My undergraduate real analysis teacher in 1954–1955.

Since $\lim_{n\to\infty} \frac{|x-c|^n}{n!} = 0$ by Exercise 9.15, we conclude $\lim_{n\to\infty} R_n(x) = 0$. ∎

Example 1
We assume the familiar differentiation properties of e^x, $\sin x$, etc.

(a) Let $f(x) = e^x$ for $x \in \mathbb{R}$. Then $f^{(n)}(x) = e^x$ for all $n = 0, 1, 2, \ldots$, so $f^{(n)}(0) = 1$ for all n. The Taylor series for e^x about 0 is

$$\sum_{k=0}^{\infty} \frac{1}{k!} x^k.$$

For any bounded interval $(-M, M)$ in $\mathbb{R}$ all the derivatives of f are bounded [by e^M, in fact], so Corollary 31.4 shows

$$e^x = \sum_{k=0}^{\infty} \frac{1}{k!} x^k \quad \text{for all} \quad x \in \mathbb{R}.$$

(b) If $f(x) = \sin x$ for $x \in \mathbb{R}$, then

$$f^{(n)}(x) = \begin{cases} \cos x & n = 1, 5, 9, \ldots \\ -\sin x & n = 2, 6, 10, \ldots \\ -\cos x & n = 3, 7, 11, \ldots \\ \sin x & n = 0, 4, 8, 12, \ldots; \end{cases}$$

thus

$$f^{(n)}(0) = \begin{cases} 1 & n = 1, 5, 9, \ldots \\ -1 & n = 3, 7, 11, \ldots \\ 0 & \text{otherwise.} \end{cases}$$

Hence the Taylor series for $\sin x$ about 0 is

$$\sum_{k=0}^{\infty} \frac{(-1)^k}{(2k+1)!} x^{2k+1}.$$

The derivatives of f are all bounded by 1, so

$$\sin x = \sum_{k=0}^{\infty} \frac{(-1)^k}{(2k+1)!} x^{2k+1} \quad \text{for all} \quad x \in \mathbb{R}. \qquad \square$$

Example 2

In Example 2 of §26 we used Abel's theorem to prove

$$\log_e 2 = 1 - \frac{1}{2} + \frac{1}{3} - \frac{1}{4} + \frac{1}{5} - \frac{1}{6} + \frac{1}{7} - \cdots. \tag{1}$$

Here is another proof, based on Taylor's Theorem. Let $f(x) = \log_e(1+x)$ for $x \in (-1, \infty)$. Differentiating, we find

$$f'(x) = (1+x)^{-1}, \qquad f''(x) = -(1+x)^{-2}, \qquad f'''(x) = 2(1+x)^{-3},$$

etc. A simple induction argument shows

$$f^{(n)}(x) = (-1)^{n+1}(n-1)!(1+x)^{-n} \quad \text{for} \quad n \geq 1. \tag{2}$$

In particular, $f^{(n)}(0) = (-1)^{n+1}(n-1)!$, so the Taylor series for f about 0 is

$$\sum_{k=1}^{\infty} \frac{(-1)^{k+1}}{k} x^k = x - \frac{x^2}{2} + \frac{x^3}{3} - \frac{x^4}{4} + \frac{x^5}{5} - \cdots.$$

We also could have obtained this Taylor series using Example 1 in §26, but we need formula (2) anyway. We now apply Theorem 31.3 with $a = -1$, $b = +\infty$, $c = 0$ and $x = 1$. Thus for each n there exists $y_n \in (0, 1)$ such that $R_n(1) = \frac{f^{(n)}(y_n)}{n!}$. Equation (2) shows

$$R_n(1) = \frac{(-1)^{n+1}(n-1)!}{(1+y_n)^n n!};$$

hence

$$|R_n(1)| = \frac{1}{(1+y_n)^n n} < \frac{1}{n} \quad \text{for all} \quad n.$$

Therefore $\lim_{n \to \infty} R_n(1) = 0$, so the Taylor series for f, evaluated at $x = 1$, converges to $f(1) = \log_e 2$, i.e., (1) holds. $\qquad\square$

The next version of Taylor's theorem gives the remainder in integral form. The proof uses results from integration theory that should be familiar from calculus; they also appear in the next chapter.

31.5 Taylor's Theorem.

Let f be defined on (a, b) where $a < c < b$, and suppose the nth derivative $f^{(n)}$ exists and is continuous on (a, b). Then for $x \in (a, b)$

we have

$$R_n(x) = \int_c^x \frac{(x-t)^{n-1}}{(n-1)!} f^{(n)}(t)\, dt. \tag{1}$$

Proof

For $n = 1$, Eq. (1) asserts

$$R_1(x) = f(x) - f(c) = \int_c^x f'(t)\, dt;$$

this holds by Theorem 34.1. For $n \geq 2$, we repeatedly apply integration by parts, i.e., we use mathematical induction. So, assume (1) above holds for some n, $n \geq 1$. We evaluate the integral in (1) by Theorem 34.2, using $u(t) = f^{(n)}(t)$, $v'(t) = \frac{(x-t)^{n-1}}{(n-1)!}$, so that $u'(t) = f^{(n+1)}(t)$ and $v(t) = -\frac{(x-t)^n}{n!}$. We obtain $R_n(x)$

$$= u(x)v(x) - u(c)v(c) - \int_c^x v(t)u'(t)\, dt$$

$$= f^{(n)}(x) \cdot 0 + f^{(n)}(c)\frac{(x-c)^n}{n!} + \int_c^x \frac{(x-t)^n}{n!} f^{(n+1)}(t)\, dt. \tag{2}$$

The definition of R_{n+1} in Definition 31.2 shows

$$R_{n+1}(x) = R_n(x) - \frac{f^{(n)}(c)}{n!}(x-c)^n; \tag{3}$$

hence from (2) we see that (1) holds for $n + 1$. ∎

31.6 Corollary.

If f is as in Theorem 31.5, then for each x in (a, b) different from c there is some y between c and x such that

$$R_n(x) = (x-c) \cdot \frac{(x-y)^{n-1}}{(n-1)!} f^{(n)}(y). \tag{1}$$

This form of R_n is known as Cauchy's form of the remainder.

Proof

We suppose $x < c$, the case $x > c$ being similar. The Intermediate Value Theorem for Integrals 33.9 shows

$$\frac{1}{c-x} \int_x^c \frac{(x-t)^{n-1}}{(n-1)!} f^{(n)}(t)\, dt = \frac{(x-y)^{n-1}}{(n-1)!} f^{(n)}(y) \tag{2}$$

for some y in (x, c). Since the integral in (2) equals $-R_n(x)$ by Theorem 31.5, formula (1) holds. ∎

Recall that, for a nonnegative integer n, the binomial theorem tells us

$$(a + b)^n = \sum_{k=0}^{n} \binom{n}{k} a^k b^{n-k}$$

where $\binom{n}{0} = 1$ and

$$\binom{n}{k} = \frac{n!}{k!(n-k)!} = \frac{n(n-1)\cdots(n-k+1)}{k!} \quad \text{for } 1 \le k \le n.$$

Let $a = x$ and $b = 1$; then

$$(1 + x)^n = 1 + \sum_{k=1}^{n} \frac{n(n-1)\cdots(n-k+1)}{k!} x^k.$$

This result holds for some values of x even if the exponent n is not an integer, provided we allow the series to be an infinite series.

We next prove this using Corollary 31.6. Our proof is motivated by that in [55].

31.7 Binomial Series Theorem.
If $\alpha \in \mathbb{R}$ and $|x| < 1$, then

$$(1 + x)^\alpha = 1 + \sum_{k=1}^{\infty} \frac{\alpha(\alpha-1)\cdots(\alpha-k+1)}{k!} x^k. \tag{1}$$

Proof
For $k = 1, 2, 3, \ldots$, let $a_k = \frac{\alpha(\alpha-1)\cdots(\alpha-k+1)}{k!}$. If α is a nonnegative integer, then $a_k = 0$ for $k > \alpha$ and (1) holds for all x as noted in our discussion prior to this theorem. Henceforth we assume α is not a nonnegative integer so that $a_k \ne 0$ for all k. Since

$$\lim_{k \to \infty} \left| \frac{a_{k+1}}{a_k} \right| = \lim_{k \to \infty} \left| \frac{\alpha - k}{k + 1} \right| = 1,$$

the series in (1) has radius of convergence 1; see Theorem 23.1 and Corollary 12.3. Lemma 26.3 shows the series $\sum ka_k x^{k-1}$ also has radius of convergence 1, so it converges for $|x| < 1$. Hence

$$\lim_{n \to \infty} na_n x^{n-1} = 0 \quad \text{for} \quad |x| < 1. \tag{2}$$

Let $f(x) = (1 + x)^\alpha$ for $|x| < 1$. For $n = 1, 2, \ldots$, we have

$$f^{(n)}(x) = \alpha(\alpha - 1)\cdots(\alpha - n + 1)(1 + x)^{\alpha-n} = n!a_n(1 + x)^{\alpha-n}.$$

Thus $f^{(n)}(0) = n!a_n$ for all $n \geq 1$, and the series in (1) is the Taylor series for f.

For the remainder of this proof, x is a fixed number satisfying $|x| < 1$. By Corollary 31.6, with $c = 0$, for each n, there is y_n between 0 and x such that

$$R_n(x) = x \cdot \frac{(x - y_n)^{n-1}}{(n-1)!} f^{(n)}(y_n) = x \cdot \frac{(x - y_n)^{n-1}}{(n-1)!} n!a_n(1 + y_n)^{\alpha-n}.$$

So

$$|R_n(x)| \leq |x| \left| \frac{x - y_n}{1 + y_n} \right|^{n-1} n|a_n|(1 + y_n)^{\alpha-1}. \tag{3}$$

We easily show that

$$\left| \frac{x - y}{1 + y} \right| \leq |x| \quad \text{if} \quad y \quad \text{is between} \quad 0 \quad \text{and} \quad x;$$

indeed, note $y = xz$ for some $z \in [0, 1]$, so

$$\left| \frac{x - y}{1 + y} \right| = \left| \frac{x - xz}{1 + xz} \right| = |x| \cdot \left| \frac{1 - z}{1 + xz} \right| \leq |x|$$

since $1 + xz \geq 1 - z$. Therefore, (3) implies

$$|R_n(x)| \leq |x|^n n|a_n|(1 + y_n)^{\alpha-1}.$$

The sequence $(1 + y_n)^{\alpha-1}$ is bounded, because the continuous function $y \to (1 + y)^{\alpha-1}$ is bounded on $[-|x|, |x|]$, so from (2) and Exercise 8.4, we conclude $\lim_n R_n(x) = 0$, so that (1) holds.   ∎

We next give an example of a function f whose Taylor series exists but does not represent the function. The function f is *infinitely differentiable on* $\mathbb{R}$, i.e., derivatives of all order exist at all points

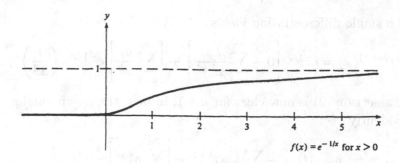

$f(x) = e^{-1/x}$ for $x > 0$

FIGURE 31.1

of $\mathbb{R}$. The example may appear artificial, but the existence of such functions [see also Exercise 31.4] is vital to the theory of distributions, an important theory related to work in differential equations and Fourier analysis.

Example 3

Let $f(x) = e^{-1/x}$ for $x > 0$ and $f(x) = 0$ for $x \leq 0$; see Fig. 31.1. Clearly f has derivatives of all orders at all $x \neq 0$. We will prove

$$f^{(n)}(0) = 0 \quad \text{for} \quad n = 0, 1, 2, 3, \ldots . \tag{1}$$

Hence the Taylor series for f is identically zero, so f does not agree with its Taylor series in any open interval containing 0. First we show for each n there is a polynomial p_n of degree $2n$ such that

$$f^{(n)}(x) = e^{-1/x} p_n(1/x) \quad \text{for} \quad x > 0. \tag{2}$$

This is obvious for $n = 0$; simply set $p_0(t) = 1$ for all t. And this is easy for $n = 1$ and $n = 2$; the reader should check (2) holds with $n = 1$ and $p_1(t) = t^2$ and (2) holds with $n = 2$ and $p_2(t) = t^4 - 2t^3$. To apply induction, we assume the result is true for n and write

$$p_n(t) = a_0 + a_1 t + a_2 t^2 + \cdots + a_{2n} t^{2n} \quad \text{where} \quad a_{2n} \neq 0.$$

Then for $x > 0$ we have

$$f^{(n)}(x) = e^{-1/x} \left[\sum_{k=0}^{2n} \frac{a_k}{x^k} \right],$$

and a single differentiation yields

$$f^{(n+1)}(x) = e^{-1/x} \left[0 - \sum_{k=1}^{2n} \frac{ka_k}{x^{k+1}} \right] + \left[\sum_{k=0}^{2n} \frac{a_k}{x^k} \right] e^{-1/x} \cdot \left(\frac{1}{x^2} \right).$$

The assertion (2) is now clear for $n+1$; in fact, the polynomial p_{n+1} is evidently

$$p_{n+1}(t) = - \sum_{k=1}^{2n} ka_k t^{k+1} + \left[\sum_{k=0}^{2n} a_k t^k \right] \cdot (t^2),$$

which has degree $2n+2$.

We next prove (1) by induction. Assume $f^{(n)}(0) = 0$ for some $n \geq 0$. We need to prove

$$\lim_{x \to 0} \frac{f^{(n)}(x) - f^{(n)}(0)}{x - 0} = \lim_{x \to 0} \frac{1}{x} f^{(n)}(x) = 0.$$

Obviously $\lim_{x \to 0^-} \frac{1}{x} f^{(n)}(x) = 0$ since $f^{(n)}(x) = 0$ for all $x < 0$. By Theorem 20.10 it suffices to verify

$$\lim_{x \to 0^+} \frac{1}{x} f^{(n)}(x) = 0.$$

In view of (2), it suffices to show

$$\lim_{x \to 0^+} e^{-1/x} q\left(\frac{1}{x} \right) = 0$$

for any polynomial q. In fact, since $q(1/x)$ is a finite sum of terms of the form $b_k(1/x)^k$, where $k \geq 0$, it suffices to show

$$\lim_{x \to 0^+} \left(\frac{1}{x} \right)^k e^{-1/x} = 0 \quad \text{for fixed} \quad k \geq 0.$$

As noted in Exercise 30.4, it suffices to show

$$\lim_{y \to \infty} y^k e^{-y} = \lim_{y \to \infty} \frac{y^k}{e^y} = 0, \tag{3}$$

and this can be verified via k applications of L'Hospital's Rule 30.2. $\square$

As we observed in Remark 2.4 on page 12, Newton's method allows one to solve some equations of the form $f(x) = 0$.

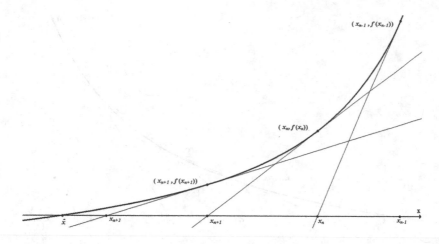

FIGURE 31.2

31.8 Newton's Method.

Newton's method for finding an approximate solution to $f(x) = 0$ is to begin with a reasonable initial guess x_0 and then compute

$$x_n = x_{n-1} - \frac{f(x_{n-1})}{f'(x_{n-1})} \quad \text{for} \quad n \geq 1.$$

Here x_n is chosen so that $(x_n, 0)$ lies on the line through $(x_{n-1}, f(x_{n-1}))$ having slope $f'(x_{n-1})$. See Fig. 31.2. Often the sequence (x_n) converges rapidly to a solution of $f(x) = 0$. ☐

31.9 Secant Method.

A similar approach to approximating solutions of $f(x) = 0$ is to start with two reasonable guesses x_0 and x_1 and then compute

$$x_n = x_{n-1} - \frac{f(x_{n-1})(x_{n-2} - x_{n-1})}{f(x_{n-2}) - f(x_{n-1})} \quad \text{for} \quad n \geq 2.$$

Here x_n is chosen so that $(x_n, 0)$ lies on the line through $(x_{n-1}, f(x_{n-1}))$ and $(x_{n-2}, f(x_{n-2}))$. See Fig. 31.3. ☐

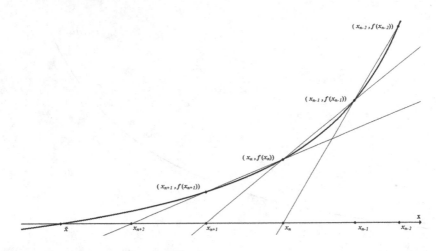

FIGURE 31.3 secant method

Example 4

Consider the equation $x^{10} + x - 1 = 0$. Let $f(x) = x^{10} + x - 1$. The equation $f(x) = 0$ will have at least two solutions, because $f(0) = -1$ and $\lim_{x \to \infty} f(x) = \lim_{x \to -\infty} f(x) = +\infty$. In fact, it has exactly two solutions, because $f'(x) = 10x^9 + 1$ is 0 at only one point, $y = -1/\sqrt[9]{10} \approx -0.774$, so that f decreases on $(-\infty, y)$ and increases on (y, ∞).

We will illustrate both approximation methods, starting with Newton's method. Since $f(0) = -1$ and $f(1) = 1$. a good initial guess for one of the solutions would be $x_0 = 1$. We omit the calculations, which we did on the spreadsheet Excel. With $x_0 = 1$ and Newton's method, we obtained $x_1 \approx 0.909090909$, $x_2 \approx 0.852873482$, $x_3 \approx 0.836193375$, $x_4 \approx 0.835083481$ and $x_5 \approx 0.835079043$. All subsequent approximations equaled x_5's, so this is surely the value to 9-place accuracy.

Since $f(-1) = -1$, the other solution is less than -1, and since $f(-2) = 1,021$, a good initial guess would be close to -1. But we started at $x_0 = -2$ to see what happens. We obtained $x_1 \approx -1.800546982$, $x_2 \approx -1.621810423$, $x_3 \approx -1.462802078$, $x_4 \approx -1.324100397$, $x_5 \approx -1.209349769$, $x_6 \approx -1.126853632$, $x_7 \approx -1.085350048$, $x_8 \approx -1.076156798$, $x_9 \approx -1.075766739$, and $x_{10} \approx -1.075766066$. All further values of x_n approximated x_{10}. Of course, $x_0 = -2$ was a poor choice, and $x_0 = -1.2$ would have been more sensible. With this choice, $x_5 \approx -1.075766066$.

As we will see, with the secant method the convergence is a bit slower. To find the positive solution, we started with $x_0 = 2$ and $x_1 = 1$. We obtained $x_9 \approx 0.835079043$, so it took about twice as many iterations to get 9-place accuracy as with Newton's method. For the other solution, we started with $x_0 = -2$ and $x_1 = -1$, and we obtained 9-place accuracy at $x_8 \approx -1.075766066$. If you look at a graph, you will see why choosing guesses on each side of the true solution can be efficient. If we had started with $x_0 = -2$ and $x_1 = -1.5$, we would not have reach 9-place accuracy until x_{12}. □

Unlike much of calculus, there isn't a single easily-found and easily-applied theorem justifying the use of either of these methods. We'll prove two such theorems, Theorems 31.12 and 31.13. They might be of value when studying the theory or when writing a program to implement a method. But for simple problems, it is much more efficient to simply assume and hope that the method will succeed, and then use common sense to assess whether the method worked.

The secant method predates Newton's method, but it does not seem as well known. In fact, the secant method is an example of the regula falsi (false position) method for approximating solutions of equations, which even appears in the Phind Papyrus attributed to the scribe Ahmes. It works in many examples, but a computer implementation would need to avoid a premature attempt to divide by 0. One advantage to this method is that the derivatives of f might be very complicated or not even exist. Another advantage to this method is that it can be used to find approximate solutions of equations, like $\cos x - x = 0$, when studying functions in algebra or trigonometry. The methodology also might be useful when motivating the sequential definition of the derivative.

We now work toward a theorem providing hypotheses that assure us that Newton's method works. We will need a technical lemma about sequences.

31.10 Lemma.

Let (a_n) be a sequence of nonnegative numbers, and let C and δ be positive numbers satisfying $C\delta < 1$.

(a) *If $a_0 \le \delta$ and $a_n \le Ca_{n-1}^2$ for $n \ge 1$, then*

$$a_n \le (C\delta)^{2^n - 1} a_0 \quad \text{for all} \quad n \ge 0.$$

(b) *If $\max\{a_0, a_1\} \le \delta$ and $a_n \le C\max\{a_{n-1}, a_{n-2}\}^2$ for $n \ge 2$, then*

$$\max\{a_{2n}, a_{2n+1}\} \le (C\delta)^{2^n - 1} \max\{a_0, a_1\} \quad \text{for all} \quad n \ge 0.$$

Proof

We prove (a) by mathematical induction, noting first that it is obvious for $n = 0$ and that $a_1 \le Ca_0^2 \le C\delta a_0$. In general, if $a_n \le (C\delta)^{2^n - 1} a_0$, then

$$a_{n+1} \le Ca_n^2 \le C[(C\delta)^{2^n - 1} a_0]^2 \le C^{2^{n+1} - 1} \delta^{2^{n+1} - 2} \delta a_0 = (C\delta)^{2^{n+1} - 1} a_0.$$

For (b), we write $M = \max\{a_0, a_1\}$, so that $M \le \delta$ and we need to show

$$\max\{a_{2n}, a_{2n+1}\} \le (C\delta)^{2^n - 1} M \quad \text{for all} \quad n \ge 0.$$

This is trivial for $n = 0$. In general, if we have $\max\{a_{2n}, a_{2n+1}\} \le (C\delta)^{2^n - 1} M$, then

$$a_{2n+2} \le C\max\{a_{2n+1}, a_{2n}\}^2 \le C[(C\delta)^{2^n - 1} M]^2$$

$$= C^{2^{n+1} - 1} \delta^{2^{n+1} - 2} M^2 \le C^{2^{n+1} - 1} \delta^{2^{n+1} - 2} \delta M = (C\delta)^{2^{n+1} - 1} M,$$

and

$$a_{2n+3} \le C\max\{a_{2n+2}, a_{2n+1}\}^2 \le C\max\{(C\delta)^{2^{n+1} - 1} M, (C\delta)^{2^n - 1} M\}^2$$

$$\le C[(C\delta)^{2^n - 1} M]^2 = C^{2^{n+1} - 1} \delta^{2^{n+1} - 2} M^2 \le C^{2^{n+1} - 1} \delta^{2^{n+1} - 2} \delta M = (C\delta)^{2^{n+1} - 1} M.$$

Thus $\max\{a_{2n+2}, a_{2n+3}\} \le (C\delta)^{2^{n+1} - 1} M$, and the induction proof of (b) is complete. ∎

31.11 Discussion.

We assume f is defined and differentiable on an open interval $J = (c, d)$. A glance at the definition

$$x_n = x_{n-1} - \frac{f(x_{n-1})}{f'(x_{n-1})}$$

suggests it is reasonable to assume $|f'|$ is bounded away from 0 on J. This implies f' is always positive or always negative [Intermediate Value Theorem for Derivatives 29.8 on page 236]. Therefore f is strictly increasing or strictly decreasing on J, and f is one-to-one [Corollary 29.7]. In addition, we assume f'' exists on J and $|f''|$ is bounded above.

Suppose there exists $\delta_0 > 0$ with the property that $f(\overline{x}) = 0$ for some $\overline{x}$ in the closed interval $I = [c + \delta_0, d - \delta_0] \subset J$. We will show that if the initial value x_0 of Newton's sequence is sufficiently close to $\overline{x}$, then the iterates x_n in Newton's method converge to $\overline{x}$. To see how this goes, we assume for now that each x_n is indeed in the interval J. From the definition, we have

$$x_n - \overline{x} = x_{n-1} - \overline{x} - \frac{f(x_{n-1})}{f'(x_{n-1})} = \frac{f'(x_{n-1}) \cdot (x_{n-1} - \overline{x}) - f(x_{n-1})}{f'(x_{n-1})}.$$

By Taylor's Theorem 31.3 on page 250, with $n = 2$, $c = x_{n-1}$ and $x = \overline{x}$, we obtain

$$0 = f(\overline{x}) = f(x_{n-1}) + f'x_{n-1}) \cdot (\overline{x} - x_{n-1}) + \frac{1}{2}f''(y_n) \cdot (\overline{x} - x_{n-1})^2$$

for some y_n between x_{n-1} and $\overline{x}$. Hence

$$f'(x_{n-1}) \cdot (x_{n-1} - \overline{x}) - f(x_{n-1}) = \frac{1}{2}f''(y_n) \cdot (\overline{x} - x_{n-1})^2,$$

so

$$x_n - \overline{x} = \frac{f''(y_n)}{2f'(x_{n-1})}(x_{n-1} - \overline{x})^2.$$

Therefore

$$|x_n - \overline{x}| \le \frac{\sup\{|f''(x)| : x \in J\}}{2\inf\{|f'(x)| : x \in J\}}|x_{n-1} - \overline{x}|^2,$$

so that

$$|x_n - \overline{x}| \le C|x_{n-1} - \overline{x}|^2,$$

where $C = \frac{\sup\{|f''(x)|:x\in J\}}{2\inf\{|f'(x)|:x\in J\}}$. Now select $\delta > 0$ so that $C\delta < 1$. By Lemma 31.10(a), with $a_n = |x_n - \overline{x}|$, we could conclude

$$|x_n - \overline{x}| \le (C\delta)^{2^n - 1}|x_0 - \overline{x}| \quad \text{for all} \quad n \ge 0, \tag{1}$$

if we could arrange for $|x_0 - \overline{x}| = a_0 \le \delta$.

The inequalities (1) are our goal, and would assure us the sequence (x_n) converges to $\overline{x}$, but we need to backtrack and determine how to be sure each x_n is in the interval J, as well as arrange for $|x_0 - \overline{x}| \le \delta$. In addition to requiring $C\delta < 1$, we will need to choose δ so that $2\delta \le \delta_0$. Finally, we select x_0 in I so that $|f(x_0)| < m\delta$ where $m = \inf\{|f'(x)| : x \in J\}$. Note $(x_0 - 2\delta, x_0 + 2\delta) \subseteq (x_0 - \delta_0, x_0 + \delta_0) \subseteq J$. We also have

$$|x_0 - \overline{x}| < \delta, \tag{2}$$

because by the Mean Value Theorem 29.3, there is y_0 between x_0 and $\overline{x}$ such that

$$|x_0 - \overline{x}| = \frac{|f(x_0) - f(\overline{x})|}{|f'(y_0)|} = \frac{|f(x_0)|}{|f'(y_0)|} \le \frac{|f(x_0)|}{m} < \delta.$$

By (1) and (2), for each n, we have $|x_n - \overline{x}| \le |x_0 - \overline{x}| < \delta$; thus $|x_n - x_0| \le |x_n - \overline{x}| + |\overline{x} - x_0| < 2\delta$, so that each x_n is in the interval J. We summarize what we have.

31.12 Theorem.
Consider a function f having a zero $\overline{x}$ on an interval $J = (c, d)$, and assume f'' exists on J. Assume $|f''|$ is bounded above on J and $|f'|$ is bounded away from 0 on J. Choose $\delta_0 > 0$ so that $I = [c + \delta_0, d - \delta_0] \subset J$ is a nondegenerate interval containing $\overline{x}$ and so that $[c + \delta_0, d - \delta_0] \subseteq J$. Let

$$C = \frac{\sup\{|f''(x)| : x \in J\}}{2\inf\{|f'(x)| : x \in J\}},$$

and select $\delta > 0$ so that $2\delta \le \delta_0$ and $C\delta < 1$. Let $m = \inf\{|f'(x)| : x \in J\}$. Consider any x_0 in I satisfying $|f(x_0)| < m\delta$. Then the sequence of iterates given by Newton's method,

$$x_n = x_{n-1} - \frac{f(x_{n-1})}{f'(x_{n-1})} \quad \text{for} \quad n \ge 1,$$

is a well-defined sequence and converges to $\overline{x}$. Also,

$$|x_n - \overline{x}| \leq C|x_{n-1} - \overline{x}|^2 \quad for \quad n \geq 1, \quad and \tag{1}$$

$$|x_n - \overline{x}| \leq (C\delta)^{2^n - 1}|x_0 - \overline{x}| \quad for \quad n \geq 0. \tag{2}$$

In view of (1), the convergence of (x_n) is said to be "quadratic convergence." Here is an analogous theorem for the secant method. Comparing its conclusion with that in Theorem 31.12, it appears that it will take about twice as many iterations using the secant method to get the same accuracy as using Newton's method. This is what happened in Example 4 on page 260.

31.13 Theorem.

Notation and hypotheses are as in Theorem 31.12, except we set

$$C = \frac{3\sup\{|f''(x)| : x \in J\}}{2\inf\{|f'(x)| : x \in J\}},$$

and we consider distinct x_0 and x_1 in I satisfying

$$\max\{|f(x_0)|, |f(x_1)|\} < m\delta. \tag{1}$$

As before $\delta > 0$ is chosen so the $2\delta \leq \delta_0$ and $C\delta < 1$.
    The sequence (x_n) of iterates given by the secant method,

$$x_n = x_{n-1} - \frac{f(x_{n-1})(x_{n-2} - x_{n-1})}{f(x_{n-2}) - f(x_{n-1})} \quad for \quad n \geq 2, \tag{2}$$

is well defined and converges to $\overline{x}$. Also,

$$|x_n - \overline{x}| \leq C \cdot \max\{|x_{n-1} - \overline{x}|, |x_{n-2} - \overline{x}|\}^2 \quad for \ all \quad n \geq 2, \tag{3}$$

and for $n \geq 0$, we have

$$\max\{|x_{2n} - \overline{x}|, |x_{2n+1} - \overline{x}|\} \leq (C\delta)^{2^n - 1}\max\{|x_0 - \overline{x}|, |x_1 - \overline{x}|\}. \tag{4}$$

Proof

As in Discussion 31.11, (1) implies

$$\max\{|x_0 - \overline{x}|, |x_1 - \overline{x}|\} < \delta. \tag{5}$$

Since

$$x_n - \overline{x} = x_{n-1} - \overline{x} - \frac{f(x_{n-1})(x_{n-2} - x_{n-1})}{f(x_{n-2}) - f(x_{n-1})} \quad for \quad n \geq 2,$$

the Mean Value Theorem shows there is w_n between x_{n-1} and x_{n-2} so that

$$x_n - \overline{x} = x_{n-1} - \overline{x} - \frac{f(x_{n-1})}{f'(w_n)}.$$

By Taylor's Theorem 31.3 on page 250, with $n = 2$, $c = \overline{x}$ and $x = x_{n-1}$, we have

$$f(x_{n-1}) = f(\overline{x}) + f'(\overline{x}) \cdot (x_{n-1} - \overline{x}) + \frac{1}{2} f''(y_n) \cdot (x_{n-1} - \overline{x})^2$$

$$= f'(\overline{x}) \cdot (x_{n-1} - \overline{x}) + \frac{1}{2} f''(y_n) \cdot (x_{n-1} - \overline{x})^2$$

for some y_n between x_{n-1} and $\overline{x}$. Hence

$$x_n - \overline{x} = (x_{n-1} - \overline{x}) - \frac{f'(\overline{x}) \cdot (x_{n-1} - \overline{x}) + f''(y_n) \cdot (x_{n-1} - \overline{x})^2/2}{f'(w_n)}$$

$$= \frac{(x_{n-1} - \overline{x})f'(w_n) - f'(\overline{x}) \cdot (x_{n-1} - \overline{x}) - f''(y_n) \cdot (x_{n-1} - \overline{x})^2/2}{f'(w_n)}$$

$$= \frac{x_{n-1} - \overline{x}}{f'(w_n)} \left\{ [f'(w_n) - f'(\overline{x})] - \frac{f''(y_n) \cdot (x_{n-1} - \overline{x})}{2} \right\}.$$

By the Mean Value Theorem, applied to f', we obtain

$$x_n - \overline{x} = \frac{x_{n-1} - \overline{x}}{f'(w_n)} \left\{ f''(z_n) \cdot (w_n - \overline{x}) - \frac{f''(y_n) \cdot (x_{n-1} - \overline{x})}{2} \right\} \quad (6)$$

for some z_n between w_n and $\overline{x}$.

In general, $a < c < b$ implies $|c - x| \leq \max\{|a - x|, |b - x|\}$ for all x in $\mathbb{R}$ [Exercise 31.12]. Applying this, with $a = \min\{x_{n-1}, x_{n-2}\}$, $b = \max\{x_{n-1}, x_{n-2}\}$, $c = w_n$ and $x = \overline{x}$, we obtain

$$|w_n - \overline{x}| \leq \max\{|x_{n-1} - \overline{x}|, |x_{n-2} - \overline{x}|\}.$$

Using this and (6), above we conclude

$$|x_n - \overline{x}| \leq \frac{3}{2} \frac{\sup\{f''(x) : x \in J\}}{\inf\{f'(x) : x \in J\}} \max\{|x_{n-1} - \overline{x}|, |x_{n-2} - \overline{x}|\}^2,$$

so that (3) holds. Substituting $a_n = |x_n - \overline{x}|$ into (3) and (5), we see that Lemma 31.10(b) implies the inequalities in (4). Inequalities (4) and (5) imply $|x_n - x_0| < 2\delta$ for all n, so each x_n is in J, as needed for the preceding argument. This completes the proof. ∎

Exercises

31.1 Find the Taylor series for $\cos x$ and indicate why it converges to $\cos x$ for all $x \in \mathbb{R}$.

31.2 Repeat Exercise 31.1 for the functions $\sinh x = \frac{1}{2}(e^x - e^{-x})$ and $\cosh x = \frac{1}{2}(e^x + e^{-x})$.

31.3 In Example 2, why did we apply Theorem 31.3 instead of Corollary 31.4?

31.4 Consider a, b in $\mathbb{R}$ where $a < b$. Show there exist infinitely differentiable functions f_a, g_b, $h_{a,b}$ and $h_{a,b}^*$ on $\mathbb{R}$ with the following properties. You may assume, without proof, that the sum, product, etc. of infinitely differentiable functions is again infinitely differentiable. The same applies to the quotient provided the denominator never vanishes.

   (a) $f_a(x) = 0$ for $x \leq a$ and $f_a(x) > 0$ for $x > a$. *Hint*: Let $f_a(x) = f(x - a)$ where f is the function in Example 3.

   (b) $g_b(x) = 0$ for $x \geq b$ and $g_b(x) > 0$ for $x < b$.

   (c) $h_{a,b}(x) > 0$ for $x \in (a, b)$ and $h_{a,b}(x) = 0$ for $x \notin (a, b)$.

   (d) $h_{a,b}^*(x) = 0$ for $x \leq a$ and $h_{a,b}^*(x) = 1$ for $x \geq b$. *Hint*: Use the function $\frac{f_a}{f_a + g_b}$.

31.5 Let $g(x) = e^{-1/x^2}$ for $x \neq 0$ and $g(0) = 0$.

   (a) Show $g^{(n)}(0) = 0$ for all $n = 0, 1, 2, 3, \ldots$. *Hint*: Use Example 3.

   (b) Show the Taylor series for g about 0 agrees with g only at $x = 0$.

31.6 An older proof of Theorem 31.3 goes as follows, which we outline for $c = 0$. Assume $x > 0$, let M be as in the proof of Theorem 31.3, and let

$$F(t) = f(t) + \sum_{k=1}^{n-1} \frac{(x-t)^k}{k!} f^{(k)}(t) + M \cdot \frac{(x-t)^n}{n!}$$

for t in $[0, x]$.

   (a) Show F is differentiable on $[0, x]$ and

$$F'(t) = \frac{(x-t)^{n-1}}{(n-1)!}[f^{(n)}(t) - M].$$

(b) Show $F(0) = F(x)$.

(c) Apply Rolle's Theorem 29.2 to F to obtain y in $(0, x)$ such that $f^{(n)}(y) = M$.

31.7 Show the sequence in Exercise 9.5 comes from Newton's method when solving $f(x) = x^2 - 2 = 0$.

31.8 (a) Show $x^4 + x^3 - 1 = 0$ has exactly two solutions.

(b) Use Newton's method or the secant method to find the solutions to $x^4 + x^3 - 1 = 0$ to six-place accuracy.

31.9 Exercise 18.6 asked for a proof that $x = \cos x$ for some x in $(0, \frac{\pi}{2})$. Use Newton's method to find x to six-place accuracy. *Hint*: Apply Newton's method to $f(x) = x - \cos x$.

31.10 Exercise 18.7 asked for a proof that $xe^x = 2$ for some x in $(0, 1)$. Use Newton's method to find x to six-place accuracy.

31.11 Suppose f is differentiable on (a, b), f' is bounded on (a, b), f' never vanishes on (a, b), and the sequence (x_n) in (a, b) converges to $\bar{x}$ in (a, b). Show that if

$$x_n = x_{n-1} - \frac{f(x_{n-1})}{f'(x_{n-1})} \quad \text{for all} \quad n \geq 1,$$

then $f(\bar{x}) = 0$.

31.12 This result will complete the proof of Theorem 31.13. Show that if $a < c < b$, then $|c - x| \leq \max\{|a - x|, |b - x|\}$ for all $x \in \mathbb{R}$. *Hint*: Consider two cases: $x \leq c$ and $x > c$.

6

CHAPTER

Integration

This chapter serves two purposes. It contains a careful development of the Riemann integral, which is the integral studied in standard calculus courses. It also contains an introduction to a generalization of the Riemann integral called the Riemann-Stieltjes integral. The generalization is easy and natural. Moreover, the Riemann-Stieltjes integral is an important tool in probability and statistics, and other areas of mathematics.

§32 The Riemann Integral

The theory of the Riemann integral is no more difficult than several other topics dealt with in this book. The one drawback is that it involves some technical notation and terminology.

32.1 Definition.
Let f be a bounded function on a closed interval $[a, b]$.[1] For $S \subseteq [a, b]$, we adopt the notation

---

[1] Here and elsewhere in this chapter, we assume $a < b$.

K.A. Ross, *Elementary Analysis: The Theory of Calculus*, Undergraduate Texts in Mathematics, DOI 10.1007/978-1-4614-6271-2_6, © Springer Science+Business Media New York 2013

$$M(f,S) = \sup\{f(x) : x \in S\} \quad \text{and} \quad m(f,S) = \inf\{f(x) : x \in S\}.$$

A *partition of* $[a,b]$ is any finite ordered subset P having the form

$$P = \{a = t_0 < t_1 < \cdots < t_n = b\}.$$

The *upper Darboux sum* $U(f,P)$ of f with respect to P is the sum

$$U(f,P) = \sum_{k=1}^{n} M(f,[t_{k-1},t_k]) \cdot (t_k - t_{k-1})$$

and the *lower Darboux sum* $L(f,P)$ is

$$L(f,P) = \sum_{k=1}^{n} m(f,[t_{k-1},t_k]) \cdot (t_k - t_{k-1}).$$

Note

$$U(f,P) \leq \sum_{k=1}^{n} M(f,[a,b]) \cdot (t_k - t_{k-1}) = M(f,[a,b]) \cdot (b-a);$$

likewise $L(f,P) \geq m(f,[a,b]) \cdot (b-a)$, so

$$m(f,[a,b]) \cdot (b-a) \leq L(f,P) \leq U(f,P) \leq M(f,[a,b]) \cdot (b-a). \quad (1)$$

The *upper Darboux integral* $U(f)$ of f over $[a,b]$ is defined by

$$U(f) = \inf\{U(f,P) : P \text{ is a partition of } [a,b]\}$$

and the *lower Darboux integral* is

$$L(f) = \sup\{L(f,P) : P \text{ is a partition of } [a,b]\}.$$

In view of (1), $U(f)$ and $L(f)$ are real numbers.

We will prove in Theorem 32.4 that $L(f) \leq U(f)$. This is not obvious from (1). [Why?] We say f is *integrable* on $[a,b]$ provided $L(f) = U(f)$. In this case, we write $\int_a^b f$ or $\int_a^b f(x)\,dx$ for this common value:

$$\int_a^b f = \int_a^b f(x)\,dx = L(f) = U(f). \quad (2)$$

Specialists call this integral the *Darboux integral*. Riemann's definition of the integral is a little different [Definition 32.8], but we will show in Theorem 32.9 that the definitions are equivalent. For this

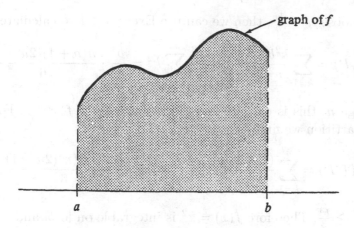
graph of f

a b

FIGURE 32.1

reason, we will follow customary usage and call the integral defined above the *Riemann integral*.

For nonnegative functions, $\int_a^b f$ is interpreted as the area of the region under the graph of f [see Fig. 32.1] for the following reason. Each lower Darboux sum represents the area of a union of rectangles inside the region, and each upper Darboux sum represents the area of a union of rectangles that contains the region. Moreover, $\int_a^b f$ is the unique number that is larger than or equal to all lower Darboux sums and smaller than or equal to all upper Darboux sums. Figure 19.2 on page 145 illustrates the situation for $[a, b] = [0, 1]$ and

$$P = \left\{ 0 < \frac{1}{n} < \frac{2}{n} < \cdots < \frac{n-1}{n} < 1 \right\}.$$

Example 1
The simplest function whose integral is not obvious is $f(x) = x^2$. Consider f on the interval $[0, b]$ where $b > 0$. For a partition

$$P = \{0 = t_0 < t_1 < \cdots < t_n = b\},$$

we have

$$U(f, P) = \sum_{k=1}^{n} \sup\{x^2 : x \in [t_{k-1}, t_k]\} \cdot (t_k - t_{k-1}) = \sum_{k=1}^{n} t_k^2 (t_k - t_{k-1}).$$

If we choose $t_k = \frac{kb}{n}$, then we can use Exercise 1.1 to calculate

$$U(f,P) = \sum_{k=1}^{n} \frac{k^2 b^2}{n^2} \left(\frac{b}{n} \right) = \frac{b^3}{n^3} \sum_{k=1}^{n} k^2 = \frac{b^3}{n^3} \cdot \frac{n(n+1)(2n+1)}{6}.$$

For large n, this is close to $\frac{b^3}{3}$, so we conclude $U(f) \leq \frac{b^3}{3}$. For the same partition we find

$$L(f,P) = \sum_{k=1}^{n} \frac{(k-1)^2 b^2}{n^2} \left(\frac{b}{n} \right) = \frac{b^3}{n^3} \cdot \frac{(n-1)(n)(2n-1)}{6},$$

so $L(f) \geq \frac{b^3}{3}$. Therefore $f(x) = x^2$ is integrable on $[0,b]$ and

$$\int_0^b x^2 \, dx = \frac{b^3}{3}.$$

Of course, any calculus student could have calculated this integral using a formula that is based on the Fundamental Theorem of Calculus; see Example 1 in §34. □

Example 2

Consider the interval $[a,b]$, where $a < b$, and let $f(x) = 1$ for rational x in $[a,b]$, and let $f(x) = 0$ for irrational x in $[a,b]$. For any partition

$$P = \{a = t_0 < t_1 < \cdots < t_n = b\},$$

we have

$$U(f,P) = \sum_{k=1}^{n} M(f, [t_{k-1}, t_k]) \cdot (t_k - t_{k-1}) = \sum_{k=1}^{n} 1 \cdot (t_k - t_{k-1}) = b - a$$

and

$$L(f,P) = \sum_{k=1}^{n} 0 \cdot (t_k - t_{k-1}) = 0.$$

It follows that $U(f) = b - a$ and $L(f) = 0$. The upper and lower Darboux integrals for f do not agree, so f is not integrable! □

We next develop some properties of the integral.

32.2 Lemma.

Let f be a bounded function on $[a, b]$. If P and Q are partitions of $[a, b]$ and $P \subseteq Q$, then

$$L(f, P) \leq L(f, Q) \leq U(f, Q) \leq U(f, P). \qquad (1)$$

Proof

The middle inequality is obvious. The proofs of the first and third inequalities are similar, so we will prove

$$L(f, P) \leq L(f, Q). \qquad (2)$$

An induction argument [Exercise 32.4] shows we may assume Q has only one more point, say u, than P. If

$$P = \{a = t_0 < t_1 < \cdots < t_n = b\},$$

then

$$Q = \{a = t_0 < t_1 < \cdots < t_{k-1} < u < t_k < \cdots < t_n = b\}$$

for some $k \in \{1, 2, \ldots, n\}$. The lower Darboux sums for P and Q are the same except for the terms involving t_{k-1} or t_k. In fact, their difference is

$$L(f, Q) - L(f, P) = m(f, [t_{k-1}, u]) \cdot (u - t_{k-1}) + m(f, [u, t_k]) \cdot (t_k - u)$$
$$-m(f, [t_{k-1}, t_k]) \cdot (t_k - t_{k-1}). \qquad (3)$$

To establish (2) it suffices to show this quantity is nonnegative. Using Exercise 4.7(a), we see

$$m(f, [t_{k-1}, t_k]) \cdot (t_k - t_{k-1})$$
$$= m(f, [t_{k-1}, t_k]) \cdot \{(t_k - u) + (u - t_{k-1})\}$$
$$\leq m(f, [u, t_k]) \cdot (t_k - u) + m(f, [t_{k-1}, u]) \cdot (u - t_{k-1}). \quad \blacksquare$$

32.3 Lemma.

If f is a bounded function on $[a, b]$, and if P and Q are partitions of $[a, b]$, then $L(f, P) \leq U(f, Q)$.

Proof

The set $P \cup Q$ is also a partition of $[a, b]$. Since $P \subseteq P \cup Q$ and $Q \subseteq P \cup Q$, we can apply Lemma 32.2 to obtain

$$L(f, P) \leq L(f, P \cup Q) \leq U(f, P \cup Q) \leq U(f, Q). \quad \blacksquare$$

32.4 Theorem.
If f is a bounded function on $[a, b]$, then $L(f) \leq U(f)$.

Proof
Fix a partition P of $[a, b]$. Lemma 32.3 shows $L(f, P)$ is a lower bound for the set

$$\{U(f, Q) : Q \text{ is a partition of } [a, b]\}.$$

Therefore $L(f, P)$ is less than or equal to the greatest lower bound [infimum!] of this set. That is

$$L(f, P) \leq U(f). \tag{1}$$

Now (1) shows that $U(f)$ is an upper bound for the set

$$\{L(f, P) : P \text{ is a partition of } [a, b]\},$$

so $U(f) \geq L(f)$. ∎

Note that Theorem 32.4 also follows from Lemma 32.3 and Exercise 4.8; see Exercise 32.5. The next theorem gives a "Cauchy criterion" for integrability.

32.5 Theorem.
A bounded function f on $[a, b]$ is integrable if and only if for each $\epsilon > 0$ there exists a partition P of $[a, b]$ such that

$$U(f, P) - L(f, P) < \epsilon. \tag{1}$$

Proof
Suppose first that f is integrable and consider $\epsilon > 0$. There exist partitions P_1 and P_2 of $[a, b]$ satisfying

$$L(f, P_1) > L(f) - \frac{\epsilon}{2} \quad \text{and} \quad U(f, P_2) < U(f) + \frac{\epsilon}{2}.$$

For $P = P_1 \cup P_2$, we apply Lemma 32.2 to obtain

$$U(f, P) - L(f, P) \leq U(f, P_2) - L(f, P_1)$$
$$< U(f) + \frac{\epsilon}{2} - \left[L(f) - \frac{\epsilon}{2}\right] = U(f) - L(f) + \epsilon.$$

Since f is integrable, $U(f) = L(f)$, so (1) holds.

Conversely, suppose for $\epsilon > 0$ the inequality (1) holds for some partition P. Then we have

$$U(f) \leq U(f,P) = U(f,P) - L(f,P) + L(f,P)$$
$$< \epsilon + L(f,P) \leq \epsilon + L(f).$$

Since ϵ is arbitrary, we conclude $U(f) \leq L(f)$. Hence we have $U(f) = L(f)$ by Theorem 32.4, i.e., f is integrable. ∎

The remainder of this section is devoted to establishing the equivalence of Riemann's and Darboux's definitions of integrability. Subsequent sections will depend only on items Definition 32.1 through Theorem 32.5. Therefore the reader who is content with the Darboux integral in Definition 32.1 can safely proceed directly to the next section.

32.6 Definition.
The *mesh* of a partition P is the maximum length of the subintervals comprising P. Thus if

$$P = \{a = t_0 < t_1 < \cdots < t_n = b\},$$

then

$$\text{mesh}(P) = \max\{t_k - t_{k-1} : k = 1, 2, \ldots, n\}.$$

Here is another "Cauchy criterion" for integrability.

32.7 Theorem.
A bounded function f on $[a,b]$ is integrable if and only if for each $\epsilon > 0$ there exists a $\delta > 0$ such that

$$\text{mesh}(P) < \delta \quad \text{implies} \quad U(f,P) - L(f,P) < \epsilon \qquad (1)$$

for all partitions P of $[a,b]$.

Proof
The ϵ–δ condition in (1) implies integrability by Theorem 32.5.

Conversely, suppose f is integrable on $[a,b]$. Let $\epsilon > 0$ and select a partition

$$P_0 = \{a = u_0 < u_1 < \cdots < u_m = b\}$$

of $[a, b]$ such that

$$U(f, P_0) - L(f, P_0) < \frac{\epsilon}{2}. \tag{2}$$

Since f is bounded, there exists $B > 0$ such that $|f(x)| \leq B$ for all $x \in [a, b]$. Let $\delta = \frac{\epsilon}{8mB}$; m is the number of intervals comprising P_0.
To verify (1), we consider any partition

$$P = \{a = t_0 < t_1 < \cdots < t_n = b\}$$

with $\text{mesh}(P) < \delta$. Let $Q = P \cup P_0$. If Q has one more element than P, then a glance at (3) in the proof of Lemma 32.2 leads us to

$$L(f, Q) - L(f, P) \leq B \cdot \text{mesh}(P) - (-B) \cdot \text{mesh}(P) = 2B \cdot \text{mesh}(P).$$

Since Q has at most m elements that are not in P, an induction argument shows

$$L(f, Q) - L(f, P) \leq 2mB \cdot \text{mesh}(P) < 2mB\delta = \frac{\epsilon}{4}.$$

By Lemma 32.2 we have $L(f, P_0) \leq L(f, Q)$, so

$$L(f, P_0) - L(f, P) < \frac{\epsilon}{4}.$$

Similarly

$$U(f, P) - U(f, P_0) < \frac{\epsilon}{4},$$

so

$$U(f, P) - L(f, P) < U(f, P_0) - L(f, P_0) + \frac{\epsilon}{2}.$$

Now (2) implies $U(f, P) - L(f, P) < \epsilon$ and we have verified (1). ∎

Now we give Riemann's definition of integrability.

32.8 Definition.
Let f be a bounded function on $[a, b]$, and let $P = \{a = t_0 < t_1 < \cdots < t_n = b\}$ be a partition of $[a, b]$. A *Riemann sum* of f associated with the partition P is a sum of the form

$$\sum_{k=1}^{n} f(x_k)(t_k - t_{k-1})$$

where $x_k \in [t_{k-1}, t_k]$ for $k = 1, 2, \ldots, n$. The choice of x_k's is quite arbitrary, so there are infinitely many Riemann sums associated with a single function and partition.

The function f is *Riemann integrable* on $[a, b]$ if there exists a number r with the following property. For each $\epsilon > 0$ there exists $\delta > 0$ such that

$$|S - r| < \epsilon \qquad (1)$$

for every Riemann sum S of f associated with a partition P having mesh$(P) < \delta$. The number r is the *Riemann integral* of f on $[a, b]$ and will be provisionally written as $\mathcal{R} \int_a^b f$.

32.9 Theorem.
A bounded function f on $[a, b]$ is Riemann integrable if and only if it is [Darboux] integrable, in which case the values of the integrals agree.

Proof
Suppose first that f is [Darboux] integrable on $[a, b]$ in the sense of Definition 32.1. Let $\epsilon > 0$, and let $\delta > 0$ be chosen so that (1) of Theorem 32.7 holds. We show

$$\left| S - \int_a^b f \right| < \epsilon \qquad (1)$$

for every Riemann sum

$$S = \sum_{k=1}^n f(x_k)(t_k - t_{k-1})$$

associated with a partition P having mesh$(P) < \delta$. Clearly we have $L(f, P) \le S \le U(f, P)$, so (1) follows from the inequalities

$$U(f, P) < L(f, P) + \epsilon \le L(f) + \epsilon = \int_a^b f + \epsilon$$

and

$$L(f, P) > U(f, P) - \epsilon \ge U(f) - \epsilon = \int_a^b f - \epsilon.$$

This proves (1); hence f is Riemann integrable and

$$\mathcal{R}\int_a^b f = \int_a^b f.$$

Now suppose f is Riemann integrable in the sense of Definition 32.8, and consider $\epsilon > 0$. Let $\delta > 0$ and r be as given in Definition 32.8. Select any partition

$$P = \{a = t_0 < t_1 < \cdots < t_n = b\}$$

with $\text{mesh}(P) < \delta$, and for each $k = 1, 2, \ldots, n$, select x_k in $[t_{k-1}, t_k]$ so that

$$f(x_k) < m(f, [t_{k-1}, t_k]) + \epsilon.$$

The Riemann sum S for this choice of x_k's satisfies

$$S \leq L(f, P) + \epsilon(b - a)$$

as well as

$$|S - r| < \epsilon.$$

It follows that

$$L(f) \geq L(f, P) \geq S - \epsilon(b - a) > r - \epsilon - \epsilon(b - a).$$

Since ϵ is arbitrary, we have $L(f) \geq r$. A similar argument shows $U(f) \leq r$. Since $L(f) \leq U(f)$, we see $L(f) = U(f) = r$. This shows f is [Darboux] integrable and

$$\int_a^b f = r = \mathcal{R}\int_a^b f.$$

■

32.10 Corollary.
Let f be a bounded Riemann integrable function on $[a, b]$. Suppose (S_n) is a sequence of Riemann sums, with corresponding partitions P_n, satisfying $\lim_n \text{mesh}(P_n) = 0$. Then the sequence (S_n) converges to $\int_a^b f$.

Proof

Let $\epsilon > 0$. There is a $\delta > 0$ so that if S is a Riemann sum with corresponding partition P, and if $\text{mesh}(P) < \delta$, then

$$\left| S - \int_a^b f \right| < \epsilon.$$

Choose N so that $\text{mesh}(P_n) < \delta$ for $n > N$. Then

$$\left| S_n - \int_a^b f \right| < \epsilon \quad \text{for} \quad n > N.$$

Since $\epsilon > 0$ is arbitrary, this shows $\lim_n S_n = \int_a^b f$. ∎

32.11 Remark.

I recently had occasion to use the following simple observation. If one ignores the end intervals of the partitions, the "almost Riemann sums" so obtained still converge to the integral; see [59]. This arose because the intervals had the form $[a, b]$, but the partition points had the form $\frac{k}{n}$. Thus the partition points were nice and equally spaced, *except* for the end ones.

Exercises

32.1 Find the upper and lower Darboux integrals for $f(x) = x^3$ on the interval $[0, b]$. *Hint*: Exercise 1.3 and Example 1 in §1 will be useful.

32.2 Let $f(x) = x$ for rational x and $f(x) = 0$ for irrational x.

    (a) Calculate the upper and lower Darboux integrals for f on the interval $[0, b]$.

    (b) Is f integrable on $[0, b]$?

32.3 Repeat Exercise 32.2 for g where $g(x) = x^2$ for rational x and $g(x) = 0$ for irrational x.

32.4 Supply the induction argument needed in the proof of Lemma 32.2.

32.5 Use Exercise 4.8 to prove Theorem 32.4. Specify the sets S and T in this case.

32.6 Let f be a bounded function on $[a, b]$. Suppose there exist sequences (U_n) and (L_n) of upper and lower Darboux sums for f such that $\lim(U_n - L_n) = 0$. Show f is integrable and $\int_a^b f = \lim U_n = \lim L_n$.

32.7 Let f be integrable on $[a, b]$, and suppose g is a function on $[a, b]$ such that $g(x) = f(x)$ except for finitely many x in $[a, b]$. Show g is integrable and $\int_a^b f = \int_a^b g$. *Hint:* First reduce to the case where f is the function identically equal to 0.

32.8 Show that if f is integrable on $[a, b]$, then f is integrable on every interval $[c, d] \subseteq [a, b]$.

§33   Properties of the Riemann Integral

In this section we establish some basic properties of the Riemann integral and we show many familiar functions, including piecewise continuous and piecewise monotonic functions, are Riemann integrable.

A function is *monotonic* on an interval if it is either increasing or decreasing on the interval; see Definition 29.6.

33.1 Theorem.
Every monotonic function f on $[a, b]$ is integrable.

Proof
We assume f is increasing on $[a, b]$ and leave the decreasing case to Exercise 33.1. We also assume $f(a) < f(b)$, since otherwise f would be a constant function. Since $f(a) \leq f(x) \leq f(b)$ for all $x \in [a, b]$, f is clearly bounded on $[a, b]$. In order to apply Theorem 32.5, let $\epsilon > 0$ and select a partition $P = \{a = t_0 < t_1 < \cdots < t_n = b\}$ with mesh less than $\frac{\epsilon}{f(b) - f(a)}$. Then

$$U(f, P) - L(f, P) = \sum_{k=1}^{n} \{M(f, [t_{k-1}, t_k]) - m(f, [t_{k-1}, t_k])\} \cdot (t_k - t_{k-1})$$

$$= \sum_{k=1}^{n} [f(t_k) - f(t_{k-1})] \cdot (t_k - t_{k-1}).$$

Since $\text{mesh}(P) < \frac{\epsilon}{f(b)-f(a)}$, we have

$$U(f,P) - L(f,P) < \sum_{k=1}^{n}[f(t_k) - f(t_{k-1})] \cdot \frac{\epsilon}{f(b) - f(a)}$$

$$= [f(b) - f(a)] \cdot \frac{\epsilon}{f(b) - f(a)} = \epsilon.$$

Theorem 32.5 now shows f is integrable. ∎

33.2 Theorem.
Every continuous function f on $[a, b]$ is integrable.

Proof
Again, in order to apply Theorem 32.5, consider $\epsilon > 0$. Since f is uniformly continuous on $[a, b]$ by Theorem 19.2, there exists $\delta > 0$ such that

$$x, y \in [a, b] \quad \text{and} \quad |x - y| < \delta \quad \text{imply} \quad |f(x) - f(y)| < \frac{\epsilon}{b - a}. \quad (1)$$

Consider any partition $P = \{a = t_0 < t_1 < \cdots < t_n = b\}$ where

$$\max\{t_k - t_{k-1} : k = 1, 2, \ldots, n\} < \delta.$$

Since f assumes its maximum and minimum on each interval $[t_{k-1}, t_k]$ by Theorem 18.1, it follows from (1) above that

$$M(f, [t_{k-1}, t_k]) - m(f, [t_{k-1}, t_k]) < \frac{\epsilon}{b - a}$$

for each k. Therefore we have

$$U(f,P) - L(f,P) < \sum_{k=1}^{n} \frac{\epsilon}{b - a}(t_k - t_{k-1}) = \epsilon,$$

and Theorem 32.5 shows f is integrable. ∎

33.3 Theorem.
Let f and g be integrable functions on $[a, b]$, and let c be a real number. Then
  (i) *cf is integrable and $\int_a^b cf = c \int_a^b f$;*
  (ii) *$f + g$ is integrable and $\int_a^b (f + g) = \int_a^b f + \int_a^b g$.*

Exercise 33.8 shows fg, $\max(f,g)$ and $\min(f,g)$ are also integrable, but there are no formulas giving their integrals in terms of $\int_a^b f$ and $\int_a^b g$.

Proof
The proof of (i) involves three cases: $c > 0$, $c = -1$, and $c < 0$. Of course, (i) is obvious for $c = 0$.

Let $c > 0$ and consider a partition

$$P = \{a = t_0 < t_1 < \cdots < t_n = b\}$$

of $[a, b]$. A simple exercise [Exercise 33.2] shows

$$M(cf, [t_{k-1}, t_k]) = c \cdot M(f, [t_{k-1}, t_k])$$

for all k, so $U(cf, P) = c \cdot U(f, P)$. Another application of the same exercise shows $U(cf) = c \cdot U(f)$. Similar arguments show $L(cf) = c \cdot L(f)$. Since f is integrable, we have $L(cf) = c \cdot L(f) = c \cdot U(f) = U(cf)$. Hence cf is integrable and

$$\int_a^b cf = U(cf) = c \cdot U(f) = c \int_a^b f, \qquad c > 0. \tag{1}$$

Now we deal with the case $c = -1$. Exercise 5.4 implies $U(-f, P) = -L(f, P)$ for all partitions P of $[a, b]$. Hence we have

$$
\begin{aligned}
U(-f) &= \inf\{U(-f, P) : P \text{ is a partition of } [a, b]\} \\
&= \inf\{-L(f, P) : P \text{ is a partition of } [a, b]\} \\
&= -\sup\{L(f, P) : P \text{ is a partition of } [a, b]\} = -L(f).
\end{aligned}
$$

Replacing f by $-f$, we also obtain $L(-f) = -U(f)$. Since f is integrable, $U(-f) = -L(f) = -U(f) = L(-f)$; hence $-f$ is integrable and

$$\int_a^b (-f) = -\int_a^b f. \tag{2}$$

The case $c < 0$ is handled by applying (2), and then (1) to $-c$:

$$\int_a^b cf = -\int_a^b (-c)f = -(-c)\int_a^b f = c \int_a^b f.$$

To prove (ii) we will again use Theorem 32.5. Let $\epsilon > 0$. By Theorem 32.5 there exist partitions P_1 and P_2 of $[a, b]$ such that

$$U(f, P_1) - L(f, P_1) < \frac{\epsilon}{2} \quad \text{and} \quad U(g, P_2) - L(g, P_2) < \frac{\epsilon}{2}.$$

Lemma 32.2 shows that if $P = P_1 \cup P_2$, then

$$U(f, P) - L(f, P) < \frac{\epsilon}{2} \quad \text{and} \quad U(g, P) - L(g, P) < \frac{\epsilon}{2}. \quad (3)$$

For any subset S of $[a, b]$, we have

$$\inf\{f(x) + g(x) : x \in S\} \geq \inf\{f(x) : x \in S\} + \inf\{g(x) : x \in S\},$$

i.e., $m(f + g, S) \geq m(f, S) + m(g, S)$. It follows that

$$L(f + g, P) \geq L(f, P) + L(g, P)$$

and similarly we have

$$U(f + g, P) \leq U(f, P) + U(g, P).$$

Therefore from (3) we obtain

$$U(f + g, P) - L(f + g, P) < \epsilon.$$

Theorem 32.5 now shows $f + g$ is integrable. Since

$$\int_a^b (f + g) = U(f + g) \leq U(f + g, P) \leq U(f, P) + U(g, P)$$

$$< L(f, P) + L(g, P) + \epsilon \leq L(f) + L(g) + \epsilon = \int_a^b f + \int_a^b g + \epsilon$$

and

$$\int_a^b (f + g) = L(f + g) \geq L(f + g, P) \geq L(f, P) + L(g, P)$$

$$> U(f, P) + U(g, P) - \epsilon \geq U(f) + U(g) - \epsilon = \int_a^b f + \int_a^b g - \epsilon,$$

we see that

$$\int_a^b (f + g) = \int_a^b f + \int_a^b g.$$

■

33.4 Theorem.
  (i) *If f and g are integrable on $[a, b]$ and if $f(x) \leq g(x)$ for x in $[a, b]$, then $\int_a^b f \leq \int_a^b g$.*
  (ii) *If g is a continuous nonnegative function on $[a, b]$ and if $\int_a^b g = 0$, then g is identically 0 on $[a, b]$.*

Proof
  (i) By Theorem 33.3, $h = g - f$ is integrable on $[a, b]$. Since $h(x) \geq 0$ for all $x \in [a, b]$, it is clear that $L(h, P) \geq 0$ for all partitions P of $[a, b]$, so $\int_a^b h = L(h) \geq 0$. Applying Theorem 33.3 again, we see

$$\int_a^b g = \int_a^b f + \int_a^b h \geq \int_a^b f.$$

  (ii) Otherwise, since g is continuous, there is a nonempty interval $(c, d) \subseteq [a, b]$ and $\alpha > 0$ satisfying $g(x) \geq \alpha/2$ for $x \in (c, d)$. Then

$$\int_a^b g \geq \int_c^d g \geq \frac{\alpha}{2}(d - c) > 0,$$

contradicting $\int_a^b g = 0$. ∎

33.5 Theorem.
If f is integrable on $[a, b]$, then $|f|$ is integrable on $[a, b]$ and

$$\left| \int_a^b f \right| \leq \int_a^b |f|. \tag{1}$$

Proof
This follows easily from Theorem 33.4(i) provided we know $|f|$ is integrable on $[a, b]$. In fact, $-|f| \leq f \leq |f|$; therefore

$$-\int_a^b |f| \leq \int_a^b f \leq \int_a^b |f|,$$

which implies (1).
  We now show $|f|$ is integrable, a point that was conveniently glossed over in Exercise 25.1. For any subset S of $[a, b]$, we have

$$M(|f|, S) - m(|f|, S) \leq M(f, S) - m(f, S) \tag{2}$$

by Exercise 33.6. From (2) it follows that

$$U(|f|, P) - L(|f|, P) \le U(f, P) - L(f, P) \qquad (3)$$

for all partitions P of $[a, b]$. By Theorem 32.5, for each $\epsilon > 0$ there exists a partition P such that

$$U(f, P) - L(f, P) < \epsilon.$$

In view of (3), the same remark applies to $|f|$, so $|f|$ is integrable by Theorem 32.5. ∎

33.6 Theorem.
Let f be a function defined on $[a, b]$. If $a < c < b$ and f is integrable on $[a, c]$ and on $[c, b]$, then f is integrable on $[a, b]$ and

$$\int_a^b f = \int_a^c f + \int_c^b f. \qquad (1)$$

Proof
Since f is bounded on both $[a, c]$ and $[c, b]$, f is bounded on $[a, b]$. In this proof we will decorate upper and lower sums so that it will be clear which intervals we are dealing with. Let $\epsilon > 0$. By Theorem 32.5 there exist partitions P_1 and P_2 of $[a, c]$ and $[c, b]$ such that

$$U_a^c(f, P_1) - L_a^c(f, P_1) < \frac{\epsilon}{2} \quad \text{and} \quad U_c^b(f, P_2) - L_c^b(f, P_2) < \frac{\epsilon}{2}.$$

The set $P = P_1 \cup P_2$ is a partition of $[a, b]$, and it is obvious that

$$U_a^b(f, P) = U_a^c(f, P_1) + U_c^b(f, P_2) \qquad (2)$$

with a similar identity for lower sums. It follows that

$$U_a^b(f, P) - L_a^b(f, P) < \epsilon,$$

so f is integrable on $[a, b]$ by Theorem 32.5. Also (1) holds because

$$\int_a^b f \le U_a^b(f, P) = U_a^c(f, P_1) + U_c^b(f, P_2)$$

$$< L_a^c(f, P_1) + L_c^b(f, P_2) + \epsilon \le \int_a^c f + \int_c^b f + \epsilon$$

and similarly $\int_a^b f > \int_a^c f + \int_c^b f - \epsilon.$ ∎

Most functions encountered in calculus and analysis are covered by the next definition. However, see Exercises 33.10–33.12.

33.7 Definition.
A function f on $[a, b]$ is *piecewise monotonic* if there is a partition

$$P = \{a = t_0 < t_1 < \cdots < t_n = b\}$$

of $[a, b]$ such that f is monotonic on each interval (t_{k-1}, t_k). The function f is *piecewise continuous* if there is a partition P of $[a, b]$ such that f is uniformly continuous on each interval (t_{k-1}, t_k).

33.8 Theorem.
If f is a piecewise continuous function or a bounded piecewise monotonic function on $[a, b]$, then f is integrable on $[a, b]$.

Proof
Let P be the partition described in Definition 33.7. Consider a fixed interval $[t_{k-1}, t_k]$. If f is piecewise continuous, then its restriction to (t_{k-1}, t_k) can be extended to a continuous function f_k on $[t_{k-1}, t_k]$ by Theorem 19.5. If f is piecewise monotonic, then its restriction to (t_{k-1}, t_k) can be extended to a monotonic function f_k on $[t_{k-1}, t_k]$; for example, if f is increasing on (t_{k-1}, t_k), simply define

$$f_k(t_k) = \sup\{f(x) : x \in (t_{k-1}, t_k)\}$$

and

$$f_k(t_{k-1}) = \inf\{f(x) : x \in (t_{k-1}, t_k)\}.$$

In either case, f_k is integrable on $[t_{k-1}, t_k]$ by Theorem 33.1 or 33.2. Since f agrees with f_k on $[t_{k-1}, t_k]$ except possibly at the endpoints, Exercise 32.7 shows f is also integrable on $[t_{k-1}, t_k]$. Now Theorem 33.6 and a trivial induction argument show f is integrable on $[a, b]$. ∎

We close this section with a simple but useful result.

33.9 Intermediate Value Theorem for Integrals.

If f is a continuous function on $[a, b]$, then for at least one x in (a, b) we have

$$f(x) = \frac{1}{b-a} \int_a^b f.$$

Proof

Let M and m be the maximum and minimum values of f on $[a, b]$. If $m = M$, then f is a constant function and $f(x) = \frac{1}{b-a} \int_a^b f$ for all $x \in [a, b]$. Otherwise, $m < M$ and by Theorem 18.1, there exist distinct x_0 and y_0 in $[a, b]$ satisfying $f(x_0) = m$ and $f(y_0) = M$. Since each function $M - f$ and $f - m$ is nonnegative and not identically 0, Theorem 33.4(ii) shows $\int_a^b m < \int_a^b f < \int_a^b M$. Thus

$$m < \frac{1}{b-a} \int_a^b f < M,$$

and by the Intermediate Value Theorem 18.2 for continuous functions, we have $f(x) = \frac{1}{b-a} \int_a^b f$ for some x between x_0 and y_0. Since x is in (a, b), this completes the proof. ∎

33.10 Discussion.

An important question concerns when one can interchange limits and integrals, i.e., when is

$$\lim_{n \to \infty} \int_a^b f_n(x) \, dx = \int_a^b \lim_{n \to \infty} f_n(x) \, dx \tag{1}$$

true? By Theorems 24.3 and 25.2, if the f_ns are continuous and converge uniformly to $f = \lim_{n \to \infty} f_n$ on $[a, b]$, then f is continuous and (1) holds. It turns out that if each f_n is just Riemann integrable and $f_n \to f$ uniformly, then f is Riemann integrable and (1) holds; see Exercise 33.9. What happens if $f_n \to f$ pointwise on $[a, b]$? One problem is that f need not be integrable even if it is bounded and each f_n is integrable.

Consider, for example, the non-integrable function f in Example 2 on page 272: $f(x) = 1$ for rational x in $[a, b]$ and $f(x) = 0$ for irrational x in $[a, b]$. Let $(x_k)_{k \in \mathbb{N}}$ be an enumeration of the rationals in $[a, b]$, and define $f_n(x_k) = 1$ for $1 \le k \le n$ and $f_n(x) = 0$ for all other x in $[a, b]$. Then $f_n \to f$ pointwise on $[a, b]$, and each f_n is integrable.

This example leaves open the possibility that (1) will hold provided all the functions f_n *and* the limit function f are integrable. However, Exercise 33.15 provides an example of a sequence (f_n) of functions on $[0, 1]$ converging pointwise to a function f, with all the functions integrable, and yet (1) does not hold. Nevertheless, there is an important theorem that does apply to sequences of functions that converge pointwise. $\qquad\square$

33.11 Dominated Convergence Theorem.
Suppose (f_n) is a sequence of integrable functions on $[a, b]$ and $f_n \to f$ pointwise where f is an integrable function on $[a, b]$. If there exists an $M > 0$ such that $|f_n(x)| \le M$ for all n and all x in $[a, b]$, then

$$\lim_{n \to \infty} \int_a^b f_n(x)\, dx = \int_a^b \lim_{n \to \infty} f_n(x)\, dx.$$

We omit the proof. An elementary proof of the Dominated Convergence Theorem is given by Jonathan W. Lewin [42]. Here is a corollary.

33.12 Monotone Convergence Theorem.
Suppose (f_n) is a sequence of integrable functions on $[a, b]$ such that $f_1(x) \le f_2(x) \le \cdots$ for all x in $[a, b]$. Suppose also that $f_n \to f$ pointwise where f is an integrable function on $[a, b]$. Then

$$\lim_{n \to \infty} \int_a^b f_n(x)\, dx = \int_a^b \lim_{n \to \infty} f_n(x)\, dx.$$

This follows from the Dominated Convergence Theorem, because there exists an $M > 0$ such that $|f_1(x)| \le M$ and also $|f(x)| \le M$ for all x in $[a, b]$. This implies $|f_n(x)| \le M$ for all n and all x in $[a, b]$, since $-M \le f_1(x) \le f_n(x) \le M$ for all x.

Our version of the Dominated Convergence Theorem is a special case of a much more general theorem, which is usually stated and proved for the family of all "Lebesgue integrable functions," not just for Riemann integrable functions. There is also a Monotone Convergence Theorem for Lebesgue integrable functions, but in that generality it does not follow immediately from the Dominated Convergence Theorem, because in that setting integrable functions need not be bounded. An elementary proof of the Monotone Convergence Theorem is proved for Riemann integrable functions, without resort to Lebesgue theory, by Brian S. Thomson [67].

Exercises

33.1 Complete the proof of Theorem 33.1 by showing that a decreasing function on $[a, b]$ is integrable.

33.2 This exercise could have appeared just as easily in §4. Let S be a nonempty bounded subset of $\mathbb{R}$. For fixed $c > 0$, let $cS = \{cs : s \in S\}$. Show $\sup(cS) = c \cdot \sup(S)$ and $\inf(cS) = c \cdot \inf(S)$.

33.3 A function f on $[a, b]$ is called a *step function* if there exists a partition $P = \{a = u_0 < u_1 < \cdots < c_m = b\}$ of $[a, b]$ such that f is constant on each interval (u_{j-1}, u_j), say $f(x) = c_j$ for x in (u_{j-1}, u_j).

(a) Show that a step function f is integrable and evaluate $\int_a^b f$.

(b) Evaluate the integral $\int_0^4 P(x)\, dx$ for the postage-stamp function P in Exercise 17.16.

33.4 Give an example of a function f on $[0, 1]$ that is *not* integrable for which $|f|$ is integrable. *Hint:* Modify Example 2 in §32.

33.5 Show $\left| \int_{-2\pi}^{2\pi} x^2 \sin^8(e^x)\, dx \right| \leq \frac{16\pi^3}{3}$.

33.6 Prove (2) in the proof of Theorem 33.5. *Hint:* For $x_0, y_0 \in S$, we have $|f(x_0)| - |f(y_0)| \leq |f(x_0) - f(y_0)| \leq M(f, S) - m(f, S)$.

33.7 Let f be a bounded function on $[a, b]$, so that there exists $B > 0$ such that $|f(x)| \leq B$ for all $x \in [a, b]$.

(a) Show

$$U(f^2, P) - L(f^2, P) \leq 2B[U(f, P) - L(f, P)]$$

for all partitions P of $[a, b]$. *Hint:* $f(x)^2 - f(y)^2 = [f(x) + f(y)] \cdot [f(x) - f(y)]$.

(b) Show that if f is integrable on $[a, b]$, then f^2 also is integrable on $[a, b]$.

33.8 Let f and g be integrable functions on $[a, b]$.

(a) Show fg is integrable on $[a, b]$. *Hint*: Use Exercise 33.7 and $4fg = (f + g)^2 - (f - g)^2$.

(b) Show $\max(f, g)$ and $\min(f, g)$ are integrable on $[a, b]$. *Hint*: Exercise 17.8.

33.9 Let (f_n) be a sequence of integrable functions on $[a, b]$, and suppose $f_n \to f$ uniformly on $[a, b]$. Prove f is integrable and

$$\int_a^b f = \lim_{n \to \infty} \int_a^b f_n.$$

Compare this result with Theorems 25.2 and 33.11.

33.10 Let $f(x) = \sin \frac{1}{x}$ for $x \neq 0$ and $f(0) = 0$. Show f is integrable on $[-1, 1]$. *Hint*: See the answer to Exercise 33.11(c).

33.11 Let $f(x) = x \, \text{sgn}(\sin \frac{1}{x})$ for $x \neq 0$ and $f(0) = 0$.

(a) Show f is not piecewise continuous on $[-1, 1]$.

(b) Show f is not piecewise monotonic on $[-1, 1]$.

(c) Show f is integrable on $[-1, 1]$.

33.12 Let f be the function described in Exercise 17.14.

(a) Show f is not piecewise continuous or piecewise monotonic on any interval $[a, b]$.

(b) Show f is integrable on every interval $[a, b]$ and $\int_a^b f = 0$.

33.13 Suppose f and g are continuous functions on $[a, b]$ such that $\int_a^b f = \int_a^b g$. Prove there exists x in (a, b) such that $f(x) = g(x)$.

33.14 (a) Prove the following generalization of the Intermediate Value Theorem for Integrals. If f and g are continuous functions on $[a, b]$ and $g(t) \geq 0$ for all $t \in [a, b]$, then there exists x in (a, b) such that

$$\int_a^b f(t)g(t) \, dt = f(x) \int_a^b g(t) \, dt.$$

(b) Show Theorem 33.9 is a special case of part (a).

(c) Does the conclusion in part (a) hold if $[a, b] = [-1, 1]$ and $f(t) = g(t) = t$ for all t?

33.15 For integers $n \geq 3$, define the function f_n on $[0,1]$ by the rules:

$$f_n(0) = f_n\left(\frac{2}{n}\right) = f_n(1) = 0 \quad \text{and} \quad f_n\left(\frac{1}{n}\right) = n,$$

and so that its graph is a straight line from $(0,0)$ to $(\frac{1}{n}, n)$, from $(\frac{1}{n}, n)$ to $(\frac{2}{n}, 0)$, and from $(\frac{2}{n}, 0)$ to $(1,0)$.

(a) Graph f_3, f_4 and f_5.

(b) Show $f_n \to 0$ pointwise on $[0,1]$.

(c) Show $\lim_n \int_0^1 f_n(x)\, dx \neq \int_0^1 0\, dx$. Why doesn't this contradict the Dominated Convergence Theorem?

§34 Fundamental Theorem of Calculus

There are two versions of the Fundamental Theorem of Calculus. Each says, roughly speaking, that differentiation and integration are inverse operations. In fact, our first version [Theorem 34.1] says "the integral of the derivative of a function is given by the function," and our second version [Theorem 34.3] says "the derivative of the integral of a continuous function is the function." It is somewhat traditional for books to prove our second version first and use it to prove our first version, although some books do avoid this approach. F. Cunningham, Jr. [18] offers some good reasons for avoiding the traditional approach:

(a) Theorem 34.3 implies Theorem 34.1 only for functions g whose derivative g' is continuous; see Exercise 34.1.
(b) Making Theorem 34.1 depend on Theorem 34.3 obscures the fact that the two theorems say different things, have different applications, and may leave the impression Theorem 34.3 is *the* fundamental theorem.
(c) The need for Theorem 34.1 in calculus is immediate and easily motivated.

In what follows, we say a function h defined on (a, b) is *integrable* on $[a, b]$ if every extension of h to $[a, b]$ is integrable. In view of

Exercise 32.7, the value $\int_a^b h$ will not depend on the values of the extensions at a or b.

34.1 Fundamental Theorem of Calculus I.
If g is a continuous function on $[a, b]$ that is differentiable on (a, b), and if g' is integrable on $[a, b]$, then

$$\int_a^b g' = g(b) - g(a). \tag{1}$$

Proof
Let $\epsilon > 0$. By Theorem 32.5, there exists a partition $P = \{a = t_0 < t_1 < \cdots < t_n = b\}$ of $[a, b]$ such that

$$U(g', P) - L(g', P) < \epsilon. \tag{2}$$

We apply the Mean Value Theorem 29.3 to each interval $[t_{k-1}, t_k]$ to obtain x_k in (t_{k-1}, t_k) for which

$$(t_k - t_{k-1})g'(x_k) = g(t_k) - g(t_{k-1}).$$

Hence we have

$$g(b) - g(a) = \sum_{k=1}^{n} [g(t_k) - g(t_{k-1})] = \sum_{k=1}^{n} g'(x_k)(t_k - t_{k-1}).$$

It follows that

$$L(g', P) \leq g(b) - g(a) \leq U(g', P); \tag{3}$$

see Definition 32.1. Since

$$L(g', P) \leq \int_a^b g' \leq U(g', P),$$

inequalities (2) and (3) imply

$$\left| \int_a^b g' - [g(b) - g(a)] \right| < \epsilon.$$

Since ϵ is arbitrary, (1) holds. ■

The integration formulas in calculus all rely in the end on Theorem 34.1.

Example 1

If $g(x) = \frac{x^{n+1}}{n+1}$, then $g'(x) = x^n$, so

$$\int_a^b x^n \, dx = \frac{b^{n+1}}{n+1} - \frac{a^{n+1}}{n+1} = \frac{b^{n+1} - a^{n+1}}{n+1}. \qquad (1)$$

In particular,

$$\int_a^b x^2 \, dx = \frac{b^3 - a^3}{3}.$$

Formula (1) is valid for any powers n for which $g(x) = \frac{x^{n+1}}{n+1}$ is defined on $[a, b]$. See Examples 3 and 4 in §28 and Exercises 29.15 and 37.5. For example,

$$\int_a^b \sqrt{x} \, dx = \int_a^b x^{1/2} \, dx = \frac{2}{3}[b^{3/2} - a^{3/2}] \quad \text{for} \quad 0 \leq a < b.$$

$\square$

34.2 Theorem [Integration by Parts].

If u and v are continuous functions on $[a, b]$ that are differentiable on (a, b), and if u' and v' are integrable on $[a, b]$, then

$$\int_a^b u(x)v'(x) \, dx + \int_a^b u'(x)v(x) \, dx = u(b)v(b) - u(a)v(a). \qquad (1)$$

Proof

Let $g = uv$; then $g' = uv' + u'v$ by Theorem 28.3. Exercise 33.8 shows g' is integrable. Now Theorem 34.1 shows

$$\int_a^b g'(x) \, dx = g(b) - g(a) = u(b)v(b) - u(a)v(a),$$

so (1) holds. ∎

Note the use of Exercise 33.8 above can be avoided if u' and v' are continuous, which is normally the case.

Example 2

Here is a simple application of integration by parts. To calculate $\int_0^\pi x \cos x \, dx$, we note the integrand has the form $u(x)v'(x)$ where $u(x) = x$ and $v(x) = \sin x$. Hence

$$\int_0^\pi x \cos x \, dx = u(\pi)v(\pi) - u(0)v(0) - \int_0^\pi 1 \cdot \sin x \, dx = - \int_0^\pi \sin x \, dx = -2.$$

□

In what follows we use the convention $\int_a^b f = - \int_b^a f$ for $a > b$.

34.3 Fundamental Theorem of Calculus II.
Let f be an integrable function on $[a,b]$. For x in $[a,b]$, let

$$F(x) = \int_a^x f(t) \, dt.$$

Then F is continuous on $[a,b]$. If f is continuous at x_0 in (a,b), then F is differentiable at x_0 and

$$F'(x_0) = f(x_0).$$

Proof
Select $B > 0$ so that $|f(x)| \leq B$ for all $x \in [a,b]$. If $x, y \in [a,b]$ and $|x - y| < \frac{\epsilon}{B}$ where $x < y$, say, then

$$|F(y) - F(x)| = \left| \int_x^y f(t) \, dt \right| \leq \int_x^y |f(t)| \, dt \leq \int_x^y B \, dt = B(y - x) < \epsilon.$$

This shows F is [uniformly] continuous on $[a,b]$.

Suppose f is continuous at x_0 in (a,b). Observe

$$\frac{F(x) - F(x_0)}{x - x_0} = \frac{1}{x - x_0} \int_{x_0}^x f(t) \, dt$$

for $x \neq x_0$. The trick is to observe

$$f(x_0) = \frac{1}{x - x_0} \int_{x_0}^x f(x_0) \, dt,$$

and therefore

$$\frac{F(x) - f(x_0)}{x - x_0} - f(x_0) = \frac{1}{x - x_0} \int_{x_0}^x [f(t) - f(x_0)] \, dt. \qquad (1)$$

Let $\epsilon > 0$. Since f is continuous at x_0, there exists $\delta > 0$ such that

$$t \in (a,b) \quad \text{and} \quad |t - x_0| < \delta \quad \text{imply} \quad |f(t) - f(x_0)| < \epsilon;$$

see Theorem 17.2. It follows from (1) that

$$\left| \frac{F(x) - F(x_0)}{x - x_0} - f(x_0) \right| \leq \epsilon$$

for x in (a, b) satisfying $|x - x_0| < \delta$; the cases $x > x_0$ and $x < x_0$ require separate arguments. We have just shown

$$\lim_{x \to x_0} \frac{F(x) - F(x_0)}{x - x_0} = f(x_0).$$

In other words, $F'(x_0) = f(x_0)$. ∎

A useful technique of integration is known as "substitution." A more accurate description of the process is "change of variable." The technique is the reverse of the chain rule.

34.4 Theorem [Change of Variable].
Let u be a differentiable function on an open interval J such that u' is continuous, and let I be an open interval such that $u(x) \in I$ for all $x \in J$. If f is continuous on I, then $f \circ u$ is continuous on J and

$$\int_a^b f \circ u(x) u'(x)\, dx = \int_{u(a)}^{u(b)} f(u)\, du \qquad (1)$$

for a, b in J.

Note $u(a)$ need not be less than $u(b)$, even if $a < b$.

Proof
The continuity of $f \circ u$ follows from Theorem 17.5. Fix c in I and let $F(u) = \int_c^u f(t)\, dt$. Then $F'(u) = f(u)$ for all $u \in I$ by Theorem 34.3. Let $g = F \circ u$. By the Chain Rule 28.4, we have

$$g'(x) = F'(u(x)) \cdot u'(x) = f(u(x)) \cdot u'(x),$$

so by Theorem 34.1

$$\int_a^b f \circ u(x) u'(x)\, dx = \int_a^b g'(x)\, dx = g(b) - g(a) = F(u(b)) - F(u(a))$$

$$= \int_c^{u(b)} f(t)\, dt - \int_c^{u(a)} f(t)\, dt = \int_{u(a)}^{u(b)} f(t)\, dt.$$

This proves (1). ∎

Example 3
Let g be a one-to-one differentiable function on an open interval I. Then $J = g(I)$ is an open interval, and the inverse function g^{-1} is differentiable on J by Theorem 29.9. We show

$$\int_a^b g(x)\,dx + \int_{g(a)}^{g(b)} g^{-1}(u)\,du = b \cdot g(b) - a \cdot g(a) \qquad (1)$$

for a, b in I.

We put $f = g^{-1}$ and $u = g$ in the change of variable formula to obtain

$$\int_a^b g^{-1} \circ g(x) g'(x)\,dx = \int_{g(a)}^{g(b)} g^{-1}(u)\,du.$$

Since $g^{-1} \circ g(x) = x$ for x in I, we obtain

$$\int_{g(a)}^{g(b)} g^{-1}(u)\,du = \int_a^b x g'(x)\,dx.$$

Now integrate by parts with $u(x) = x$ and $v(x) = g(x)$:

$$\int_{g(a)}^{g(b)} g^{-1}(u)\,du = b \cdot g(b) - a \cdot g(a) - \int_a^b g(x)\,dx.$$

This is formula (1). ☐

Exercises

34.1 Use Theorem 34.3 to prove Theorem 34.1 for the case g' is continuous. *Hint*: Let $F(x) = \int_a^x g'$; then $F' = g'$. Apply Corollary 29.5.

34.2 Calculate
(a) $\lim_{x \to 0} \frac{1}{x} \int_0^x e^{t^2}\,dt$ (b) $\lim_{h \to 0} \frac{1}{h} \int_3^{3+h} e^{t^2}\,dt$.

34.3 Let f be defined as follows: $f(t) = 0$ for $t < 0$; $f(t) = t$ for $0 \le t \le 1$; $f(t) = 4$ for $t > 1$.

(a) Determine the function $F(x) = \int_0^x f(t)\,dt$.

(b) Sketch F. Where is F continuous?

(c) Where is F differentiable? Calculate F' at the points of differentiability.

34.4 Repeat Exercise 34.3 for f where $f(t) = t$ for $t < 0$; $f(t) = t^2 + 1$ for $0 \le t \le 2$; $f(t) = 0$ for $t > 2$.

34.5 Let f be a continuous function on $\mathbb{R}$ and define

$$F(x) = \int_{x-1}^{x+1} f(t)\, dt \quad \text{for} \quad x \in \mathbb{R}.$$

Show F is differentiable on $\mathbb{R}$ and compute F'.

34.6 Let f be a continuous function on $\mathbb{R}$ and define

$$G(x) = \int_0^{\sin x} f(t)\, dt \quad \text{for} \quad x \in \mathbb{R}.$$

Show G is differentiable on $\mathbb{R}$ and compute G'.

34.7 Use change of variables to integrate $\int_0^1 x\sqrt{1 - x^2}\, dx$.

34.8 **(a)** Use integration by parts to evaluate

$$\int_0^1 x \arctan x\, dx.$$

Hint: Let $u(x) = \arctan x$, so that $u'(x) = \frac{1}{1+x^2}$.

(b) If you used $v(x) = \frac{x^2}{2}$ in part (a), do the computation again with $v(x) = \frac{x^2+1}{2}$. This interesting example is taken from J.L. Borman [10].

34.9 Use Example 3 to show $\int_0^{1/2} \arcsin x\, dx = \frac{\pi}{12} + \frac{\sqrt{3}}{2} - 1$.

34.10 Let g be a strictly increasing continuous function mapping $[0, 1]$ onto $[0, 1]$. Give a geometric argument showing $\int_0^1 g(x)dx + \int_0^1 g^{-1}(u)du = 1$.

34.11 Suppose f is a continuous function on $[a, b]$. Show that if $\int_a^b f(x)^2 dx = 0$, then $f(x) = 0$ for all x in $[a, b]$. *Hint:* See Theorem 33.4.

34.12 Show that if f is a continuous real-valued function on $[a, b]$ satisfying $\int_a^b f(x)g(x)\, dx = 0$ for every continuous function g on $[a, b]$, then $f(x) = 0$ for all x in $[a, b]$.

§35 * Riemann-Stieltjes Integrals

In this long section we introduce a useful generalization of the Riemann integral. In the Riemann integral, all intervals of the same length are given the same weight. For example, in our definition of upper sums

$$U(f, P) = \sum_{k=1}^{n} M(f, [t_{k-1}, t_k]) \cdot (t_k - t_{k-1}), \qquad (*)$$

the factors $(t_k - t_{k-1})$ are the lengths of the intervals involved. In applications such as probability and statistics, it is desirable to modify the definition so as to weight the intervals according to some increasing function F. In other words, the idea is to replace the factors $(t_k - t_{k-1})$ in (*) by $[F(t_k) - F(t_{k-1})]$. The Riemann integral is, then, the special case where $F(t) = t$ for all t.

It is also desirable to allow some *points* to have positive weight. This corresponds to the situation where F has jumps, i.e., where the left-hand and right-hand limits of F differ. In fact, if (c_k) is a sequence in $\mathbb{R}$, then the sums

$$\sum_{k=1}^{\infty} c_k f(u_k)$$

can be viewed as a generalized integral for a suitable F [see Examples 1 and 3 on pages 301 and 309]. In this case, F has a jump at each u_k.

The traditional treatment, in all books that I am aware of, replaces the factors $(t_k - t_{k-1})$ in (*) by $[F(t_k) - F(t_{k-1})]$ and develops the theory from there, though some authors emphasize upper and lower sums while others stress generalized Riemann sums. In this section, we offer a slightly different treatment, so

Warning. Theorems in this section do not necessarily correspond to theorems in other texts.

We deviate from tradition because:

(a) Our treatment is more general. Functions that are Riemann-Stieltjes integrable in the traditional sense are integrable in our sense [Theorem 35.20].

(b) In the traditional theory, if f and F have a common discontinuity, then f is not integrable using F. Such unfortunate results disappear in our approach. We will show piecewise continuous and piecewise monotonic functions are always integrable using F [Theorem 35.17]. We also will observe that if F is a step function, then *all* bounded functions are integrable; see Example 1.

(c) We will give a definition involving Riemann-Stieltjes sums that is equivalent to our definition involving upper and lower sums [Theorem 35.25]. The corresponding standard definitions are not equivalent.

As just explained, our development of Riemann-Stieltjes integrals has several positive features. However, this section is long with lots of technical details. Therefore, we recommend readers omit the proofs on first reading, and then decide whether to go through the details.

Many of the results in this section are straightforward generalizations of results in §§32 and 33. Accordingly, many proofs will be brief or omitted.

35.1 Notation.
We assume throughout this section that F is an increasing function on a closed interval $[a, b]$. To avoid trivialities we assume $F(a) < F(b)$. All left-hand and right-hand limits exist; see Definition 20.3 and Exercise 35.1. We use the notation

$$F(t^-) = \lim_{x \to t^-} F(x) \quad \text{and} \quad F(t^+) = \lim_{x \to t^+} F(x).$$

For the endpoints we decree

$$F(a^-) = F(a) \quad \text{and} \quad F(b^+) = F(b).$$

Note that $F(t^-) \leq F(t^+)$ for all $t \in [a, b]$. If F is continuous at t, then $F(t^-) = F(t) = F(t^+)$. Otherwise $F(t^-) < F(t^+)$ and the difference $F(t^+) - F(t^-)$ is called the *jump* of F at t. The actual value of $F(t)$ at jumps t will play no role in what follows. □

In the next definition we employ some of the notation established in Definition 32.1.

35.2 Definition.

For a bounded function f on $[a, b]$ and a partition $P = \{a = t_0 < t_1 < \cdots < t_n = b\}$ of $[a, b]$, we write

$$J_F(f, P) = \sum_{k=0}^{n} f(t_k) \cdot [F(t_k^+) - F(t_k^-)]. \tag{1}$$

The *upper Darboux-Stieltjes sum* is

$$U_F(f, P) = J_F(f, P) + \sum_{k=1}^{n} M(f, (t_{k-1}, t_k)) \cdot [F(t_k^-) - F(t_{k-1}^+)] \tag{2}$$

and the *lower Darboux-Stieltjes sum* is

$$L_F(f, P) = J_F(f, P) + \sum_{k=1}^{n} m(f, (t_{k-1}, t_k)) \cdot [F(t_k^-) - F(t_{k-1}^+)]. \tag{3}$$

These definitions explicitly take into account the possible jump effects of F at the points t_k, though F may have jumps at other points. The other terms focus on the effect of F on the open intervals (t_{k-1}, t_k). Observe

$$[a, b] = \bigcup_{k=0}^{n} \{t_k\} \cup \bigcup_{k=1}^{n} (t_{k-1}, t_k)$$

represents $[a, b]$ as a disjoint union.[2] See Remarks 35.26 for a discussion about the choice of open intervals versus closed intervals in the definitions of Riemann integrals and Riemann-Stieltjes integrals.

Note

$$U_F(f, P) - L_F(f, P)$$
$$= \sum_{k=1}^{n} [M(f, (t_{k-1}, t_k)) - m(f, (t_{k-1}, t_k))][F(t_k^-) - F(t_{k-1}^+)] \tag{4}$$

and

$$m(f, [a, b]) \cdot [F(b) - F(a)] \leq L_F(f, P) \leq U_F(f, P)$$
$$\leq M(f, [a, b]) \cdot [F(b) - F(a)]. \tag{5}$$

---

[2]In measure theory this is an example of a measurable partition, that is to say, a family of measurable sets with disjoint union $[a, b]$.

In checking (5), note

$$\sum_{k=0}^{n}[F(t_k^+) - F(t_k^-)] + \sum_{k=1}^{n}[F(t_k^-) - F(t_{k-1}^+)]$$
$$= F(t_0^+) - F(t_0^-) + \sum_{k=1}^{n}[F(t_k^+) - F(t_{k-1}^+)] \tag{6}$$
$$= F(a^+) - F(a^-) + F(b^+) - F(a^+) = F(b) - F(a),$$

since the last sum is a telescoping sum. The *upper Darboux-Stieltjes integral* is

$$U_F(f) = \inf\{U_F(f, P) : P \text{ is a partition of } [a, b]\} \tag{7}$$

and the *lower Darboux-Stieltjes integral* is

$$L_F(f) = \sup\{L_F(f, P) : P \text{ is a partition of } [a, b]\}. \tag{8}$$

Theorem 35.5 below states that $L_F(f) \leq U_F(f)$. Accordingly, we say f is *Darboux-Stieltjes integrable on* $[a, b]$ with respect to F or, more briefly, *F-integrable on* $[a, b]$, provided $L_F(f) = U_F(f)$; in this case we write

$$\int_a^b f\,dF = \int_a^b f(x)\,dF(x) = L_F(f) = U_F(f).$$

Example 1
For each u in $[a, b]$, let

$$J_u(t) = \begin{cases} 0 & \text{for } t < u, \\ 1 & \text{for } t \geq u, \end{cases}$$

for $u > a$, and let

$$J_a(t) = \begin{cases} 0 & \text{for } t = a, \\ 1 & \text{for } t > a. \end{cases}$$

Then J_u is an increasing step function with jump 1 at u. Also, every bounded function f on $[a, b]$ is J_u-integrable and

$$\int_a^b f\,dJ_u = f(u).$$

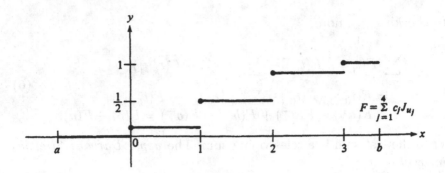

FIGURE 35.1

More generally, if $u_1, u_2, \ldots, u_m$ are distinct points in $[a, b]$ and if $c_1, c_2, \ldots, c_m$ are positive numbers, then

$$F = \sum_{j=1}^{m} c_j J_{u_j}$$

is an increasing step function with jumps c_j at u_j. See Fig. 35.1 for a special case. Every bounded function f on $[a, b]$ is F-integrable and

$$\int_a^b f \, dF = \sum_{j=1}^{m} c_j \cdot f(u_j). \tag{1}$$

To check (1), first note the validity of (1) will not change if some of the u_j are not jump points so that $c_j = 0$. So we may assume a and b are in $\{u_j : 1 \le j \le m\}$, and $a = u_1 < u_2 < \cdots < u_m = b$. Let P be the partition $\{a = u_1 < u_2 < \cdots < u_m = b\}$. Then we have $F(u_j^+) - F(u_j^-) = c_j$ for $j = 1, 2, \ldots, m$ and $F(u_j^-) - F(u_{j-1}^+) = 0$ for $j = 2, 3, \ldots, m$. Therefore

$$U_F(f, P) = L_F(f, P) = J_F(f, P) = \sum_{j=1}^{m} f(u_j) \cdot c_j$$

for any bounded function f on $[a, b]$. From (7) and (8) in Definition 35.2 and $L_F(f) \leq U_F(f)$, it follows that

$$U_F(f) = L_F(f) = \sum_{j=1}^{m} f(u_j) \cdot c_j.$$

Hence f is F-integrable and (1) holds. $\qquad\square$

Example 2

We specialize Example 1 to the case $u_1 = 0$, $u_2 = 1$, $u_3 = 2$, $u_4 = 3$; $c_1 = c_4 = \frac{1}{8}$, $c_2 = c_3 = \frac{3}{8}$. Thus we must have $a \leq 0$ and $b \geq 3$; see Fig. 35.1. For any bounded function f on $[a, b]$, we have

$$\int_a^b f\, dF = \frac{1}{8} f(0) + \frac{3}{8} f(1) + \frac{3}{8} f(2) + \frac{1}{8} f(3).$$

$\qquad\square$

35.3 Lemma.

Let f be a bounded function on $[a, b]$, and let P and Q be partitions of $[a, b]$ such that $P \subseteq Q$. Then

$$L_F(f, P) \leq L_F(f, Q) \leq U_F(f, Q) \leq U_F(f, P). \tag{1}$$

Proof

We imitate the proof of Lemma 32.2 on page 273 down to, but not including, formula (3). In the present case, the difference $L_F(f, Q) - L_F(f, P)$ equals

$$f(u) \cdot [F(u^+) - F(u^-)] + m(f, (t_{k-1}, u)) \cdot [F(u^-) - F(t_{k-1}^+)]$$
$$+ m(f, (u, t_k)) \cdot [F(t_k^-) - F(u^+)] \tag{3}$$
$$- m(f, (t_{k-1}, t_k)) \cdot [F(t_k^-) - F(t_{k-1}^+)],$$

and this is nonnegative because

$$m(f, (t_{k-1}, t_k)) \cdot [F(t_k^-) - F(t_{k-1}^+)]$$
$$= m(f, (t_{k-1}, t_k)) \cdot [F(t_k^-) - F(u^+) + F(u^+) - F(u^-)$$
$$+ F(u^-) - F(t_{k-1}^+)]$$
$$\leq m(f, (u, t_k)) \cdot [F(t_k^-) - F(u^+)] + f(u) \cdot [F(u^+) - F(u^-)]$$
$$+ m(f, (t_{k-1}, u)) \cdot [F(u^-) - F(t_{k-1}^+)].$$

$\qquad\blacksquare$

35.4 Lemma.
If f is a bounded function on $[a, b]$ and if P and Q are partitions of $[a, b]$, then $L_F(f, P) \leq U_F(f, Q)$.

Proof
Imitates the proof of Lemma 32.3 on page 273.   ∎

35.5 Theorem.
For every bounded function f on $[a, b]$, we have $L_F(f) \leq U_F(f)$.

Proof
Imitates the proof of Theorem 32.4 on page 274.   ∎

35.6 Theorem.
A bounded function f on $[a, b]$ is F-integrable if and only if for each $\epsilon > 0$ there exists a partition P such that

$$U_F(f, P) - L_F(f, P) < \epsilon.$$

Proof
Imitates the proof of Theorem 32.5 on page 274.   ∎

We next develop analogues of results in §33; we return later to generalizations of items Definition 32.6 through Theorem 32.9. We begin with the analogue of Theorem 33.2. The analogue of Theorem 33.1 is true, but its proof requires some preparation, so we defer it to Theorem 35.16.

35.7 Theorem.
Every continuous function f on $[a, b]$ is F-integrable.

Proof
To apply Theorem 35.6, let $\epsilon > 0$. Since f is uniformly continuous, there exists $\delta > 0$ such that

$$x, y \in [a, b] \quad \text{and} \quad |x - y| < \delta \quad \text{imply} \quad |f(x) - f(y)| < \frac{\epsilon}{F(b) - F(a)}.$$

Just as in the proof of Theorem 33.2, there is a partition P of $[a, b]$ such that

$$M(f, (t_{k-1}, t_k)) - m(f, (t_{k-1}, t_k)) < \frac{\epsilon}{F(b) - F(a)}$$

for each k. Hence by (4) of Definition 35.2 we have

$$U_F(f, P) - L_F(f, P) \le \sum_{k=1}^{n} \frac{\epsilon}{F(b) - F(a)} [F(t_k^-) - F(t_{k-1}^+)] \le \epsilon.$$

Theorem 35.6 now shows f is F-integrable. ∎

35.8 Theorem.
Let f and g be F-integrable functions on $[a, b]$, and let c be a real number. Then
  (i) *cf is F-integrable and $\int_a^b (cf) \, dF = c \int_a^b f \, dF$;*
  (ii) *$f + g$ is F-integrable and $\int_a^b (f + g) \, dF = \int_a^b f \, dF + \int_a^b g \, dF$.*

Proof
Imitates the proof of Theorem 33.3, using Theorem 35.6 instead of Theorem 32.5. ∎

35.9 Theorem.
If f and g are F-integrable on $[a, b]$ and if $f(x) \le g(x)$ for $x \in [a, b]$, then $\int_a^b f \, dF \le \int_a^b g \, dF$.

Proof
Imitates the proof of Theorem 33.4(i). ∎

35.10 Theorem.
If f is F-integrable on $[a, b]$, then $|f|$ is F-integrable and

$$\left| \int_a^b f \, dF \right| \le \int_a^b |f| \, dF.$$

Proof
Imitates the proof of Theorem 33.5 and uses formula (4) of Definition 35.2. ∎

35.11 Theorem.
Let f be a function defined on $[a, b]$. If $a < c < b$ and f is F-integrable on $[a, c]$ and on $[c, b]$, then f is F-integrable on $[a, b]$ and

$$\int_a^b f \, dF = \int_a^c f \, dF + \int_c^b f \, dF. \tag{1}$$

Proof
Imitates the proof of Theorem 33.6. Note that an upper or lower sum on $[a, c]$ will include the term $f(c)[F(c) - F(c^-)]$ while an upper or lower sum on $[c, b]$ will include the term $f(c)[F(c^+) - F(c)]$.   ∎

The next result clearly has no analogue in §32 or §33.

35.12 Theorem.

(a) *Let F_1 and F_2 be increasing functions on $[a, b]$. If f is F_1-integrable and F_2-integrable on $[a, b]$ and if $c > 0$, then f is cF_1-integrable, f is $(F_1 + F_2)$-integrable,*

$$\int_a^b f \, d(cF_1) = c \int_a^b f \, dF_1, \tag{1}$$

and

$$\int_a^b f \, d(F_1 + F_2) = \int_a^b f \, dF_1 + \int_a^b f \, dF_2. \tag{2}$$

(b) *Let (F_j) be a sequence of increasing functions on $[a, b]$ such that $F = \sum_{j=1}^{\infty} F_j$ defines an (automatically increasing) function on $[a, b]$. Thus the series converges on the entire interval $[a, b]$, and $F(a)$ and $F(b)$ are finite. If a bounded function f on $[a, b]$ is F_j-integrable for each j, then f is F-integrable on $[a, b]$ and*

$$\int_a^b f \, dF = \sum_{j=1}^{\infty} \int_a^b f \, dF_j. \tag{3}$$

Proof
From Theorem 20.4 we see

$$(F_1 + F_2)(t^+) = \lim_{x \to t^+} [F_1(x) + F_2(x)] = \lim_{x \to t^+} F_1(x) + \lim_{x \to t^+} F_2(x)$$
$$= F_1(t^+) + F_2(t^+)$$

with similar identities for $(F_1 + F_2)(t^-)$, $(cF_1)(t^+)$ and $(cF_1)(t^-)$. Hence for any partition P of $[a, b]$, we have

$$\begin{aligned} U_{F_1 + F_2}(f, P) &= U_{F_1}(f, P) + U_{F_2}(f, P) \\ L_{F_1 + F_2}(f, P) &= L_{F_1}(f, P) + L_{F_2}(f, P), \end{aligned} \tag{4}$$

$U_{cF_1}(f, P) = cU_{F_1}(f, P)$ and $L_{cF_1}(f, P) = cL_{F_1}(f, P)$. It is now clear that f is cF_1-integrable and that (1) holds. To check (2), let $\epsilon > 0$. By Theorem 35.6 and Lemma 35.3, there is a single partition P of $[a, b]$ so that both

$$U_{F_1}(f, P) - L_{F_1}(f, P) < \frac{\epsilon}{2} \quad \text{and} \quad U_{F_2}(f, P) - L_{F_2}(f, P) < \frac{\epsilon}{2}.$$

Hence by (4) we have

$$U_{F_1+F_2}(f, P) - L_{F_1+F_2}(f, P) < \epsilon.$$

This and Theorem 35.6 imply f is $(F_1 + F_2)$-integrable. The identity (2) follows from

$$\int_a^b f \, d(F_1 + F_2) \leq U_{F_1+F_2}(f, P) < L_{F_1+F_2}(f, P) + \epsilon$$

$$= L_{F_1}(f, P) + L_{F_2}(f, P) + \epsilon \leq \int_a^b f dF_1 + \int_a^b f dF_2 + \epsilon$$

and the similar inequality

$$\int_a^b f \, d(F_1 + F_2) > \int_a^b f \, dF_1 + \int_a^b f \, dF_2 - \epsilon.$$

Now we prove part (b). We write S_m for $\sum_{j=1}^m F_j$ and T_m for the difference (or tail sum) $F - S_m = \sum_{j=m+1}^\infty F_j$. Let B be a positive bound for $|f|$ on $[a, b]$, and consider $\epsilon > 0$. Select a positive integer m_0 so that

$$|T_m(b) - T_m(a)| < \frac{\epsilon}{5B} \quad \text{for} \quad m \geq m_0. \tag{5}$$

We will show

$$U_F(f) - L_F(f) < \frac{3\epsilon}{5}. \tag{6}$$

Since $\epsilon > 0$ is arbitrary, this will imply $U_F(f) = L_F(f)$, so that f is F-integrable. We will also show

$$\left| \int_a^b f dF - \int_a^b f dS_m \right| < \epsilon \quad \text{for} \quad m \geq m_0. \tag{7}$$

Since $\epsilon > 0$ is arbitrary, and $\int_a^b f \, dS_m = \sum_{j=1}^m \int_a^b f \, dF_j$ by (2), this will confirm (3).

We now prove (6). First we claim

$$|U_F(f,P)-U_{S_m}(f,P)| < \frac{\epsilon}{5} \text{ for all partitions } P \text{ of } [a,b] \text{ and } m \geq m_0,$$
$$(8)$$

$$|L_F(f,P)-L_{S_m}(f,P)| < \frac{\epsilon}{5} \text{ for all partitions } P \text{ of } [a,b] \text{ and } m \geq m_0.$$
$$(9)$$

Let $P = \{a = t_0 < t_1 < \cdots < t_n = b\}$. Note $T_m = \sum_{j=m+1}^{\infty} F_j$ is an increasing function on $[a,b]$, so using Eq. (2) in Definition 35.2 on page 300, the left-hand side of inequality (8) is bounded by

$$\sum_{k=0}^{n} B[T_m(t_k^+) - T_m(t_k^-)] + \sum_{k=1}^{n} B[T_m(t_k^-) - T_m(t_{k-1}^+)].$$

This sum is equal to $B[T_m(b^+)-T_m(a^-)] = B[T_m(b)-T_m(a)]$ which, by (5), is less than $\frac{\epsilon}{5}$, so (8) holds. The verification of (9) is similar.

In view of (2), f is S_m-integrable for each m, so for each m there is a partition P_m of $[a,b]$ satisfying

$$U_{S_m}(f,P_m) - L_{S_m}(f,P_m) < \frac{\epsilon}{5}. \qquad (10)$$

Now we apply the triangle inequality to (8) and (9) [with $P = P_m$] and (10) to obtain

$$U_F(f,P_m) - L_F(f,P_m) < \frac{3\epsilon}{5} \quad \text{for} \quad m \geq m_0. \qquad (11)$$

This implies (6); therefore f is F-integrable.

Now (11) and (10) imply

$$U_F(f,P_m) - \int_a^b f\,dF < \frac{3\epsilon}{5} \quad \text{and} \quad U_{S_m}(f,P_m) - \int_a^b f\,dS_m < \frac{\epsilon}{5}$$

for $m \geq m_0$. Applying these inequalities and (8), we obtain

$$\left| \int_a^b f\,dF - \int_a^b f\,dS_m \right| < \epsilon \quad \text{for} \quad m \geq m_0,$$

i.e., (7) holds. ∎

Example 3

(a) Let (u_j) be a sequence of distinct points in $[a, b]$, and let (c_j) be a sequence of nonnegative numbers[3] such that $\sum c_j < \infty$. Using the notation of Example 1 on page 301, we define

$$F = \sum_{j=1}^{\infty} c_j J_{u_j}.$$

Then F is an increasing function on $[a, b]$; note $F(a) = 0$ and $F(b) = \sum_{j=1}^{\infty} c_j < \infty$. Every bounded function f on $[a, b]$ is F-integrable and

$$\int_a^b f \, dF = \sum_{j=1}^{\infty} c_j f(u_j). \tag{1}$$

This follows from Theorem 35.12(b) with $F_j = c_j J_{u_j}$, since every bounded function is F_j-integrable for every j, as shown in Example 1.

(b) The function F in part (a) satisfies

$$F(t) = \sum \{c_j : u_j \leq t\} \quad \text{for} \quad t \in (a, b], \tag{2}$$

since $J_{u_j}(t) = 1$ if and only if $u_j \leq t$. Also $F(a) = 0$. Moreover,

$$F(t^-) = \sum \{c_j : u_j < t\} \quad \text{for} \quad t \in (a, b], \tag{3}$$

since $F(t^-) = \lim_{x \to t-} \sum \{c_j : u_j \leq x\}$. We also have

$$F(t^+) = F(t) \quad \text{for} \quad t \in (a, b], \tag{4}$$

because $F(b^+) = F(b)$ by decree, and for $t \in (a, b)$ and $x > t$, we have

$$F(x) - F(t) = \sum \{c_j : t < u_j \leq x\}, \tag{5}$$

so that $F(t^+) = \lim_{x \to t+} F(x) = F(t)$. [Details: Given $\epsilon > 0$, there is N so that $\sum_{j=N+1}^{\infty} c_j < \epsilon$. For $x > t$ and x sufficiently close to t, the sets $\{u_j : t < u_j \leq x\}$ and $\{u_1, \ldots, u_N\}$ are disjoint, so the sum in (5) is less than ϵ.]

---

[3]Allowing some c_j's to be 0 won't change the value of F, of course, but allowing c_j's to be 0 is sometimes convenient.

(c) A function F is said to be right continuous at t if $F(t) = F(t^+)$. Equation (4) shows F is right continuous except possibly at a. By (4), (2) and (3), we have

$$F(t^+) - F(t^-) = \sum \{c_j : u_j = t\}. \qquad (6)$$

Equation (6) holds for all t in $[a, b]$, even $t = a$. Therefore, F is continuous at t unless $t = u_j$ for some j. Finally, F is clearly discontinuous at each u_j unless $c_j = 0$. $\qquad\qquad\qquad\qquad\qquad\qquad\qquad$ □

The next theorem shows F-integrals can often be calculated using ordinary Riemann integrals. In fact, most F-integrals encountered in practice are either covered by Example 3 or this theorem.

35.13 Theorem.
Suppose F is differentiable on $[a, b]$ and F' is continuous on $[a, b]$. If f is continuous on $[a, b]$, then

$$\int_a^b f \, dF = \int_a^b f(x) F'(x) \, dx. \qquad (1)$$

Proof
Note fF' is Riemann integrable by Theorem 33.2, and f is F-integrable by Theorem 35.7. By Theorems 32.5 and 35.6, there is a partition $P = \{a = t_0 < t_1 < \cdots < t_n = b\}$ such that

$$U(fF', P) - L(fF', P) < \frac{\epsilon}{2} \quad \text{and} \quad U_F(f, P) - L_F(f, P) < \frac{\epsilon}{2}. \qquad (2)$$

By the Mean Value Theorem 29.3 on page 233 applied to F on each interval $[t_{k-1}, t_k]$, there exist x_k in (t_{k-1}, t_k) so that

$$F(t_k) - F(t_{k-1}) = F'(x_k)(t_k - t_{k-1});$$

hence

$$\sum_{k=1}^n f(x_k) \cdot [F(t_k) - F(t_{k-1})] = \sum_{k=1}^n f(x_k) F'(x_k) \cdot (t_k - t_{k-1}). \qquad (3)$$

Since F is continuous, it has no jumps and (3) implies

$$L_F(f, P) \le U(fF', P) \quad \text{and} \quad L(fF', P) \le U_F(f, P).$$

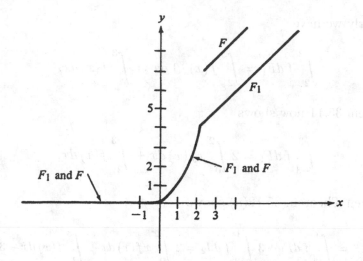

FIGURE 35.2

Now by (2) we have

$$\int_a^b f\, dF \le U_F(f,P) < \frac{\epsilon}{2} + L_F(f,P) \le \frac{\epsilon}{2} + U(fF',P)$$

$$< \frac{\epsilon}{2} + \frac{\epsilon}{2} + L(fF',P) \le \epsilon + \int_a^b f(x)F'(x)\, dx$$

and similarly $\int_a^b f\, dF > \int_a^b f(x)F'(x)\, dx - \epsilon$. Since $\epsilon > 0$ is arbitrary, (1) holds. ∎

An extension of Theorem 35.13 appears in Theorem 35.29.

Example 4
Let $F(t) = 0$ for $t < 0$, $F(t) = t^2$ for $0 \le t < 2$, and $F(t) = t + 5$ for $t \ge 2$; see Fig. 35.2. We can write $F = F_1 + 3J_2$ where F_1 is continuous and J_2 is the jump function at 2. The function F_1 is differentiable except at $t = 2$; the differentiability of F_1 at $t = 0$ is shown in Exercise 28.7. Let f be continuous on $[-3,3]$, say. Clearly $\int_{-3}^0 f\, dF_1 = 0$. Since F_1 agrees with the differentiable function t^2 on $[0,2]$, we can apply Theorem 35.13 to obtain

$$\int_0^2 f\, dF_1 = \int_0^2 f(x) \cdot 2x\, dx = 2\int_0^2 xf(x)\, dx.$$

Similarly we have

$$\int_2^3 f\,dF_1 = \int_2^3 f(x)\cdot 1\,dx = \int_2^3 f(x)\,dx.$$

Theorem 35.11 now shows

$$\int_{-3}^3 f\,dF_1 = 2\int_0^2 xf(x)\,dx + \int_2^3 f(x)\,dx,$$

and then Theorem 35.12(a) shows

$$\int_{-3}^3 f\,dF = \int_{-3}^3 f\,dF_1 + 3\int_{-3}^3 f\,dJ_2 = 2\int_0^2 xf(x)\,dx + \int_2^3 f(x)\,dx + 3f(2).$$

As a specific example, if $f(x) = x^3$, then

$$\int_{-3}^3 f\,dF = 2\int_0^2 x^4\,dx + \int_2^3 x^3\,dx + 3\cdot 8 = \frac{1061}{20} = 53.05.$$

$\square$

For the proof of Lemma 35.15, we will use the next theorem, which is interesting in its own right.

35.14 Theorem.
Let F be a right-continuous increasing function on $[a,b]$. Then we have $F = F_c + F_d$, where F_c is a continuous increasing function on $[a,b]$, and $F_d = \sum c_j J_{u_j}$ where $\{u_1, u_2, \ldots\}$ are the jump points of F and c_j is the jump at u_j, i.e., $c_j = F(u_j^+) - F(u_j^-)$ for each j. (If there are no jump points, then F is continuous, $F = F_c$ and $F_d = 0$.) In general, there might be finitely many or infinitely many such u_j, so we will not decorate the sums until we need to.

See Examples 1 and 3 on pages 301 and 309 for the notation. The function F_c is called the "continuous part" of F, and F_d is called the "discrete part" of F.

Proof

Let $F_c = F - F_d$. To show F_c is an increasing function, it suffices to show $s < t$ implies $F_c(s) \leq F_c(t)$, i.e.,

$$s < t \quad \text{implies} \quad F_d(t) - F_d(s) \leq F(t) - F(s).$$

By (2) in Example 3(b) on page 309, it suffices to show

$$s < t \quad \text{implies} \quad \sum \{c_j : s < u_j \leq t\} \leq F(t) - F(s). \qquad (1)$$

To show this, it suffices to consider a finite subset E of $\{u_j : s < u_j \leq t\}$, which we write as $\{t_0 < t_1 < \cdots < t_n\}$ where $s < t_0$ and $t_n \leq t$. Then

$$\sum \{c_j : u_j \in E\} = \sum_{k=0}^{n} [F(t_k^+) - F(t_k^-)] \leq F(t^+) - F(s) = F(t) - F(s)$$

by (6) in Definition 35.2 on page 300 with $a = s$ and $b = t$. Since E is an arbitrary finite subset of $\{u_j : s < u_j \leq t\}$, (1) holds.

To show F_c is continuous, first note that if $t \neq u_j$ for all j, then F is continuous at t, and F_d is continuous at t as shown in Example 3(c). So $F_c = F - F_d$ is continuous at t in this case. Otherwise, $t = u_j$ for some j. Then $F(u_j^+) - F(u_j^-) = c_j$ by the definition of c_j, and $F_d(u_j^+) - F_d(u_j^-) = c_j$ by (6) of Example 3(b). Therefore

$$F_c(u_j^+) - F_c(u_j^-) = F(u_j^+) - F_d(u_j^+) - F(u_j^-) + F_d(u_j^-)$$

$$= [F(u_j^+) - F(u_j^-)] - [F_d(u_j^+) - F_d(u_j^-)] = c_j - c_j = 0,$$

so that F_c is continuous at u_j, as claimed. ∎

It is easy to check that the representation of F as $F_c + F_d$ is unique in the following sense: If $F_c + F_d = G_c + G_d$ where G is another right-continuous increasing function, then $F_c = G_c$ and $F_d = G_d$.

35.15 Lemma.

If F is an increasing function on $[a, b]$ and if $\epsilon > 0$, then there exists a partition

$$P = \{a = t_0 < t_1 < \cdots < t_n = b\}$$

such that

$$F(t_k^-) - F(t_{k-1}^+) < \epsilon \quad for \quad k = 1, 2, \ldots, n. \tag{1}$$

Proof

The conclusion (1) does not depend on the exact values of F at any of its jump points, though of course they satisfy $F(t^-) \leq F(t) \leq F(t^+)$. Hence it suffices to prove (1) after F has been redefined at its jump points. We choose to define $F(t) = F(t^+)$ at all jump points, so that F is right continuous on $[a, b]$.

Now that F is right continuous, we can apply Theorem 35.14: $F = F_c + \sum c_j J_{u_j}$ where F_c is continuous, each c_j is positive, and $\sum c_j < \infty$. If F is continuous, then the sum disappears and the proof below can be simplified.

Given $\epsilon > 0$, select N so that $\sum_{j=N+1}^{\infty} c_j < \frac{\epsilon}{2}$. Again, if there are only finitely many u_j's, this part of the proof simplifies. Since F_c is uniformly continuous on $[a, b]$ [by Theorem 19.2 on page 143], there is $\delta > 0$ so that

$$|s - t| < \delta \quad \text{implies} \quad |F_c(s) - F_c(t)| < \frac{\epsilon}{2}. \tag{2}$$

Let $P = \{a = t_0 < t_1 < \cdots < t_n\}$ be a partition that includes $\{u_1, \ldots, u_N\}$ and satisfies $|t_k - t_{k-1}| < \delta$ for $k = 1, 2, \ldots, n$. Let

$$F_1 = \sum_{j=1}^{N} c_j J_{u_j} \quad \text{and} \quad F_2 = \sum_{j=N+1}^{\infty} c_j J_{u_j},$$

so that $F = F_c + F_1 + F_2$. For each $k = 1, 2, \ldots, n$, we have

$$F_c(t_k^-) - F_c(t_{k-1}^+) = F_c(t_k) - F_c(t_{k-1}) < \frac{\epsilon}{2} \quad \text{by (2)},$$

$$F_1(t_k^-) - F_1(t_{k-1}^+) = 0$$

because F_1 is constant on the open interval (t_{k-1}, t_k), and

$$F_2(t_k^-) - F_2(t_{k-1}^+) \leq F_2(b) - F_2(a) \leq \sum_{j=N+1}^{\infty} c_j < \frac{\epsilon}{2}.$$

Since $F = F_c + F_1 + F_2$, summing yields $F(t_k^-) - F(t_{k-1}^+) < \epsilon$. ∎

35.16 Theorem.

Every monotonic function f on $[a, b]$ is F-integrable.

Proof
We may assume f is increasing and $f(a) < f(b)$. Since $f(a) \leq f(x) \leq f(b)$ for all x in $[a, b]$, f is bounded on $[a, b]$. For $\epsilon > 0$ we apply Lemma 35.15 to obtain $P = \{a = t_0 < t_1 < \cdots < t_n = b\}$ where

$$F(t_k^-) - F(t_{k-1}^+) < \frac{\epsilon}{f(b) - f(a)}$$

for $k = 1, 2, \ldots, n$. Since $M(f, (t_{k-1}, t_k)) = f(t_k^-) \leq f(t_k)$ and $m(f, (t_{k-1}, t_k)) = f(t_{k-1}^+) \geq f(t_{k-1})$, we have

$$U_F(f, P) - L_F(f, P) \leq \sum_{k=1}^{n} [f(t_k) - f(t_{k-1})] \cdot [F(t_k^-) - F(t_{k-1}^+)]$$

$$< \sum_{k=1}^{n} [f(t_k) - f(t_{k-1})] \cdot \frac{\epsilon}{f(b) - f(a)} = \epsilon.$$

Since ϵ is arbitrary, Theorem 35.6 shows f is F-integrable. ∎

35.17 Theorem.
If f is piecewise continuous, or bounded and piecewise monotonic, on $[a, b]$, then f is F-integrable.

Proof
Just as in the proof of Theorem 33.8, this follows from Theorems 35.7, 35.16 and 35.11, provided we have the following generalization of Exercise 32.7. ∎

35.18 Proposition.
If f is F-integrable on $[a, b]$ and $g(x) = f(x)$ except for finitely many points, then g is F-integrable. We do not claim $\int_a^b f \, dF = \int_a^b g \, dF$.

Proof
It suffices to show $g - f$ is F-integrable, because then the sum $(g - f) + f$ would be F-integrable by Theorem 35.8(ii). Thus it suffices to assume $g(x) = 0$ except for finitely many points. Consider any partition $P = \{a = t_0 < t_1 < \cdots < t_n = b\}$ satisfying $g(x) = 0$ for all x in $[a, b] \setminus \{t_0, t_1, \ldots, t_n\}$. Then from Eqs. (2) and (3) in Definition 35.2 on page 300 we have $U_F(g, P) = J_F(g, P) = L_F(g, P)$, so g is clearly F-integrable; see Theorem 35.6, for example. ∎

If F_1 and F_2 are increasing functions with continuous derivatives, then Theorem 35.13 allows the formula on integration by parts [Theorem 34.2] to be recast as

$$\int_a^b F_1 \, dF_2 + \int_a^b F_2 \, dF_1 = F_1(b)F_2(b) - F_1(a)F_2(a).$$

There is no hope to prove this in general because if $F(t) = 0$ for $t < 0$ and $F(t) = 1$ for $t \geq 0$, then

$$\int_{-1}^1 F \, dF + \int_{-1}^1 F \, dF = 2 \neq 1 = F(1)F(1) - F(-1)F(-1).$$

The generalization does hold provided the functions in the integrands take the middle values at each of their jumps, as we next prove. The result is a special case of a theorem given by Edwin Hewitt [29].

35.19 Theorem [Integration by Parts].
Suppose F_1 and F_2 are increasing functions on $[a, b]$ and define

$$F_1^*(t) = \frac{1}{2}[F_1(t^-) + F_1(t^+)] \quad and \quad F_2^*(t) = \frac{1}{2}[F_2(t^-) + F_2(t^+)]$$

for all $t \in [a, b]$. Then

$$\int_a^b F_1^* \, dF_2 + \int_a^b F_2^* \, dF_1 = F_1(b)F_2(b) - F_1(a)F_2(a). \qquad (1)$$

As usual, we decree $F_1(b^+) = F_1(b)$, $F_1(a^-) = F_1(a)$, etc.

Proof
Both integrals in (1) exist in view of Theorem 35.16. For an $\epsilon > 0$, there exists a partition $P = \{a = t_0 < t_1 < \cdots < t_n = b\}$ such that

$$U_{F_1}(F_2^*, P) - L_{F_1}(F_2^*, P) < \epsilon.$$

Some algebraic manipulation [discussed in the next paragraph] shows

$$U_{F_2}(F_1^*, P) + L_{F_1}(F_2^*, P) = F_1(b)F_2(b) - F_1(a)F_2(a), \qquad (2)$$

so that also

$$U_{F_1}(F_2^*, P) + L_{F_2}(F_1^*, P) = F_1(b)F_2(b) - F_1(a)F_2(a). \qquad (3)$$

It follows from (2) that

$$\int_a^b F_1^* \, dF_2 + \int_a^b F_2^* \, dF_1 \le U_{F_2}(F_1^*, P) + U_{F_1}(F_2^*, P)$$
$$< U_{F_2}(F_1^*, P) + L_{F_1}(F_2^*, P) + \epsilon$$
$$= F_1(b)F_2(b) - F_1(a)F_2(a) + \epsilon,$$

while (3) leads to

$$\int_a^b F_1^* \, dF_2 + \int_a^b F_2^* \, dF_1 > F_1(b)F_2(b) - F_1(a)F_2(a) - \epsilon.$$

Since ϵ is arbitrary, (1) holds.

It remains to verify (2). Observe

$$U_{F_2}(F_1^*, P) + L_{F_1}(F_2^*, P) = \sum_{k=0}^n F_1^*(t_k) \cdot [F_2(t_k^+) - F_2(t_k^-)]$$
$$+ \sum_{k=1}^n M(F_1^*, (t_{k-1}, t_k)) \cdot [F_2(t_k^-) - F_2(t_{k-1}^+)]$$
$$+ \sum_{k=0}^n F_2^*(t_k) \cdot [F_1(t_k^+) - F_1(t_k^-)]$$
$$+ \sum_{k=1}^n m(F_2^*, (t_{k-1}, t_k)) \cdot [F_1(t_k^-) - F_1(t_{k-1}^+)]$$
$$= \sum_{k=0}^n \frac{1}{2}[F_1(t_k^-) + F_1(t_k^+)] \cdot [F_2(t_k^+) - F_2(t_k^-)]$$
$$+ \sum_{k=1}^n F_1(t_k^-) \cdot [F_2(t_k^-) - F_2(t_{k-1}^+)]$$
$$+ \sum_{k=0}^n \frac{1}{2}[F_2(t_k^-) + F_2(t_k^+)] \cdot [F_1(t_k^+) - F_1(t_k^-)]$$
$$+ \sum_{k=1}^n F_2(t_{k-1}^+) \cdot [F_1(t_k^-) - F_1(t_{k-1}^+)].$$

The first and third sums add to

$$\sum_{k=0}^n [F_1(t_k^+)F_2(t_k^+) - F_1(t_k^-)F_2(t_k^-)], \qquad (4)$$

while the second and fourth sums add to

$$\sum_{k=1}^{n} [F_1(t_k^-)F_2(t_k^-) - F_1(t_{k-1}^+)F_2(t_{k-1}^+)]. \tag{5}$$

Since the sums in (4) and (5) add to $F_1(b)F_2(b) - F_1(a)F_2(a)$, by Eq. (6) in Definition 35.2 [on page 300] applied to F_1F_2, the equality (2) holds. The preceding algebra simplifies considerably if F_1 and F_2 are continuous. ∎

We next compare our approach to Riemann-Stieltjes integration to the usual approach. For a bounded function f on $[a, b]$, the usual Darboux-Stieltjes integral is defined via the upper sums

$$\tilde{U}_F(f, P) = \sum_{k=1}^{n} M(f, [t_{k-1}, t_k]) \cdot [F(t_k) - F(t_{k-1})]$$

and the lower sums

$$\tilde{L}_F(f, P) = \sum_{k=1}^{n} m(f, [t_{k-1}, t_k]) \cdot [F(t_k) - F(t_{k-1})].$$

The expressions $\tilde{U}_F(f)$, $\tilde{L}_F(f)$ and $\tilde{\int}_a^b f \, dF$ are defined in analogy to those in Definition 35.2 on page 300. The usual Riemann-Stieltjes integral is defined via the sums

$$\tilde{S}_F(f, P) = \sum_{k=1}^{n} f(x_k)[F(t_k) - F(t_{k-1})],$$

where x_k is in $[t_{k-1}, t_k]$, and the mesh is defined in Definition 32.6; compare Definition 35.24.

The usual Riemann-Stieltjes integrability criterion implies the usual Darboux-Stieltjes integrability criterion; these criteria are not equivalent in general, but they are equivalent if F is continuous. See, for example, [53, §12.2], [55, Chap. 8], or [62, Chap. 6], the most complete treatment being in [55].

35.20 Theorem.
If f is Darboux-Stieltjes integrable on $[a, b]$ with respect to F in the usual sense, then f is F-integrable and the integrals agree.

Proof

For any partition P, $\tilde{L}_F(f, P)$ equals

$$\sum_{k=1}^{n} m(f, [t_{k-1}, t_k]) \cdot [F(t_k) - F(t_k^-) + F(t_k^-) - F(t_{k-1}^+)$$

$$+ F(t_{k-1}^+) - F(t_{k-1})]$$

$$\leq \sum_{k=1}^{n} f(t_k)[F(t_k) - F(t_k^-)]$$

$$+ \sum_{k=1}^{n} m(f, (t_{k-1}, t_k)) \cdot [F(t_k^-) - F(t_{k-1}^+)]$$

$$+ \sum_{k=1}^{n} f(t_{k-1})[F(t_{k-1}^+) - F(t_{k-1})].$$

The first and third sums add to

$$\sum_{k=1}^{n} f(t_k)[F(t_k) - F(t_k^-)] + \sum_{k=0}^{n-1} f(t_k)[F(t_k^+) - F(t_k)]$$

$$= f(t_n)[F(t_n) - F(t_n^-)] + \sum_{k=1}^{n-1} f(t_k)[F(t_k^+) - F(t_k^-)]$$

$$+ f(t_0)[F(t_0^+) - F(t_0)]$$

$$= \sum_{k=0}^{n} f(t_k)[F(t_k^+) - F(t_k^-)] = J_F(f, P).$$

These observations and a glance at the definition of $L_F(f, P)$ now show $\tilde{L}_F(f, P) \leq L_F(f, P)$. Likewise we have $\tilde{U}_F(f, P) \geq U_F(f, P)$, so

$$U_F(f, P) - L_F(f, P) \leq \tilde{U}_F(f, P) - \tilde{L}_F(f, P). \qquad (1)$$

If $\epsilon > 0$, the usual theory [Theorem 32.5] shows there exists a partition P such that $\tilde{U}_F(f, P) - \tilde{L}_F(f, P) < \epsilon$. By (1) we see that we also have $U_F(f, P) - L_F(f, P) < \epsilon$, so f is F-integrable by Theorem 35.6.

To see equality of the integrals, simply observe

$$\tilde{\int_a^b} f \, dF \leq \tilde{U}_F(f, P) < \tilde{L}_F(f, P) + \epsilon \leq L_F(f, P) + \epsilon \leq \int_a^b f \, dF + \epsilon$$

and similarly

$$\tilde{\int}_a^b f \, dF > \int_a^b f \, dF - \epsilon.$$

        ■

We will define Riemann-Stieltjes integrals using a mesh defined in terms of F instead of the usual mesh in Definition 32.6.

35.21 Definition.
The F-*mesh* of a partition P is

$$F\text{-mesh}(P) = \max\{F(t_k^-) - F(t_{k-1}^+) : k = 1, 2, \ldots, n\}.$$

It is convenient to restate Lemma 35.15:

35.22 Lemma.
If $\delta > 0$, there exists a partition P such that F-mesh$(P) < \delta$.

35.23 Theorem.
A bounded function f on $[a, b]$ is F-integrable if and only if for each $\epsilon > 0$ there exists $\delta > 0$ such that

$$F\text{-mesh}(P) < \delta \quad \text{implies} \quad U_F(f, P) - L_F(f, P) < \epsilon \qquad (1)$$

for all partitions P of $[a, b]$.

Proof
Suppose the ϵ–δ condition stated in the theorem holds. If we have $\epsilon > 0$, then (1) applies to some partition P by Lemma 35.22 and hence $U_F(f, P) - L_F(f, P) < \epsilon$. Since this remark applies to all $\epsilon > 0$, Theorem 35.6 implies f is F-integrable.

    The converse is proved just as in Theorem 32.7 with "mesh" replaced by "F-mesh" and references to Lemma 32.2 replaced by references to Lemma 35.3.   ■

35.24 Definition.
Let f be bounded on $[a, b]$, and let

$$P = \{a = t_0 < t_1 < \cdots < t_n = b\}.$$

A *Riemann-Stieltjes sum* of f associated with P and F is a sum of the form

$$J_F(f, P) + \sum_{k=1}^{n} f(x_k)[F(t_k^-) - F(t_{k-1}^+)]$$

where x_k is in (t_{k-1}, t_k) for $k = 1, 2, \ldots, n$.

The function f is *Riemann-Stieltjes integrable on* $[a, b]$ if there exists r in $\mathbb{R}$ with the following property. For each $\epsilon > 0$ there exists $\delta > 0$ such that

$$|S - r| < \epsilon \tag{1}$$

for every Riemann-Stieltjes sum S of f associated with a partition P having F-mesh$(P) < \delta$. We call r the *Riemann-Stieltjes integral of f* and temporarily write it as

$$\mathcal{RS} \int_a^b f \, dF.$$

35.25 Theorem.
A bounded function f on $[a, b]$ is F-integrable if and only if it is Riemann-Stieltjes integrable, in which case the integrals are equal.

Proof
The proof that F-integrability implies Riemann-Stieltjes integrability imitates the corresponding proof in Theorem 32.9. The proof of the converse also imitates the corresponding proof, but a little care is needed, so we give it.

Let f be a Riemann-Stieltjes integrable function, and let r be as in Definition 35.24. Consider $\epsilon > 0$, and let $\delta > 0$ be as provided in Definition 35.24. By Lemma 35.22 there exists a partition $P = \{a = t_0 < t_1 < \cdots < t_n = b\}$ with F-mesh$(P) < \delta$. For each $k = 1, 2, \ldots, n$, select x_k in (t_{k-1}, t_k) so that $f(x_k) < m(f, (t_{k-1}, t_k)) + \epsilon$. The Riemann-Stieltjes sum S for this choice of x_k's satisfies

$$S \leq L_F(f, P) + \epsilon[F(b) - F(a)]$$

and also

$$|S - r| < \epsilon;$$

hence $L_F(f) \geq L_F(f, P) > r - \epsilon - \epsilon[F(b) - F(a)]$. Since $\epsilon > 0$ is arbitrary, it follows that $L_F(f) \geq r$ and similarly $U_F(f) \leq r$. Therefore $L_F(f) = U_F(f) = r$. Thus f is F-integrable and

$$\int_a^b f \, dF = r = \mathcal{RS} \int_a^b f \, dF.$$

$\blacksquare$

35.26 Remarks.

In the definition of upper and lower Darboux sums, we used closed intervals $[t_{k-1}, t_k]$ for Riemann integrals, in §32, and we used open intervals (t_{k-1}, t_k) in this section. The reason we used closed intervals in §32 is because that is completely standard. The reason we used open intervals in this section is because our development was motivated by the desire to have Riemann-Stieltjes integrals more compatible with their measure-theoretic generalization, Lebesgue-Stieltjes integrals, where the "measurable partition"

$$[a, b] = \bigcup_{k=0}^{n} \{t_k\} \cup \bigcup_{k=1}^{n} (t_{k-1}, t_k)$$

is a natural starting point. Moreover, the arguments in Examples 1 and 3 and in Proposition 35.18 would be somewhat more complicated if we had used closed intervals.

The reader may wonder whether the theories would have changed if we had used closed intervals in this section or open intervals in §32. It turns out there would be no essential changes, as we now explain. We do so using increasing functions F, as in this section; see Corollary 35.28 for the case of Riemann integrals. We write $\overline{L}_F(f, P)$ and $\overline{U}_F(f, P)$ for the lower and upper Darboux-Stieltjes sums using closed intervals $[t_{k-1}, t_k]$ instead of open intervals (t_{k-1}, t_k). These will sometimes be different from the corresponding Darboux-Stieltjes sums defined in Definition 35.2 on page 300. But if we define

$$\overline{U}_F(f) = \inf\{\overline{U}_F(f, P) : P \text{ partitions } [a, b]\}$$

and we define $\overline{L}_F(f)$ similarly, we have the following fact.

35.27 Proposition.

If f is bounded on $[a, b]$, then $\overline{U}_F(f) = U_F(f)$ and $\overline{L}_F(f) = L_F(f)$. Thus f is integrable using closed intervals in the Darboux-Stieltjes definitions if and only if it is integrable using open intervals, and in this case the integrals are equal.

Proof

We prove $\overline{U}_F(f) = U_F(f)$; the proof of $\overline{L}_F(f) = L_F(f)$ is similar. For each partition P of $[a, b]$, it is clear that $U_F(f, P) \leq \overline{U}_F(f, P)$. Therefore $U_F(f) \leq \overline{U}_F(f)$ and it suffices to show

$$\overline{U}_F(f) \leq U_F(f). \tag{1}$$

Let $\epsilon > 0$. There is a partition P of $[a, b]$ such that

$$U_F(f, P) < U_F(f) + \frac{\epsilon}{4}. \tag{2}$$

Note (2) holds for any refinement of P by Lemma 35.3.

As before, if there are any jumps we list them as $u_1, u_2, \ldots$ and define $c_j = F(u_j^+) - F(u_j^-)$. Since $\sum c_j$ is finite, there is an integer N so that $\sum_{j=N+1}^{\infty} c_j < \frac{\epsilon}{4}$. Of course, this step is trivial if F has only finitely many jumps. We first refine P so that for every $j \leq N$, the jump u_j is equal to some t_k in P. Then if P_1 and P_2 are such refinements of P, we have

$$|J_F(f, P_1) - J_F(f, P_2)| \leq \sum_{j=N+1}^{\infty} c_j < \frac{\epsilon}{4}. \tag{3}$$

Now we further refine P as follows. Using the definitions of $F(t_k^-)$ and $F(t_k^+)$, we see there are points s_k and u_k for $k = 0, 1, \ldots, n$ so that

$$a = s_0 = t_0 < u_0 < s_1 < t_1 < u_1 < s_2 < t_2 < u_2 < s_3 < \cdots$$

$$\cdots < s_{n-1} < t_{n-1} < u_{n-1} < s_n < t_n = u_n = b,$$

and all $F(u_k^-) - F(t_k^+)$ and $F(t_k^-) - F(s_k^+)$ are less than $\frac{\epsilon}{4(n+1)B}$, where B is a positive bound for the absolute value $|f|$. We estimate $\overline{U}_F(f, P^\sharp)$ for the new partition $P^\sharp$ consisting of all s_k, t_k and u_k.

Observe:

$$\sum_{k=0}^{n} M(f, [s_k, t_k]) \cdot [F(t_k^-) - F(s_k^+)] \leq \sum_{k=0}^{n} B \cdot [F(t_k^-) - F(s_k^+)]) < \frac{\epsilon}{4};$$

$$\sum_{k=0}^{n} M(f, [t_k, u_k]) \cdot [F(u_k^-) - F(t_k^+)] \leq \sum_{k=0}^{n} B \cdot [F(u_k^-) - F(t_k^+)] < \frac{\epsilon}{4};$$

$$\text{and} \quad \sum_{k=1}^{n} M(f, [u_{k-1}, s_k]) \cdot [F(s_k^-) - F(u_{k-1}^+)]$$

$$\leq \sum_{k=1}^{n} M(f, (t_{k-1}, t_k)) \cdot [F(t_k^-) - F(t_{k-1}^+)] = U_F(f, P) - J_F(f, P).$$

Summing these three inequalities, we see

$$\overline{U}_F(f, P^\sharp) - \overline{J}_F(f, P^\sharp) < U_F(f, P) - J_F(f, P) + \frac{2\epsilon}{4}.$$

Now we invoke (3) to conclude

$$\overline{U}_F(f, P^\sharp) < U_F(f, P) - J_F(f, P) + \overline{J}_F(f, P^\sharp) + \frac{2\epsilon}{4} < U_F(f, P) + \frac{3\epsilon}{4}.$$

Finally, using (2), we obtain $\overline{U}_F(f, P^\sharp) < U_F(f) + \epsilon$; hence

$$\overline{U}_F(f) \leq \overline{U}_F(f, P^\sharp) < U_F(f) + \epsilon.$$

Since ϵ is arbitrary, (1) is established. ∎

Proposition 35.27 is true for $F(t) = t$, so it is true for Riemann integrals. However, for this important case, we use closed intervals as is traditional. Thus our notation $U(f, P)$, $U(f)$ refers to the theory using closed intervals, and we do not want to confuse the situation here by switching to $\overline{U}(f, P$, $\overline{U}(f)$, etc. To distinguish the theory using open intervals, we introduce new notation $U^\flat(f, P)$, $U^\flat(f)$, etc.

35.28 Corollary.
In §32, on the Riemann integral, the definitions of $L(f, P)$ and $U(f, P)$ for upper and lower Darboux sums could have used open intervals (t_{k-1}, t_k) instead of closed intervals $[t_{k-1}, t_k]$. The resulting

Darboux sums $L^b(f, P)$ and $U^b(f, P)$ might be different. But if we define

$$U^b(f) = \inf\{U^b(f, P) : P \text{ partitions } [a, b]\}$$

and we define $L^b(f)$ similarly, then we have $U^b(f) = U(f)$ and $L^b(f) = L(f)$. Thus the two approaches to Riemann integration give the same integrals.

Proof
This is Proposition 35.27 for $F(t) = t$; only the notation here is different. ■

We next give a generalization of Theorem 35.13. The proof is somewhat complicated and uses Corollary 35.28. It was originally worked out by me and my colleague Theodore W. Palmer.

35.29 Theorem.
Suppose F is differentiable on $[a, b]$ and F' is Riemann integrable on $[a, b]$. A bounded function f on $[a, b]$ is F-integrable if and only if fF' is Riemann integrable, in which case

$$\int_a^b f \, dF = \int_a^b f(x)F'(x) \, dx.$$

We will give the proof after Lemma 35.30 and its corollary. First look at Definition 35.2 on page 300. Being differentiable, F is continuous. So the J_F terms in the definition are 0, and we have

$$U_F(f, P) = \sum_{k=1}^n M(f, (t_{k-1}, t_k)) \cdot [F(t_k) - F(t_{k-1})]$$

and

$$L_F(f, P) = \sum_{k=1}^n m(f, (t_{k-1}, t_k)) \cdot [F(t_k) - F(t_{k-1})].$$

We will use these formulas in the proof of Lemma 35.30 below. We also use the notation $U^b(f)$, etc., from Corollary 35.28.

35.30 Lemma.
Assume F is differentiable on $[a,b]$, F' is Riemann integrable on $[a,b]$, and f is a bounded function on $[a,b]$. Let B be a positive bound for $|f|$. If we have $U^b(F',P) - L^b(F',P) < \epsilon/B$, then

$$|U_F(f,P) - U^b(fF',P)| \le \epsilon \tag{1}$$

and

$$|L_F(f,P) - L^b(fF',P)| \le \epsilon. \tag{2}$$

Proof
Let $P = \{a = t_0 < t_1 < \cdots < t_n = b\}$. By the Mean Value Theorem 29.3, for each k there exists x_k in (t_{k-1}, t_k) so that

$$F(t_k) - F(t_{k-1}) = F'(x_k) \cdot (t_k - t_{k-1}). \tag{3}$$

Now consider an arbitrary y_k in (t_{k-1}, t_k) and observe

$$\sum_{k=1}^{n} |F'(x_k) - F'(y_k)| \cdot (t_k - t_{k-1}) \le U^b(F',P) - L^b(F',P) < \frac{\epsilon}{B} \tag{4}$$

since the sum is bounded by $\sum_{k=1}^{n}[M(F',(t_{k-1},t_k)) - m(F',(t_{k-1},t_k))] \cdot (t_k - t_{k-1})$. Applying this and (3), we obtain

$$\begin{aligned}
&|\textstyle\sum_{k=1}^{n} f(y_k)[F(t_k) - F(t_{k-1})] - \sum_{k=1}^{n} f(y_k) \cdot F'(y_k) \cdot (t_k - t_{k-1})| \\
&= |\textstyle\sum_{k=1}^{n} f(y_k) \cdot F'(x_k) \cdot (t_k - t_{k-1}) - \sum_{k=1}^{n} f(y_k) \cdot F'(y_k) \cdot (t_k - t_{k-1})| \\
&\le \textstyle\sum_{k=1}^{n} |f(y_k)| \cdot |F'(x_k) - F'(y_k)| \cdot (t_k - t_{k-1}) \\
&\le B \cdot \textstyle\sum_{k=1}^{n} |F'(x_k) - F'(y_k)| \cdot (t_k - t_{k-1}) < \epsilon.
\end{aligned}$$

Therefore for y_k in (t_{k-1}, t_k), we have

$$\begin{aligned}
\textstyle\sum_{k=1}^{n} f(y_k)[F(t_k) - F(t_{k-1})] &< \epsilon + \textstyle\sum_{k=1}^{n} f(y_k) \cdot F'(y_k) \cdot (t_k - t_{k-1}) \\
&\le \epsilon + \textstyle\sum_{k=1}^{n} M(fF',(t_{k-1},t_k)) \cdot (t_k - t_{k-1}) = \epsilon + U^b(fF',P)
\end{aligned} \tag{5}$$

and

$$\begin{aligned}
\textstyle\sum_{k=1}^{n} f(y_k) \cdot F'(y_k) \cdot (t_k - t_{k-1}) &< \epsilon + \textstyle\sum_{k=1}^{n} f(y_k)[F(t_k) - F(t_{k-1})] \\
&\le \epsilon + \textstyle\sum_{k=1}^{n} M(f,(t_{k-1},t_k))[F(t_k) - F(t_{k-1})] = \epsilon + U_F(f,P).
\end{aligned} \tag{6}$$

Since (5) and (6) hold for arbitrary y_k in (t_{k-1}, t_k), for each k, we conclude[4]

$$U_F(f, P) = \sum_{k=1}^{n} M(f, (t_{k-1}, t_k)) \cdot [F(t_k) - F(t_{k-1})] \le \epsilon + U^b(fF', P).$$

(7)

and

$$U^b(fF', P) = \sum_{k=1}^{n} M(fF', (t_{k-1}, t_k)) \cdot (t_k - t_{k-1}) \le \epsilon + U_F(f, P). \quad (8)$$

Now (1) follows from (7) and (8); (2) has a similar proof. ∎

35.31 Corollary.

With F and f as in Lemma 35.30, we have $U_F(f) = U(fF')$ and $L_F(f) = L(fF')$.

Proof
Consider $\epsilon > 0$. Select partitions P_1 and P_2 of $[a, b]$ so that

$$|U_F(f, P_1) - U_F(f)| < \epsilon \qquad \text{and} \qquad |U(fF') - U^b(fF', P_2)| < \epsilon.$$

Such a partition P_2 exists because $U(fF') = U^b(fF')$ by Corollary 35.28. Since F' is Riemann integrable by hypothesis, $U^b(F') = U(F') = L(F') = L^b(F')$. So there exists a partition P_3 so that

$$U^b(F', P_3) - L^b(F', P_3) < \epsilon/B.$$

The last three inequalities still hold if we replace P_1, P_2, P_3 by a common refinement P. Now by Lemma 35.30, we have

$$|U_F(f, P) - U^b(fF', P)| \le \epsilon,$$

so by the triangle inequality, we have

$$|U_F(f) - U(fF')| < 3\epsilon.$$

Since ϵ is arbitrary, $U_F(f) = U(fF')$. Likewise $L_F(f) = L(fF')$. ∎

---

[4]Some readers may appreciate more explanation. To avoid breaking the flow of the proofs, this will be provided after the proof of Theorem 35.29 is completed.

Proof of Theorem 35.29

Suppose f is F-integrable. Then $U_F(f) = L_F(f)$. Hence $U(fF') = L(fF')$ by Corollary 35.31. Thus fF' is Riemann integrable and

$$\int_a^b f\,dF = U_F(f) = U(fF') = \int_a^b f(x)F'(x)dx.$$

Now suppose fF' is Riemann integrable. Then $U(fF') = L(fF')$; hence $U_F(f) = L_F(f)$. Thus f is F-integrable and

$$\int_a^b f(x)F'(x)dx = U(fF') = U_F(f) = \int_a^b f\,dF.$$

$\blacksquare$

How did we so quickly conclude inequalities (5) and (6) in Lemma 35.30? Intuitively, we knew we could select each y_k so that $f(y_k)$ is very close to $M(f, (t_{k-1}, t_k))$, which appears in the definition of $U_F(f, P)$ given after the statement of Theorem 35.29. Similarly, we concluded inequality (6) because we could select each y_k so that $(fF')(y_k)$ is very close to $M(fF', (t_{k-1}, t_k))$, which appears in the definition of $U^b(fF', P)$. In fact, these conclusions follow from the following lemma.

35.32 Lemma.

Suppose $a_1, \ldots, a_n$ are (fixed) nonnegative numbers and $B_1, \ldots, B_n$ are nonempty bounded subsets of $\mathbb{R}$. If for some constant K, we have

$$\sum_{k=1}^n a_k b_k \leq K$$

for all choices of b_k in B_k, $k = 1, 2, \ldots, n$, then

$$\sum_{k=1}^n a_k \sup B_k \leq K.$$

Proof

We prove the following by induction on $m = 1, 2, \ldots, n$:

$$P_m : \text{``}a_m b_m \leq K - \sum_{k=1}^{m-1} a_k \sup B_k - \sum_{k=m+1}^n a_k b_k,$$

for all choices of b_k in B_k, $k = m, \ldots, n$." As always, sums $\sum_{k=1}^{0}$ and $\sum_{k=n+1}^{n}$ are taken to be 0. Assertion P_1 holds by the hypothesis of the lemma. Assume P_m holds for some $m = 1, \ldots, n-1$. Since b_m in B_m is arbitrary, P_m implies

$$Q_m : \text{``} a_m \sup B_m \leq K - \sum_{k=1}^{m-1} a_k \sup B_k - \sum_{k=m+1}^{n} a_k b_k,$$

for all choices of b_k in B_k, $k = m + 1, \ldots, n$," but this is exactly assertion P_{m+1}. By induction, all P_n and Q_n hold. Since Q_n is the conclusion of the lemma, the proof is complete. ∎

The nontrivial part of Exercise 4.14, $\sup A + \sup B \leq \sup(A + B)$, is a special case of Lemma 35.32 where $n = 2$, $a_1 = a_2 = 1$, $B_1 = A$, $B_2 = B$, and $K = \sup(A + B)$.

Exercise 35.6 asserts $\int_a^b f\,dF = \lim_{n \to \infty} \int_a^b f_n\,dF$ provided the sequence (f_n) of F-integrable functions converges uniformly to f on $[a, b]$. Moreover, all of the comments in Discussion 33.10 carry over to this setting.

Exercises

35.1 Let F be an increasing function on $[a, b]$.

    (a) Show $\lim_{x \to t-} F(x)$ exists for t in $(a, b]$ and is equal to $\sup\{F(x) : x \in (a, t)\}$.

    (b) Show $\lim_{x \to t+} F(x)$ exists for t in $[a, b)$ and is equal to $\inf\{F(x) : x \in (t, b)\}$.

35.2 Calculate $\int_0^3 x^2\,dF(x)$ for the function F in Example 4.

35.3 Let F be the step function such that $F(t) = n$ for $t \in [n, n+1)$, n an integer. Calculate
    (a) $\int_0^6 x\,dF(x)$,   (b) $\int_0^3 x^2\,dF(x)$,
    (c) $\int_{1/4}^{\pi/4} x^2\,dF(x)$.

35.4 Let $F(t) = \sin t$ for $t \in [-\frac{\pi}{2}, \frac{\pi}{2}]$. Calculate
    (a) $\int_0^{\pi/2} x\,dF(x)$ (b) $\int_{-\pi/2}^{\pi/2} x\,dF(x)$.

35.5 Let $f(x) = 1$ for rational x and $f(x) = 0$ for irrational x.

(a) Show that if F is continuous on $[a, b]$ and $F(a) < F(b)$, then f is not F-integrable on $[a, b]$.

(b) Observe f is F-integrable if F is as in Example 1 or 3.

35.6 Let (f_n) be a sequence of F-integrable functions on $[a, b]$, and suppose $f_n \to f$ uniformly on $[a, b]$. Show f is F-integrable and

$$\int_a^b f \, dF = \lim_{n \to \infty} \int_a^b f_n \, dF.$$

35.7 Let f and g be F-integrable functions on $[a, b]$. Show

(a) f^2 is F-integrable.

(b) fg is F-integrable.

(c) $\max(f, g)$ and $\min(f, g)$ are F-integrable.

35.8 Let g be continuous on $[a, b]$ where $g(x) \geq 0$ for all $x \in [a, b]$ and define $F(t) = \int_a^t g(x) \, dx$ for $t \in [a, b]$. Show that if f is continuous, then

$$\int_a^b f \, dF = \int_a^b f(x)g(x) \, dx.$$

35.9 Let f be continuous on $[a, b]$.

(a) Show $\int_a^b f \, dF = f(x)[F(b) - F(a)]$ for some x in $[a, b]$.

(b) Show Exercise 33.14 is a special case of part (a).

35.10 Let $F(x) = f(x) = x^n$ for $x \in [a, b]$ and some positive odd integer n. Show $\int_a^b f \, dF = \frac{1}{2}[b^{2n} - a^{2n}]$.

35.11 Here is a "change of variable" formula. Let f be F-integrable on $[a, b]$. Let ϕ be a continuous, strictly increasing function on an interval $[c, d]$ such that $\phi(c) = a$ and $\phi(d) = b$. Define

$$g(u) = f(\phi(u)) \quad \text{and} \quad G(u) = F(\phi(u)) \quad \text{for} \quad u \in [c, d].$$

Show g is G-integrable and $\int_c^d g \, dG = \int_a^b f \, dF$.

35.12 Let (u_j) be an enumeration of the rationals in $[a, b]$ and let (c_j) be a sequence of positive integers such that $\sum c_j < \infty$.

(a) Show $F = \sum c_j J_{u_j}$ defines a strictly increasing function on $[a, b]$.

(b) At what points is F continuous?

§36 * Improper Integrals

The Riemann integral in §32 has been defined only for functions that are bounded on a closed interval $[a, b]$. It is convenient to be able to integrate some functions that are unbounded or are defined on an unbounded interval.

36.1 Definition.

Consider an interval $[a, b)$ where b is finite or $+\infty$. Suppose f is a function on $[a, b)$ that is integrable on each $[a, d]$ for $a < d < b$, and suppose the limit

$$\lim_{d \to b^-} \int_a^d f(x)\, dx$$

exists either as a finite number, $+\infty$ or $-\infty$. Then we define

$$\int_a^b f(x)\, dx = \lim_{d \to b^-} \int_a^d f(x)\, dx. \tag{1}$$

If b is finite and f is integrable on $[a, b]$, this definition agrees with that in Definition 32.1 [Exercise 36.1]. If $b = +\infty$ or if f is not integrable on $[a, b]$, but the limit in (1) exists, then (1) defines an *improper integral*.

An analogous definition applies if f is defined on $(a, b]$ where a is finite or $-\infty$ and if f is integrable on each $[c, b]$ for $a < c < b$. Then we define

$$\int_a^b f(x)\, dx = \lim_{c \to a^+} \int_c^b f(x)\, dx \tag{2}$$

whenever the limit exists.

If f is defined on (a, b) and integrable on all closed subintervals $[c, d]$, then we fix α in (a, b) and define

$$\int_a^b f(x)\, dx = \int_a^\alpha f(x)\, dx + \int_\alpha^b f(x)\, dx \tag{3}$$

provided the integrals on the right exist and the sum is not of the form $+\infty + (-\infty)$. Here we agree $\infty + L = \infty$ if $L \neq -\infty$ and $(-\infty) + L = -\infty$ if $L \neq \infty$. It is easy [Exercise 36.2] to see this definition does not depend on the choice of α.

Whenever the improper integrals defined above exist and are finite, the integrals are said to *converge*. Otherwise they *diverge* to $+\infty$ or to $-\infty$.

Example 1
Consider $f(x) = \frac{1}{x}$ for $x \in (0, \infty)$. For $d > 1$, we have $\int_1^d \frac{1}{x}\,dx = \log_e d$, so

$$\int_1^\infty \frac{1}{x}\,dx = \lim_{d \to \infty} \log_e d = +\infty.$$

This improper integral diverges to $+\infty$. For $0 < c < 1$, we have $\int_c^1 \frac{1}{x}\,dx = -\log_e c$, so

$$\int_0^1 \frac{1}{x}\,dx = \lim_{c \to 0^+} [-\log_e c] = +\infty.$$

Also we have

$$\int_0^\infty \frac{1}{x} = +\infty. \qquad \square$$

Example 2
Consider $f(x) = x^{-p}$ for $x \in [1, \infty)$ and a fixed positive number $p \neq 1$. For $d > 1$,

$$\int_1^d x^{-p}\,dx = \frac{1}{1-p}[d^{1-p} - 1].$$

It follows that

$$\int_1^\infty x^{-p}\,dx = \frac{1}{1-p}[0-1] = \frac{1}{p-1} \quad \text{if } p > 1$$

and

$$\int_1^\infty x^{-p}\,dx = +\infty \quad \text{if } 0 < p < 1. \qquad \square$$

Example 3
We have $\int_0^d \sin x\,dx = 1 - \cos d$ for all d. The value $(1 - \cos d)$ oscillates between 0 and 2, as $d \to \infty$, and therefore the limit

$$\lim_{d \to \infty} \int_0^d \sin x\,dx \quad \text{does not exist.}$$

Thus the symbol $\int_0^\infty \sin x\, dx$ has no meaning and is not an improper integral. Similarly, $\int_{-\infty}^0 \sin x\, dx$ and $\int_{-\infty}^\infty \sin x\, dx$ have no meaning. Note the limit

$$\lim_{a \to \infty} \int_{-a}^a \sin x\, dx$$

clearly exists and equals 0. When such a "symmetric" limit exists even though the improper integral $\int_{-\infty}^\infty$ does not, we have what is called a *Cauchy principal value* of $\int_{-\infty}^\infty$. Thus 0 is the Cauchy principal value of $\int_{-\infty}^\infty \sin x\, dx$, but this is not an improper integral. $\qquad\qquad\square$

It is especially valuable to extend Riemann-Stieltjes integrals to infinite intervals; see the discussion after Theorem 36.4 below. Let F be a bounded increasing function on some interval I. The function F can be extended to all of $\mathbb{R}$ by a simple device: if I is bounded below, define

$$F(t) = \inf\{F(u) : u \in I\} \quad \text{for} \quad t \le \inf I;$$

if I is bounded above, define

$$F(t) = \sup\{F(u) : u \in I\} \quad \text{for} \quad t \ge \sup I.$$

For this reason, we will henceforth assume F is an increasing function on all of $\mathbb{R}$. We will use the notations

$$F(-\infty) = \lim_{t \to -\infty} F(t) \quad \text{and} \quad F(\infty) = \lim_{t \to \infty} F(t).$$

Improper Riemann-Stieltjes integrals are defined in analogy to improper Riemann integrals.

36.2 Definition.
Suppose f is F-integrable on each interval $[a, b]$ in $\mathbb{R}$. We make the following definitions whenever the limits exist:

$$\int_0^\infty f\, dF = \lim_{b \to \infty} \int_0^b f\, dF; \qquad \int_{-\infty}^0 f\, dF = \lim_{a \to -\infty} \int_a^0 f\, dF.$$

If both limits exist and their sum does not have the form $\infty + (-\infty)$, we define

$$\int_{-\infty}^{\infty} f\, dF = \int_{-\infty}^{0} f\, dF + \int_{0}^{\infty} f\, dF.$$

If this sum is finite, we say f is F-integrable on $\mathbb{R}$. If f is F-integrable on $\mathbb{R}$ for $F(t) = t$ [i.e., the integrals are Riemann integrals], we say f is *integrable* on $\mathbb{R}$.

36.3 Theorem.
If f is F-integrable on each interval $[a, b]$ and if $f(x) \geq 0$ for all $x \in \mathbb{R}$, then f is F-integrable on $\mathbb{R}$ or else $\int_{-\infty}^{\infty} f\, dF = +\infty$.

Proof
We indicate why $\lim_{a \to -\infty} \int_{a}^{0} f\, dF$ exists, and leave the case of $\lim_{b \to \infty} \int_{0}^{b} f\, dF$ to the reader. Let $h(a) = \int_{a}^{0} f\, dF$ for $a < 0$, and note $a' < a < 0$ implies $h(a') \geq h(a)$. This property implies $\lim_{a \to -\infty} h(a)$ exists and

$$\lim_{a \to -\infty} h(a) = \sup\{h(a) : a \in (-\infty, 0)\}.$$

We omit the simple argument. ∎

36.4 Theorem.
Suppose $-\infty < F(-\infty) < F(\infty) < \infty$. Let f be a bounded function on $\mathbb{R}$ that is F-integrable on each interval $[a, b]$. Then f is F-integrable on $\mathbb{R}$.

Proof
Select a constant B such that $|f(x)| \leq B$ for all $x \in \mathbb{R}$. Since we have $F(\infty) - F(-\infty) < \infty$, constant functions are F-integrable. Since $0 \leq f + B \leq 2B$, Theorem 36.3 shows $f + B$ is F-integrable. It follows [Exercise 36.10] that $f = (f + B) + (-B)$ is also F-integrable. ∎

Increasing functions F defined on $\mathbb{R}$ come up naturally in probability and statistics. In these disciplines, F is called a *distribution function* if we also have $F(-\infty) = 0$ and $F(\infty) = 1$. Of course, the function $F(t) = t$ that corresponds to the Riemann integral is not a distribution function. Here is how a distribution function comes up in probability. Consider a random experiment with numerical outcomes; then $F(t)$ can represent the probability the numerical value will be $\leq t$. Thus F will be right continuous. As a very simple example, suppose the experiment involves tossing three fair coins and counting the number of heads. The numerical values 0, 1, 2, and 3 will result with probabilities $\frac{1}{8}$, $\frac{3}{8}$, $\frac{3}{8}$, and $\frac{1}{8}$, respectively. The corresponding distribution function is defined in Example 2 of §35, page 303, and sketched in Fig. 35.1.

Frequently a distribution function F has the form

$$F(t) = \int_{-\infty}^{t} g(x)\,dx$$

for an integrable function g satisfying $g(x) \geq 0$ for all $x \in \mathbb{R}$. Then g is called a *density* for F. Note we must have

$$\int_{-\infty}^{\infty} g(x)\,dx = 1.$$

If g is continuous, then $g(t) = F'(t)$ for all t by Theorem 34.3.

Example 4
It turns out that $\int_{-\infty}^{\infty} e^{-x^2}\,dx = \sqrt{\pi}$ [Exercise 36.7] and hence

$$\int_{-\infty}^{\infty} e^{-x^2/2}\,dx = \sqrt{2\pi}.$$

The most important density in probability is the *normal density*

$$g(x) = \frac{1}{\sqrt{2\pi}} e^{-x^2/2}$$

which gives rise to the *normal distribution*

$$F(t) = \frac{1}{\sqrt{2\pi}} \int_{-\infty}^{t} e^{-x^2/2}\,dx;$$

see Fig. 36.1. □

Exercises 36.1–36.8 below deal only with Riemann integrals.

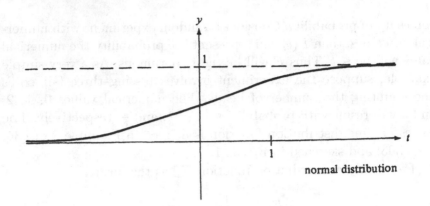

normal distribution

FIGURE 36.1

Exercises

36.1 Show that if f is integrable on $[a, b]$ as in Definition 32.1, then

$$\lim_{d \to b^-} \int_a^d f(x)\, dx = \int_a^b f(x)\, dx.$$

36.2 Show the definition (3) in Definition 36.1 does not depend on the choice of α.

36.3 **(a)** Show

$$\int_0^1 x^{-p}\, dx = \frac{1}{1-p} \quad \text{if} \quad 0 < p < 1 \quad \text{and} \quad \int_0^1 x^{-p}\, dx = +\infty \quad \text{if} \quad p > 1.$$

(b) Show $\int_0^\infty x^{-p}\, dx = +\infty$ for all $p > 0$.

36.4 Calculate
(a) $\int_0^1 \log_e x\, dx$,   **(b)** $\int_2^\infty \frac{\log_e x}{x}\, dx$,
(c) $\int_0^\infty \frac{1}{1+x^2}\, dx$.

36.5 Let f be a continuous function on (a, b) such that $f(x) \geq 0$ for all $x \in (a, b)$; a can be $-\infty$, b can be $+\infty$. Show the improper integral $\int_a^b f(x)\, dx$ exists and equals

$$\sup \left\{ \int_c^d f(x)\, dx : [c, d] \subseteq (a, b) \right\}.$$

36.6 Prove the following *comparison tests*. Let f and g be continuous functions on (a, b) such that $0 \leq f(x) \leq g(x)$ for all x in (a, b); a can be $-\infty$, b can be $+\infty$.

(a) If $\int_a^b g(x)\,dx < \infty$, then $\int_a^b f(x)\,dx < \infty$.

(b) If $\int_a^b f(x)\,dx = +\infty$, then $\int_a^b g(x)\,dx = +\infty$.

36.7 (a) Use Exercise 36.6 to show $\int_{-\infty}^{\infty} e^{-x^2}\,dx < \infty$.

(b) Show this integral equals $\sqrt{\pi}$. *Hint:* Calculate the double integral $\int_{-\infty}^{\infty}\int_{-\infty}^{\infty} e^{-x^2} e^{-y^2}\,dx\,dy$ using polar coordinates.

36.8 Suppose f is continuous on (a, b) and $\int_a^b |f(x)|\,dx < \infty$; again a can be $-\infty$, b can be $+\infty$. Show the integral $\int_a^b f(x)\,dx$ exists and is finite.

36.9 Let F be the normal distribution function in Example 4.

(a) Show that if f is continuous on $\mathbb{R}$ and if the improper integral $\int_{-\infty}^{\infty} f(x)e^{-x^2/2}\,dx$ exists, then the improper integral $\int_{-\infty}^{\infty} f\,dF$ exists and

$$\int_{-\infty}^{\infty} f\,dF = \frac{1}{\sqrt{2\pi}} \int_{-\infty}^{\infty} f(x)e^{-x^2/2}\,dx.$$

Calculate

(b) $\int_{-\infty}^{\infty} x^2\,dF(x)$,

(c) $\int_{-\infty}^{\infty} e^{x^2}\,dF(x)$,

(d) $\int_{-\infty}^{\infty} |x|\,dF(x)$,

(e) $\int_{-\infty}^{\infty} x\,dF(x)$.

36.10 Let f and g be F-integrable functions on $\mathbb{R}$. Show $f+g$ is F-integrable on $\mathbb{R}$ and

$$\int_{-\infty}^{\infty} (f + g)\,dF = \int_{-\infty}^{\infty} f\,dF + \int_{-\infty}^{\infty} g\,dF.$$

36.11 Show that if f and g are F-integrable on $\mathbb{R}$ and if $f(x) \leq g(x)$ for x in $\mathbb{R}$, then $\int_{-\infty}^{\infty} f\,dF \leq \int_{-\infty}^{\infty} g\,dF$.

36.12 Generalize Exercise 36.6 to F-integrals on $\mathbb{R}$.

36.13 Generalize Exercise 36.8 to F-integrals on $\mathbb{R}$.

36.14 Let (u_j) be a sequence of distinct points in $\mathbb{R}$, and let (c_j) be a sequence of nonnegative numbers such that $\sum c_j < \infty$.

(a) Observe $F = \sum_{j=1}^{\infty} c_j J_{u_j}$ is an increasing function on $\mathbb{R}$. See Example 3 on page 309.

(b) Show every bounded function f on $\mathbb{R}$ is F-integrable and

$$\int_{-\infty}^{\infty} f\,dF = \sum_{j=1}^{\infty} c_j f(u_j).$$

(c) When will F be a distribution function?

(d) As in Exercise 35.12, if (u_j) is an enumeration of the rationals, then F is strictly increasing on $\mathbb{R}$.

36.15 (a) Give an example of a sequence (f_n) of integrable functions on $\mathbb{R}$ where $\int_{-\infty}^{\infty} f_n(x)\,dx = 1$ for all n and yet $f_n \to 0$ uniformly on $\mathbb{R}$.

(b) Suppose F is a distribution function on $\mathbb{R}$. Show that if (f_n) is a sequence of F-integrable functions on $\mathbb{R}$ and if $f_n \to f$ uniformly on $\mathbb{R}$, then f is F-integrable on $\mathbb{R}$ and

$$\int_{-\infty}^{\infty} f\,dF = \lim_{n \to \infty} \int_{-\infty}^{\infty} f_n\,dF.$$

7

Capstone

CHAPTER

§37 * A Discussion of Exponents and Logarithms

In this book we have carefully developed the theory, but have been casual about using the familiar exponential, logarithmic and trigonometric functions in examples and exercises. Most readers probably found this an acceptable approach, since they are comfortable with these basic functions. In this section, we indicate three ways to develop the exponential and logarithmic functions assuming only the axioms in Chap. 1 and the theoretical results in later chapters. We will provide proofs for the third approach.

Recall, for x in $\mathbb{R}$ and a positive integer n, x^n is the product of x by itself n times. For $x \neq 0$, we have the convention $x^0 = 1$. And for $x \neq 0$ and negative integers $-n$ where $n \in \mathbb{N}$, we define x^{-n} to be the reciprocal of x^n, i.e., $x^{-n} = (x^n)^{-1}$.

37.1 Piecemeal Approach.
This approach starts with Example 2 on page 238 and Exercise 29.15 where it is shown that x^r is meaningful whenever $x > 0$ and r is

K.A. Ross, *Elementary Analysis: The Theory of Calculus*, 339
Undergraduate Texts in Mathematics, DOI 10.1007/978-1-4614-6271-2_7,
© Springer Science+Business Media New York 2013

rational, i.e., $r \in \mathbb{Q}$. Moreover,

$$\text{if} \quad h(x) = x^r, \quad \text{then} \quad h'(x) = rx^{r-1}.$$

The algebraic properties $x^r x^s = x^{r+s}$ and $(xy)^r = x^r y^r$ can be verified for $r, s \in \mathbb{Q}$ and positive x and y. For any $t \in \mathbb{R}$ and $x > 0$, we *define*

$$x^t = \sup\{x^r : r \in \mathbb{Q} \text{ and } r \le t\}$$

and

$$\left(\frac{1}{x}\right)^t = (x^t)^{-1}.$$

This defines x^t for $x > 0$. It can be shown that with this definition x^t is finite and the algebraic properties mentioned above still hold. Further, it can be shown that $h(x) = x^t$ is differentiable and $h'(x) = tx^{t-1}$.

Next we can consider a fixed $b > 0$ and the function B defined by $B(x) = b^x$ for $x \in \mathbb{R}$. The function B is differentiable and $B'(x) = c_b B(x)$ for some constant c_b. We elaborate on this last claim. In view of Exercise 28.14, we can write

$$B'(x) = \lim_{h \to 0} \frac{b^{x+h} - b^x}{h} = b^x \cdot \lim_{h \to 0} \frac{b^h - 1}{h}$$

provided these limits exist. Some analysis shows that the last limit does exist, so

$$B'(x) = c_b B(x) \quad \text{where} \quad c_b = \lim_{h \to 0} \frac{1}{h}[b^h - 1].$$

It turns out that $c_b = 1$ for a certain b, known universally as e. Since B is one-to-one if $b \ne 1$, B has an inverse function L which is named $L(y) = \log_b y$. Since B is differentiable, Theorem 29.9 on page 237 can be applied to show L is differentiable and

$$L'(y) = \frac{1}{c_b y}.$$

Finally, the familiar properties of $\log_b$ can be established for L. $\qquad \square$

When all the details are supplied, the above approach is very tedious. It has one, and only one, merit; it is direct without any tricks. One could call it the "brute force approach." The next two approaches begin with some well defined mathematical object [either a power series or an integral] and then work backwards to develop the familiar properties of exponentials and logarithms. In both instances, for motivation we will draw on more advanced facts we believe but which have *not* been established in this book.

37.2 Exponential Power Series Approach.

This approach is adopted in two of our favorite books: [16, §4.9] and [62, Chap. 8]. As noted in Example 1 of §31, we believe

$$e^x = \sum_{k=0}^{\infty} \frac{1}{k!} x^k$$

though we have *not* proved this, since we have not even defined exponentials yet. In this approach, we *define*

$$E(x) = \sum_{k=0}^{\infty} \frac{1}{k!} x^k, \tag{1}$$

and we *define* $e = E(1)$. The series here has radius of convergence $+\infty$ [Example 1, §23], and E is differentiable on $\mathbb{R}$ [Theorem 26.5]. It is easy [Exercise 26.5] to show $E' = E$. The fundamental property

$$E(x + y) = E(x)E(y) \tag{2}$$

can be established using only the facts observed above. Actually [62] uses a theorem on multiplication of absolutely convergent series, but [16] avoids this. Other properties of E can be quickly established. In particular, E is strictly increasing on $\mathbb{R}$ and has an inverse L. Theorem 29.9 on page 237 assures us L is differentiable and $L'(y) = \frac{1}{y}$. For rational r and $x > 0$, x^r was defined in Exercise 29.15. Applying that exercise and the chain rule to $g(x) = L(x^r) - rL(x)$, we find $g'(x) = 0$ for $x > 0$. Hence g is a constant function. Since $g(1) = 0$, we conclude

$$L(x^r) = rL(x) \quad \text{for} \quad r \in \mathbb{Q} \quad \text{and} \quad x > 0. \tag{3}$$

For $b > 0$ and rational r, (3) implies

$$b^r = E(L(b^r)) = E(rL(b)).$$

Because of this, we *define*

$$b^x = E(xL(b)) \quad \text{for} \quad x \in \mathbb{R}.$$

The familiar properties of exponentials and their inverses [logarithms!] are now easy to prove. □

The choice between the approach just outlined and the next approach, which we will present in some detail, is really a matter of taste and depends on the appeal of power series. One genuine advantage to the exponential approach is that the series in (1) defining E is equally good for defining $E(z) = e^z$ for complex numbers z.

37.3 Logarithmic Integral Approach.

Let us attempt to solve $f' = f$ where f never vanishes; we expect to obtain $E(x) = e^x$ as one of the solutions. This simple differential equation can be written

$$\frac{f'}{f} = 1. \tag{1}$$

In view of the chain rule, if we could find L satisfying $L'(y) = \frac{1}{y}$, then Eq. (1) would simplify to

$$(L \circ f)' = 1,$$

so one of the solutions would satisfy

$$L \circ f(y) = y.$$

In other words, one solution f of (1) would be an inverse to L where $L'(y) = \frac{1}{y}$. But by the Fundamental Theorem of Calculus II [Theorem 34.3 on page 294], we know such a function L exists. Since we also expect $L(1) = 0$, we *define*

$$L(y) = \int_1^y \frac{1}{t}\,dt \quad \text{for} \quad y \in (0, \infty).$$

We use this definition to prove the basic facts about logarithms and exponentials. □

37.4 Theorem.

(i) *The function L is strictly increasing, continuous and differentiable on $(0, \infty)$. We have*

$$L'(y) = \frac{1}{y} \quad for \quad y \in (0, \infty).$$

(ii) $L(yz) = L(y) + L(z)$ *for* $y, z \in (0, \infty)$.
(iii) $L(\frac{y}{z}) = L(y) - L(z)$ *for* $y, z \in (0, \infty)$.
(iv) $\lim_{y \to \infty} L(y) = +\infty$ *and* $\lim_{y \to 0+} L(y) = -\infty$.

Proof
It is trivial to show the function $f(t) = t$ is continuous on $\mathbb{R}$, so its reciprocal $\frac{1}{t}$ is continuous on $(0, \infty)$ by Theorem 17.4. It is easy to see L is strictly increasing, and the rest of (i) follows immediately from Theorem 34.3.

Assertion (ii) can be proved directly [Exercise 37.1]. Alternatively, fix z and consider $g(y) = L(yz) - L(y) - L(z)$. Since $g(1) = 0$, it suffices to show $g'(y) = 0$ for $y \in (0, \infty)$ [Corollary 29.4]. But since z is fixed, we have

$$g'(y) = \frac{z}{yz} - \frac{1}{y} - 0 = 0.$$

To check (iii), note $L(\frac{1}{z}) + L(z) = L(\frac{1}{z} \cdot z) = L(1) = 0$, so that $L(\frac{1}{z}) = -L(z)$ and

$$L\left(\frac{y}{z}\right) = L\left(y \cdot \frac{1}{z}\right) = L(y) + L\left(\frac{1}{z}\right) = L(y) - L(z).$$

To see (iv), first observe $L(2) > 0$ and $L(2^n) = n \cdot L(2)$ in view of (ii). Thus $\lim_{n \to \infty} L(2^n) = +\infty$. Since L is increasing, it follows that $\lim_{y \to \infty} L(y) = +\infty$. Likewise $L(\frac{1}{2}) < 0$ and $L((\frac{1}{2})^n) = n \cdot L(\frac{1}{2})$, so $\lim_{y \to 0+} L(y) = -\infty$. ∎

The Intermediate Value Theorem 18.2 on page 134 shows L maps $(0, \infty)$ *onto* $\mathbb{R}$. Since L is a strictly increasing function, it has a continuous strictly increasing inverse [Theorem 18.4 on page 137] and the inverse has domain $\mathbb{R}$.

37.5 Definition.

We denote the function inverse to L by E. Thus

$$E(L(y)) = y \quad \text{for} \quad y \in (0, \infty)$$

and

$$L(E(x)) = x \quad \text{for} \quad x \in \mathbb{R}.$$

We also define $e = E(1)$ so that $\int_1^e \frac{1}{t}\, dt = 1$.

37.6 Theorem.

(i) *The function E is strictly increasing, continuous and differentiable on $\mathbb{R}$. Then*

$$E'(x) = E(x) \quad \text{for} \quad x \in \mathbb{R}.$$

(ii) $E(u + v) = E(u)E(v)$ *for $u, v \in \mathbb{R}$.*
(iii) $\lim_{x \to \infty} E(x) = +\infty$ *and* $\lim_{x \to -\infty} E(x) = 0$.

Proof

All of (i) follows from Theorem 37.4 in conjunction with Theorem 29.9 on page 237. In particular,

$$E'(x) = \frac{1}{L'(E(x))} = \frac{1}{\frac{1}{E(x)}} = E(x).$$

If $u, v \in \mathbb{R}$, then $u = L(y)$ and $v = L(z)$ for some $y, z \in (0, \infty)$. Then $u + v = L(yz)$ by (ii) of Theorem 37.4, so

$$E(u + v) = E(L(yz)) = yz = E(L(y))E(L(z)) = E(u)E(v).$$

Assertion (iii) follows from (iv) of Theorem 37.4 [Exercise 37.2]. ∎

Consider $b > 0$ and $r \in \mathbb{Q}$, say $r = \frac{m}{n}$ where $m, n \in \mathbb{Z}$ and $n > 0$. It is customary to write b^r for that positive number a such that $a^n = b^m$. By (ii) of Theorem 37.4, we have $nL(a) = mL(b)$; hence

$$b^r = a = E(L(a)) = E\left(\frac{1}{n} \cdot nL(a)\right) = E\left(\frac{1}{n} \cdot mL(b)\right) = E(rL(b)).$$

This motivates our next definition and also shows the definition is compatible with the usage of fractional powers in algebra.

37.7 Definition.
For $b > 0$ and x in $\mathbb{R}$, we define

$$b^x = E(xL(b)).$$

Since $L(e) = 1$, we have $e^x = E(x)$ for all $x \in \mathbb{R}$.

37.8 Theorem.
Fix $b > 0$.
  (i) *The function $B(x) = b^x$ is continuous and differentiable on $\mathbb{R}$.*
  (ii) *If $b > 1$, then B is strictly increasing; if $b < 1$, then B is strictly decreasing.*
  (iii) *If $b \neq 1$, then B maps $\mathbb{R}$ onto $(0, \infty)$.*
  (iv) *$b^{u+v} = b^u b^v$ for $u, v \in \mathbb{R}$.*

Proof
Exercise 37.3. ∎

If $b \neq 1$, the function B has an inverse function.

37.9 Definition.
For $b > 0$ and $b \neq 1$, the inverse of $B(x) = b^x$ is written $\log_b$. The domain of $\log_b$ is $(0, \infty)$ and

$$\log_b y = x \quad \text{if and only if} \quad b^x = y.$$

Note $\log_e y = L(y)$ for $y > 0$.

37.10 Theorem.
Fix $b > 0$, $b \neq 1$.
  (i) *The function $\log_b$ is continuous and differentiable on $(0, \infty)$.*
  (ii) *If $b > 1$, $\log_b$ is strictly increasing; if $b < 1$, $\log_b$ is strictly decreasing.*
  (iii) *$\log_b(yz) = \log_b y + \log_b z$ for $y, z \in (0, \infty)$.*
  (iv) *$\log_b(\frac{y}{z}) = \log_b y - \log_b z$ for $y, z \in (0, \infty)$.*

Proof
This follows from Theorem 37.4 and the identity $\log_b y = \frac{L(y)}{L(b)}$ [Exercise 37.4]. Note $L(b)$ is negative if $b < 1$. ∎

The function $E(x) = e^x$ has now been rigorously developed and, as explained in Example 1 of §31, we have

$$E(x) = \sum_{k=0}^{\infty} \frac{1}{k!} x^k.$$

In particular, $e = \sum_{k=0}^{\infty} \frac{1}{k!}$. Also

37.11 Theorem.

$$e = \lim_{h \to 0} (1+h)^{1/h} = \lim_{n \to \infty} \left(1 + \frac{1}{n}\right)^n.$$

Proof
We only need to verify the first equality. Since $L'(1) = 1$ by Theorem 37.4(i), and since $L(1) = 0$, we have

$$1 = \lim_{h \to 0} \frac{L(1+h) - L(1)}{h} = \lim_{h \to 0} \frac{1}{h} L(1+h) = \lim_{h \to 0} L\left((1+h)^{1/h}\right).$$

Since E is continuous, we can apply Theorem 20.5 on page 158 to the function $f(h) = L((1+h)^{1/h})$ to obtain

$$\lim_{h \to 0} (1+h)^{1/h} = \lim_{h \to 0} E\left(L\left((1+h)^{1/h}\right)\right) = E(1) = e. \qquad \blacksquare$$

37.12 Trigonometric Functions.
Either approach 37.2 or 37.3 can be modified to rigorously develop the trigonometric functions. They can also be developed using the exponential functions for *complex* values since

$$\sin x = \frac{1}{2i}[e^{ix} - e^{-ix}], \quad \text{etc.};$$

see [62, Chap. 8]. A development of the trigonometric functions analogous to approach 37.3 can proceed as follows. Since we believe

$$\arcsin x = \int_0^x (1 - t^2)^{-1/2} \, dt,$$

we can define $A(x)$ as this integral and obtain $\sin x$ from this. Then $\cos x$ and $\tan x$ are easy to obtain. In this development, the number π is defined to be $2 \int_0^1 (1 - t^2)^{-1/2} \, dt$. $\qquad \square$

In the exercises, use results proved in Theorem 37.4 and subsequent theorems but not the material discussed in 37.1 and 37.2.

Exercises

37.1 Prove directly that
$$\int_1^{yz} \frac{1}{t}\, dt = \int_1^y \frac{1}{t}\, dt + \int_1^z \frac{1}{t}\, dt \quad \text{for} \quad y, z \in (0, \infty).$$

37.2 Prove (iii) of Theorem 37.6.

37.3 Prove Theorem 37.8.

37.4 Consider $b > 0$, $b \neq 1$. Prove $\log_b y = \frac{L(y)}{L(b)}$ for $y \in (0, \infty)$.

37.5 Let p be any real number and define $f(x) = x^p$ for $x > 0$. Show f is differentiable and $f'(x) = px^{p-1}$; compare Exercise 29.15. *Hint*: $f(x) = E(pL(x))$.

37.6 Show $x^p y^p = (xy)^p$ for $p \in \mathbb{R}$ and positive x, y.

37.7 **(a)** Show that if $B(x) = b^x$, then $B'(x) = (\log_e b)b^x$.

   (b) Find the derivative of $\log_b$.

37.8 For $x > 0$, let $f(x) = x^x$. Show $f'(x) = [1 + \log_e x] \cdot x^x$.

37.9 **(a)** Show $\log_e y < y$ for $y > 1$.

   (b) Show
$$\frac{\log_e y}{y} < \frac{2}{\sqrt{y}} \quad \text{for} \quad y > 1. \quad \textit{Hint:} \quad \log_e y = 2\log_e \sqrt{y}.$$

   (c) Use part (b) to prove $\lim_{y\to\infty} \frac{1}{y}\log_e y = 0$. This neat little exercise is based on the paper [28].

§38 * Continuous Nowhere-Differentiable Functions

The history of continuous nowhere-differentiable functions is long and fascinating. Examples of such functions were constructed as early as 1831, by Bernard Bolzano. But a version wasn't published until

Karl Weierstrass did so in 1875 [69]; he had given examples in lectures before that and formally presented his result on July 18, 1872. These early examples tended to be of the form $\sum_{n=0}^{\infty} b^n \cos(a^n \pi x)$ or $\sum_{n=1}^{\infty} \frac{\sin(n^2 \pi x)}{n^2}$. The delay in publicizing these examples is due to the suspicion of mathematicians of the era that the definition of "function" should not be so general as to allow such monstrosities.

Excellent brief accounts are described in William Dunham's book [21, Chap. 9] and David Bressoud's book [13, §6.4]. Dunham asserts that Hermite, Poincare and Picard were appalled by Weierstrass's "pathological example." Bressoud has a quotation from Henri Poincaré that refers to these abnormal functions as counterexamples to the "flaws in our forefathers' reasoning...". There is also a quote of Hermite referring to the "plague of functions that do not have derivatives". He had no idea how bad a plague; see Theorem 38.3 on page 350 and the paragraph preceding it.

An especially thorough and detailed history can be found in Johan Thim's Master's thesis [66].

Karl Stromberg [65, Example (4.8), page 174] provides an elegant variant of van der Waerden's example [68], published in 1930. We repeat Stromberg's example almost verbatim because we don't see how to improve it.

38.1 Van der Waerden's Example.
We provide an example of a continuous function on $\mathbb{R}$ that is nowhere differentiable. Let $\phi(x) = |x|$ for $|x| \leq 2$ and extend it to be 4-periodic, i.e., $\phi(x + 4q) = \phi(x)$ for all $x \in \mathbb{R}$ and $q \in \mathbb{Z}$. Since the graph of ϕ has slope ± 1 on intervals between even integers, we have

$$|\phi(s) - \phi(t)| = |s - t| \tag{1}$$

provided there is no even integer in the open interval (s, t). The graph of ϕ is similar to that of g in Fig. 25.1 on page 204; in fact, $\phi(x) = 2g(\frac{x}{2})$ for all x.

The continuous nowhere-differentiable function f is defined by

$$f(x) = \sum_{n=1}^{\infty} f_n(x) \quad \text{where} \quad f_n(x) = 4^{-n} \phi(4^n x).$$

Since each f_n is continuous and $\sum_{n=1}^{\infty} 4^{-n} \cdot 2 < \infty$, the Weierstrass M-test 25.7 on page 205 and Theorem 25.5 show f is continuous on $\mathbb{R}$.

Fix a in $\mathbb{R}$; we will show f is not differentiable at a by showing there is a sequence (h_k) of nonzero numbers satisfying

$$h_k \to 0 \quad \text{and} \quad \lim_{k \to \infty} \frac{f(a + h_k) - f(a)}{h_k} \quad \text{does not exist.} \qquad (2)$$

Until Eq. (5), the integer $k \geq 1$ is fixed. There exists $\epsilon_k = \pm 1$ so that no even integers lie between $4^k a$ and $4^k a + \epsilon_k$. From (1), we have

$$|\phi(4^k a + \epsilon_k) - \phi(4^k a)| = 1.$$

In fact, for $n \leq k$, there are no even integers between $4^n a$ and $4^n a + 4^{n-k} \epsilon_k$, because if $4^n a < 2q < 4^n a + 4^{n-k} \epsilon_k$ for some $q \in \mathbb{Z}$, multiplying through by the integer 4^{k-n} would yield $4^k a < 4^{k-n} \cdot 2q < 4^k a + \epsilon_k$. Thus, by (1) again,

$$|\phi(4^n a + 4^{n-k} \epsilon_k) - \phi(4^n a)| = 4^{n-k} \quad \text{for} \quad 1 \leq n \leq k. \qquad (3)$$

Since ϕ is 4-periodic, we also have

$$|\phi(4^n a + 4^{n-k} \epsilon_k) - \phi(4^n a)| = 0 \quad \text{for} \quad n > k. \qquad (4)$$

Combining (3) and (4) with the definition of f_n, we conclude

$$|f_n(a + 4^{-k} \epsilon_k) - f_n(a)| = 4^{-k} \quad \text{for} \quad n \leq k \quad \text{and} = 0 \text{ for} \quad n > k.$$

We now let $h_k = 4^{-k} \epsilon_k$ so that

$$|f_n(a + h_k) - f_n(a)| = 4^{-k} \quad \text{for} \quad n \leq k \quad \text{and} = 0 \text{ for} \quad n > k.$$

Therefore the difference quotient,

$$\frac{f(a + h_k) - f(a)}{h_k} = \sum_{n=1}^{k} \frac{f_n(a + h_k) - f_n(a)}{h_k}$$

$$= \sum_{n=1}^{k} \frac{f_n(a + h_k) - f_n(a)}{4^{-k} \epsilon_k} = \sum_{n=1}^{k} (\pm 1), \qquad (5)$$

is an even integer for even k and an odd integer for odd k. Therefore the difference quotients in (5) cannot converge, as $k \to \infty$. Since $\lim_k h_k = 0$, this proves (2) so that f is not differentiable at a. $\qquad \square$

38.2 Remark.
For later use, note the function ϕ in the preceding proof satisfies $\sup\{|\phi(y)| : y \in \mathbb{R}\} = 2$, and for any y in $\mathbb{R}$ we have

$$|\phi(y+h) - \phi(y)| = h \quad \text{for sufficiently small} \quad h > 0. \tag{1}$$

We've seen that constructing continuous nowhere-differentiable functions can be a challenge. Now we take a more abstract approach. In the discussion after Definition 22.6 on page 183, it is observed that the space $C(\mathbb{R})$ of bounded continuous functions on $\mathbb{R}$ is a complete metric space using the metric

$$d(f,g) = \sup\{|f(y) - g(y)| : y \in \mathbb{R}\} \quad \text{for all} \quad f, g \in C(\mathbb{R}).$$

Recall from Example 7(b) on page 174 that the irrational numbers are much more prevalent than the rational numbers, because the set $\mathbb{Q}$ is of first category in the complete metric space $\mathbb{R}$. Similarly, as we will see in the next theorem, the continuous nowhere-differentiable functions are more prevalent than the differentiable ones. One might say mathematicians of the 19th century viewed the continuous nowhere-differentiable functions as the "irrational functions."

As just noted, the next theorem shows "most" functions in $C(\mathbb{R})$ are nowhere differentiable. The proof uses the Baire Category Theorem 21.8 in the enrichment Sect. 21; see page 173. In 1931, S. Banach and S. Mazurkiewicz independently published this result in Studia Mathematica.

38.3 Theorem.
Let $\mathcal{D}$ consist of the functions f in $C(\mathbb{R})$ that have a derivative at at least one point. Then $\mathcal{D}$ is first category in $C(\mathbb{R})$.

Proof
Consider f in $\mathcal{D}$. Then for some x in $\mathbb{R}$, the derivative $f'(x)$ exists. Choose m so that $|x| \leq m$ and $|f'(x)| < m$. There exists δ in $(0,1)$ so that

$$\left| \frac{f(x+h) - f(x)}{h} - f'(x) \right| < m - |f'(x)| \quad \text{for all} \quad h \in (0, \delta).$$

Select a positive integer $n \geq m$ satisfying $\frac{1}{n} < \delta$. Then

$$x \in [-n, n] \quad \text{and} \quad |f(x + h) - f(x)| \leq hn \quad \text{for} \quad h \in \left(0, \frac{1}{n}\right). \quad (1)$$

Thus f belongs to $\mathcal{F}_n$, which is defined to be all f in $C(\mathbb{R})$ such that some x in $\mathbb{R}$ satisfies (1) above. We have just shown $\mathcal{D} \subseteq \cup_{n=1}^{\infty} \mathcal{F}_n$, so by the Baire Category Theorem 21.8 on page 173, it suffices to show each $\mathcal{F}_n$ is closed in $C(\mathbb{R})$ and is nowhere dense in $C(\mathbb{R})$.

To see $\mathcal{F}_n$ is closed, consider a sequence (f_k) in $\mathcal{F}_n$ converging to f in $C(\mathbb{R})$. As noted after Definition 22.6, this means $f_k \to f$ uniformly on $\mathbb{R}$. We will show f is in $\mathcal{F}_n$. For each k, there exists x_k so that

$$x_k \in [-n, n] \quad \text{and} \quad |f_k(x_k + h) - f_k(x_k)| \leq hn \text{ for } h \in \left(0, \frac{1}{n}\right). \quad (2)$$

By the Bolzano-Weierstrass Theorem 11.5, a subsequence of (x_k) converges to some x in $[-n, n]$; to avoid subsequence notation, we will assume the sequence (x_k) itself converges to x. Since $x_k \to x$ and $f_k \to f$ uniformly, we have $f_k(x_k + h) \to f(x + h)$ for each $h \in [0, \frac{1}{n})$; see Exercise 24.17 on page 200. For each h in $(0, \frac{1}{n})$, we take the limit in (2) as $k \to \infty$ to obtain (1). Hence f belongs to $\mathcal{F}_n$, and $\mathcal{F}_n$ is closed.

To show $\mathcal{F}_n$ is nowhere dense, it suffices to show no ball $B_r(g) = \{f \in C(\mathbb{R}) : d(f, g) < r\}$ is a subset of any $\mathcal{F}_n$, i.e., every such ball contains a function in $C(\mathbb{R}) \setminus \mathcal{F}_n$. By the corollary to the Weierstrass Approximation Theorem 27.5 on page 220, there is a polynomial p such that $\sup\{|g(y) - p(y)| : y \in [-n - 1, n + 1]\} < \frac{r}{2}$, $p(-n - 1) = g(-n - 1)$ and $p(n + 1) = g(n + 1)$. The function $\bar{p}$ on $\mathbb{R}$, defined by $\bar{p}(y) = p(y)$ for $|y| \leq n + 1$ and $\bar{p}(y) = g(y)$ for $|y| \geq n + 1$, is in $C(\mathbb{R})$ and satisfies $d(g, \bar{p}) < \frac{r}{2}$. We choose $K \geq n$ so that it also satisfies $K > \sup\{|\bar{p}'(y)| : y \in [-n - 1, n + 1]\}$. Then for each x in $[-n, n]$, the Mean Value Theorem 29.3 on page 233 implies

$$|\bar{p}(x + h) - \bar{p}(x)| \leq hK \quad \text{for sufficiently small} \quad h > 0. \quad (3)$$

A modification ψ of the sawtooth function ϕ in van der Waerden's Example 38.1 on page 348 satisfies

$$\sup\{|\psi(y)| : y \in \mathbb{R}\} \leq \frac{r}{2} \quad (4)$$

and

$$|\psi(x+h) - \psi(x)| > 2Kh \quad \text{for sufficiently small} \quad h > 0, \qquad (5)$$

for all x in $\mathbb{R}$. In fact, using Remark 38.2, it is easy to check $\psi(y) = \frac{r}{4}\phi(\frac{12Ky}{r})$ works. Since $d(\overline{p}, g) < \frac{r}{2}$, inequality (4) shows $\overline{p}+\psi$ belongs to $B_r(g)$. From (5) and (3), we see

$$|(\overline{p} + \psi)(x+h) - (\overline{p} + \psi)(x)| > hK \geq hn$$

for each $x \in [-n, n]$ and sufficiently small $h > 0$, which shows $\overline{p} + \psi$ cannot be in $\mathcal{F}_n$. ∎

In Theorem 38.5 below, we give another interesting application of the Baire Category Theorem 21.8. It is due to Corominas and Sunyer Balaguer [17, 1954]. See also Boas [9, Chap. 1, §10], Phillips [52, Chap. 4, §3, Theorem 18], and Donoghue [20, §I.11]. Interesting generalizations can be found in [17] and Krüppel [38]. There are relatively straightforward versions for holomorphic complex-valued functions; see, e.g., Burckel [14, Example 5.64].

First, we give an easy result. Henceforth, when we refer to an open interval (α, β), we allow $\alpha = -\infty$ and $\beta = \infty$, so that in particular (α, β) may be $\mathbb{R}$. As in Sect. 31, we write $f^{(n)}$ for the nth derivative of f.

38.4 Proposition.
Let f be a function on an open interval (α, β), and suppose there is an integer n such that $f^{(n)}(x) = 0$ for all x. Then f agrees with a polynomial on (α, β) of degree at most $n - 1$.

Proof
This can be seen by integrating the function $f^{(n)}$ n times. At each step, one obtains a polynomial of one higher degree; see Corollary 29.5 and Exercise 29.7. ∎

We write $C^\infty((\alpha, \beta))$ for the space of functions f infinitely differentiable on (α, β), i.e., all the derivatives $f^{(n)}$ exist on (α, β). Here we agree $f^{(0)} = f$.

38.5 Theorem.
Suppose f is in $C^\infty((\alpha, \beta))$, and suppose for each x in (α, β), there is an integer n_x (depending on x) such that $f^{(n_x)}(x) = 0$. Then f is a polynomial on (α, β).

We begin with a lemma.

38.6 Lemma.
Suppose x_0 is in $\mathbb{R}$, p and q are polynomials, and f is a C^∞ function defined on an interval containing x_0 that agrees with p on an interval $(x_0 - h, x_0]$ and agrees with q on an interval $[x_0, x_0 + h)$. Then $p = q$.

Proof
Assume $p \neq q$, and let m be the degree of the polynomial $p - q$. Then $(p - q)^{(m)}(x_0) \neq 0$, contradicting $p^{(m)}(x_0) = f^{(m)}(x_0) = q^{(m)}(x_0)$. ∎

Proof of Theorem 38.5
It suffices to show that f agrees with a polynomial on every closed subinterval $[c, d]$ of (α, β). To see this is sufficient, consider a decreasing sequence (c_n) with limit α and an increasing sequence (d_n) with limit β, where $c_1 < d_1$. For each n, let p_n be a polynomial that agrees with f on $[c_n, d_n]$. Then each p_n agrees with p_1 on $[c_1, d_1]$. Hence $p_n = p_1$ for all n.[1] It follows that $f = p_1$ on (α, β).

Step 1. In this step, we show that every nondegenerate closed subinterval[2] I of $[c, d]$ contains an open subinterval J on which f is a polynomial. For each $k \geq 0$, we let $E_k = \{x \in I : f^{(k)}(x) = 0\}$. Then, by hypothesis $I = \bigcup_{k \geq 0} E_k$, and by the continuity of the function f and all of its derivatives, each E_k is closed. By the Baire Category Theorem 21.8 on page 173, for some index n, E_n has nonempty interior. Thus, for such an index, there is an open subinterval $J \subseteq E_n$ of I such that $f^{(n)}(x) = 0$ for all $x \in J$. By Proposition 38.4, f agrees with a polynomial on J, as claimed.

---

[1]Polynomials that agree on an infinite set are equal; otherwise their difference would be a nonzero polynomial with infinitely many zeros.
[2]Nondegenerate here means that its endpoints are not equal.

Step 2. In this step, we identify maximal nondegenerate closed subintervals of $[c, d]$ on which the restriction of f is a polynomial. For any polynomial p that agrees with f on an open interval containing a point x_0 in $[c, d]$, let

$$a_p(x_0) = \inf\{x \in [c, x_0] : f = p \quad \text{on} \quad [x, x_0]\}$$

and

$$b_p(x_0) = \sup\{x \in [x_0, d] : f = p \quad \text{on} \quad [x_0, x]\}.$$

Let $I_p(x_0) = [a_p(x_0), b_p(x_0)]$, and note $a_p(x_0) < b_p(x_0)$ and $I_p(x_0) \subseteq [c, d]$. Then

(a) $I_p(x_0)$ is the maximal closed subinterval of $[c, d]$ containing x_0 on which f agrees with p;

(b) $(a_p(x_0), b_p(x_0))$ is the maximal open subinterval of $[c, d]$ containing x_0 on which f agrees with p;

(c) The closed intervals $I_p(x_0)$ are either disjoint or identical.

To check assertion (c), assume $I_{p_0}(x_0) \cap I_{p_1}(x_1) \neq \varnothing$, where $x_0 \leq x_1$. Clearly $a_{p_1}(x_1) \leq b_{p_0}(x_0)$. If $b_{p_0}(x_0) = a_{p_1}(x_1)$, then f agrees with p_0 to the left of this value, and f agrees with p_1 to the right of this value. But then Lemma 38.6 shows $p_0 = p_1$, so the closed intervals involved aren't maximal. Therefore $a_{p_1}(x_1) < b_{p_0}(x_0)$, and the polynomials p_1 and p_0 agree on the open interval $(a_{p_1}(x_1), b_{p_0}(x_0))$. Hence $p_0 = p_1$. Now f agrees with $p_0 = p_1$ on the closed interval $[a_{p_0}(x_0), b_{p_1}(x_1)]$, so by maximality,

$$[a_{p_0}(x_0), b_{p_1}(x_1)] = [a_{p_0}(x_0), b_{p_0}(x_0)] = [a_{p_1}(x_1), b_{p_1}(x_1)].$$

Therefore $I_{p_0}(x_0) = [a_{p_0}(x_0), b_{p_0}(x_0)] = [a_{p_1}(x_1), b_{p_1}(x_1)] = I_{p_1}(x_1)$, and (c) holds.

Step 3. Our goal is to show all the closed intervals in Step 2 are the same, i.e., all equal $[c, d]$, which would prove the theorem. Assume, on the contrary, that there exist x_1, x_2 in $[c, d]$ and polynomials p_1, p_2 so that $I_{p_1}(x_1) \neq I_{p_2}(x_2)$. We assume $x_1 < x_2$. Then $b_{p_1}(x_1) < a_{p_2}(x_2)$ and we will focus on $[b_{p_1}(x_1), a_{p_2}(x_2)]$, which we will write as $[b, a]$, so

$$I_{p_1}(x_1) \neq I_{p_2}(x_2) \quad \text{and} \quad [b, a] = [b_{p_1}(x_1), a_{p_2}(x_2)]. \tag{1}$$

Consider the nondegenerate closed intervals from Step 2 that are subintervals of (b, a). Since these intervals are disjoint, there are only countably many of them. Moreover, there cannot be finitely many of them, because if there were, there would be an open interval in (b, a) that is disjoint from the finite union of them. By Step 1, there would be another closed interval as described in Step 2, a contradiction.

We relabel the countably infinite disjoint family of maximal *open* intervals in Step 2, that are subsets of (b, a), by (a_k, b_k). Observe $\bigcup_k (a_k, b_k)$ is dense in $[b, a]$; otherwise there is a nonempty open interval in $[b, a] \setminus \bigcup_k (a_k, b_k)$ and by applying Step 1 we could expand our family. Let $K = [b, a] \setminus \bigcup_k (a_k, b_k)$, which is clearly closed. Let E be the set of endpoints of the intervals $[b, a]$ and all (a_k, b_k). Observe that $E \subseteq K$. Note that, by (1), b and a are right and left endpoints, respectively, of maximal open intervals.

Step 4. In this step, we show that K is a perfect set, as defined in Discussion 21.10 on page 174, i.e., every point x in K is the limit of a sequence of points in $K \setminus \{x\}$. Otherwise, for some x in K and $h > 0$ we have $K \cap (x - h, x + h) = \{x\}$. The point x cannot be both a left and a right endpoint in E, so we may assume, say, that x is not a left endpoint in E. Then $x \neq a$ and we may suppose that $(x, x+h) \subseteq (b, a)$. Since $(x, x+h)$ intersects the dense set $\bigcup_k [a_k, b_k]$, we have $(x, x + h) \cap [a_k, b_k] \neq \varnothing$ for some k. Then $a_k < x + h$ and $x < b_k$. Since $x \neq a_k$, x is not in $\{a_k\} \cup (a_k, b_k) = [a_k, b_k)$. Since $x < b_k$, we have $x < a_k < x + h$ and a_k belongs to $(x, x + h)$. Thus $a_k \in K \cap (x, x + h)$, a contradiction. Thus K is perfect.

Step 5. In this step, we obtain a contradiction to our hypotheses by showing there is $x_0 \in K$ satisfying $f^{(n)}(x_0) \neq 0$ for all $n \geq 0$. This will prove the theorem.

Claim 1. Let J be an open interval in $[b, a]$ satisfying $J \cap K \neq \varnothing$, and consider an integer $n \geq 0$. Then exists an $x \in J \cap K$ such that $f^{(n)}(x) \neq 0$.

Assume that J is as specified, but that n_0 satisfies $f^{(n_0)}(x) = 0$ for all $x \in J \cap K$. We show by induction

$$f^{(n)}(x) = 0 \quad \text{for} \quad n \geq n_0 \quad \text{and all} \quad x \in J \cap K. \tag{2}$$

Suppose (2) holds for some $n \geq n_0$, and consider $x \in J \cap K$. Since K is perfect, there is a sequence (x_k) in $K \setminus \{x\}$ such that $\lim_k x_k = x$.

Since J is open, we may suppose that all x_k are also in J. Hence

$$f^{(n+1)}(x) = \lim_{k \to \infty} \frac{f^{(n)}(x_k) - f^{(n)}(x)}{x_k - x} = \lim_{k \to \infty} \frac{0 - 0}{x_k - x} = 0;$$

i.e., (2) holds for $n + 1$. By induction, we conclude that (2) holds as stated.

On the other hand, if $x \in J \setminus K$ then x belongs to one of the maximal open intervals (a_k, b_k). Since $J \cap K \neq \varnothing$, there exists y in $J \setminus (a_k, b_k)$, so that $y \leq a_k$ or $y \geq b_k$, say $y \geq b_k$. Since $x < b_k$ and J is an interval, J also contains b_k. Thus b_k belongs to $J \cap K$. Let p_k be the polynomial that agrees with f on (a_k, b_k). Since x is in (a_k, b_k), we have $f^{(j)}(x) = p_k^{(j)}(x)$ for all $j \geq 0$. If p_k is the zero polynomial, then clearly $f^{(n_0)}(x) = p_k^{(n_0)}(x) = 0$. Otherwise, let n_k be the degree of the polynomial p_k, so that $p^{(n_k)}$ is the nonzero constant function $p^{(n_k)}(b_k)$. Since b_k is in $J \cap K$, (2) implies that $n_k < n_0$ and $f^{(n_0)}(x) = p^{(n_0)}(x) = 0$. Summarizing, $f^{(n_0)}(x) = 0$ for all x in the open interval J, so f agrees with a polynomial on J by Proposition 38.4 and $J \subseteq \bigcup_k (a_k, b_k)$, contradicting $J \cap K \neq \varnothing$. This proves Claim 1.

Since each $f^{(n)}$ is continuous, Claim 1 implies:

Claim 2. If $n \geq 0$ and J is an open interval in $[b, a]$ satisfying $J \cap K \neq \varnothing$, then there is an open interval J' so that $\overline{J'} \subseteq J$, $J' \cap K \neq \varnothing$ and $f^{(n)}(x) \neq 0$ for all $x \in J' \cap K$.[3]

Applying Claim 2 to $J = (b, a)$, we obtain an open interval J_0 so that $J_0 \cap K \neq \varnothing$ and $f(x) \neq 0$ for $x \in J_0 \cap K$. Repeated use of the claim yields a sequence (J_n) of open intervals so that

$$J_0 \supseteq \overline{J_1} \supseteq J_1 \supseteq \overline{J_2} \supseteq J_2 \supseteq \cdots, \tag{3}$$

each $J_n \cap K \neq \varnothing$ and $f^{(n)}(x) \neq 0$ for all $x \in J_n \cap K$. Since $\bigcap_n J_n \cap K = \bigcap_n \overline{J_n} \cap K \neq \varnothing$, by Theorem 13.10 there is x_0 in K that belongs to all the intervals J_n. Therefore $f^{(n)}(x_0) \neq 0$ for all $n \geq 0$; this completes Step 5 and the proof of the theorem. ∎

---

[3]The notation $\overline{J'}$ represents the closure of the interval J'.

38.7 Corollary.
Suppose $-\infty < \alpha < \beta < \infty$. If f is continuous on $[\alpha, \beta]$, and if f satisfies the hypotheses of Theorem 38.5 on (α, β), then f agrees with a polynomial on $[\alpha, \beta]$.

Even though continuous nowhere-differentiable functions are plentiful, by Theorem 38.3, the next theorem shows every continuous function on $\mathbb{R}$ is arbitrarily close to nice functions.

38.8 Theorem.
Given a continuous function f on $\mathbb{R}$ (bounded or not) and $\epsilon > 0$, there is a piecewise linear function[4] g and a function ψ in $C^\infty(\mathbb{R})$ such that $\sup\{|f(x) - g(x)| : x \in \mathbb{R}\} < \epsilon$ and $\sup\{|f(x) - \psi(x)| : x \in \mathbb{R}\} < \epsilon$.

Proof
If you look at pictures of a couple of continuous functions f, as well as $f + \epsilon$ and $f - \epsilon$, as in Fig. 24.1 on page 195, you might conclude the existence of such a piecewise linear function is obvious. The only flaw in this thinking is that, as we know, general continuous functions can be quite complicated, and impossible to graph. For this reason, we include a proof.

Step 1. If f is continuous on $[a, b]$ and $\epsilon > 0$, then there is a piecewise linear function g on $[a, b]$ satisfying $\sup\{|f(x) - g(x)| : x \in [a, b]\} < \epsilon$. To see this, we use uniform continuity of f to obtain $\delta > 0$ so that

$$|s - t| < \delta, \quad s, t \in [a, b] \quad \text{implies} \quad |f(s) - f(t)| < \frac{\epsilon}{3}.$$

Select a partition $\{a = t_0 < t_1 < \cdots < t_n = b\}$ of $[a, b]$ with mesh $< \delta$. For each $k = 1, 2, \ldots, n$, let ℓ_k be the linear function satisfying $\ell_k(t_{k-1}) = f(t_{k-1})$ and $\ell_k(t_k) = f(t_k)$. We write m_k for $\min\{f(t_{k-1}), f(t_k)\}$ and M_k for $\max\{f(t_{k-1}), f(t_k)\}$. Then for x in $[t_{k-1}, t_k]$, we have

$$m_k - \frac{\epsilon}{3} < f(x) < M_k + \frac{\epsilon}{3} \quad \text{and} \quad M_k - m_k < \frac{\epsilon}{3}.$$

---

[4]A piecewise linear function on an interval is a continuous function that is linear on subintervals.

Thus $f(x)$ lies between $m_k - \frac{\epsilon}{3}$ and $m_k + \frac{2\epsilon}{3}$. The same is true for $\ell_k(x)$, and therefore

$$|f(x) - \ell_k(x)| < \epsilon \quad \text{for} \quad x \in [t_{k-1}, t_k].$$

The function g on $[a, b]$, defined by $g(x) = \ell_k(x)$ for x in $[t_{k-1}, t_k]$, satisfies $\sup\{|f(x) - g(x)| : x \in [a, b]\} < \epsilon$.

Step 2. Let ℓ be a linear function on an interval $[a, b]$. There is a monotonic function ψ in $C^\infty(\mathbb{R})$ so that $\psi(a) = \ell(a)$, $\psi(b) = \ell(b)$, all derivatives of ψ at a and b are 0, and

$$|\psi(x) - \ell(x)| \leq |\ell(b) - \ell(a)| \quad \text{for all} \quad x \in [a, b]. \tag{1}$$

We begin with the function $h = h_{a,b}^*$ in Exercise 31.4 on page 267. Then $h(x) = 0$ for $x \leq a$, $h(x) = 1$ for $x \geq b$, h is increasing on $[a, b]$, and all its derivative are 0 at a and at b. In fact,

$$h(x) = \frac{f(x-a)}{f(x-a) + f(b-x)}$$

where f is the function in Example 3 and Fig. 31.1 on page 257. From the quotient rule, and writing D for the denominator obtained using the quotient rule, we see

$$h'(x) = [f(b-x)f'(x-a) + f(x-a)f'(b-x)]/D(x).$$

This is nonnegative, so h is increasing. By looking at the right-hand derivatives of h at a, and the left-hand derivatives at b, we see $h^{(n)}(a) = h^{(n)}(b) = 0$ for all n in $\mathbb{N}$.[5] Let

$$\psi(x) = \ell(a) + [\ell(b) - \ell(a)] \cdot h(x) \quad \text{for} \quad x \in [a, b].$$

Condition (1) holds because, for x in $[a, b]$, both $\psi(x)$ and $\ell(x)$ lie between $\ell(a)$ and $\ell(b)$.

Step 3. Given a piecewise linear function g on $[a, b]$ and $\epsilon > 0$, there is a function ψ in $C^\infty(\mathbb{R})$ so that $\psi(a) = g(a)$, $\psi(b) = g(b)$, all derivatives of ψ at a and b are 0, and

$$\sup\{|\psi(x) - g(x)| : x \in [a, b]\} < \epsilon. \tag{2}$$

---

[5] Regarding why h, the quotient of two infinitely differentiable functions, also is infinitely differentiable, see Exercise 38.6.

As in Step 1, there is a partition $\{a = t_0 < t_1 < \cdots < t_n = b\}$ of $[a, b]$ such that $|g(t_k) - g(t_{k-1})| < \epsilon$ and such that g agrees on each $[t_{k-1}, t_k]$ with a linear function ℓ_k, $k = 1, 2, \ldots, n$. Apply Step 2 to each $[t_{k-1}, t_k]$ to obtain a function ψ_k in $C^\infty(\mathbb{R})$ so that $\psi_k(t_{k-1}) = \ell_k(t_{k-1}) = g(t_{k-1})$ and similarly $\psi_k(t_k) = g(t_k)$, all derivatives at t_{k-1} and t_k are 0, and

$$|\psi_k(x) - g(x)| \le |\ell_k(t_k) - \ell_k(t_{k-1})| < \epsilon \quad \text{for all} \quad x \in [t_{k-1}, t_k].$$

The desired ψ is defined by $\psi(x) = \psi_k(x)$ for $x \in [t_{k-1}, t_k]$, $k = 1, 2, \ldots, n$.

Step 4. Consider a continuous function f on $\mathbb{R}$ and $\epsilon > 0$. By Step 1, for each k in $\mathbb{Z}$, there is a piecewise linear function g_k on the interval $[k, k + 1]$ satisfying

$$\sup\{|g_k(x) - f(x)| : x \in [k, k+1]\} < \frac{\epsilon}{3}. \qquad (3)$$

We apply Step 3 to each interval $[k, k+1]$ to obtain ψ_k in $C^\infty(\mathbb{R})$ so that $\psi_k(k) = f(k)$, $\psi_k(k+1) = f(k+1)$, all derivatives of ψ_k are 0 at k and $k+1$, and

$$\sup\{|\psi_k(x) - g_k(x)| : x \in [k, k+1]\} < \frac{\epsilon}{3}. \qquad (4)$$

We obtain ψ in $C^\infty(\mathbb{R})$ by defining $\psi(x) = \psi_k(x)$ for x in $[k, k+1]$, $k \in \mathbb{Z}$. From (3) and (4) we conclude

$$\sup\{|\psi(x) - f(x)| : x \in \mathbb{R}\} \le \frac{2\epsilon}{3} < \epsilon. \qquad \blacksquare$$

38.9 Corollary.
The space of all bounded piecewise linear functions and the space $C^\infty(\mathbb{R}) \cap C(\mathbb{R})$ of all bounded infinitely differentiable functions on $\mathbb{R}$ are dense in $C(\mathbb{R})$, using the distance $d(f, g) = \sup\{|f(y) - g(y)| : y \in \mathbb{R}\}$ defined prior to Theorem 38.3.

The next result is both a corollary to Theorem 38.8 and a lemma for the next result.

38.10 Lemma.
Consider continuous f on $[a, b]$, $\epsilon > 0$, and a constant $c > 0$. Then there is a piecewise linear function g on $[a, b]$ such that

$\sup\{|f(x) - g(x)| : x \in [a,b]\} < \epsilon$ *and the absolute value of the slope of each of its linear segments exceeds* c.

Proof

Before continuing, you might wish to reread the first paragraph of the proof of Theorem 38.8. Since, by that theorem, it now suffices to consider the case when f itself is a linear function ℓ, it will be easy to see why this lemma is true.

It suffices to consider the case when ℓ is increasing, and we may assume c exceeds the slope of ℓ. There is a $\delta > 0$ sufficiently small that one can replace any line segment from $(x, \ell(x))$ to $(x+\delta, \ell(x+\delta))$ by two line segments, one rising from $(x, \ell(x))$, with slope exceeding c, to the line $\ell + \frac{\epsilon}{2}$, and the other dropping down to $(x+\delta, f(x+\delta))$, at a slope less than $-c$. By replacing δ by a smaller δ, if necessary, we may assume δ divides $b - a$. Then there is partition $\{t_0 = a < t_1 < \cdots < t_n = b\}$ of $[a,b]$, with $t_k = t_{k-1} + \delta$ for each k, so that each line segment from $(t_{k-1}, \ell(t_{k-1}))$ to $(t_k, \ell(t_k))$ can be replaced by two line segments each with slope exceeding c in absolute value. The resulting piecewise linear function[6] g on $[a,b]$ is continuous because it agrees with $f = \ell$ at each t_k, and it satisfies

$$\sup\{|g(x) - f(x)| : x \in [a,b]\} = \frac{\epsilon}{2} < \epsilon.$$

This establishes the lemma.   ∎

We now return to continuous nowhere-differentiable functions. So far as we know, until 1992 the construction of such functions has involved summing certain continuous functions, i.e., building them up from simpler functions. The next construction, due to Mark Lynch [44] in 1992, is quite different, using some topology and no infinite series. Lynch starts with certain subsets of the plane and shrinks them down to the graph of a continuous nowhere-differentiable function. He uses Example 6 on page 171, which states that a function on a compact subset of $\mathbb{R}$ is continuous if and only if its graph is compact. Later Lynch [45] generalized his technique to construct continuous functions that are differentiable only at the rationals.

---

[6]It is tempting to call this a "stegosaurus linear function."

Below, we simplify the technique in [45] just to obtain a continuous nowhere-differentiable function.

38.11 Mark Lynch's Construction.

Step 1. Let ℓ be a linear function on an interval $[a, b]$. Then its graph L is a line segment in the plane. We assume the slope of ℓ exceeds $2n$ in absolute value. We will construct a (closed) rhomboid[7] R about L with the following property:

> **Property R_n:** If f is a function on $[a, b]$ and if the graph of f is contained in R, then for each $x \in [a, b]$ there is $y \in [a, b]$ such that $x \neq y$ and $\left| \frac{f(y) - f(x)}{y - x} \right| > n$.

First, we suppose the slope exceeds $2n$. Let $m = \frac{a+b}{2}$ be the midpoint of the interval $[a, b]$. Then we have

$$\frac{\ell(b) - \ell(m)}{b - a} = \frac{\ell(m) - \ell(a)}{b - a} > n;$$

thus there exists $\epsilon > 0$ so that

$$\frac{\ell(b) - \epsilon - [\ell(m) + \epsilon]}{b - a} > n \quad \text{and} \quad \frac{\ell(m) - \epsilon - [\ell(a) + \epsilon]}{b - a} > n. \quad (1)$$

The desired rhomboid R is bounded above and below by the graphs of $\ell + \epsilon$ and $\ell - \epsilon$, and bounded on the left and right by the vertical line segments connecting $(x, \ell(x) - \epsilon)$ and $(x, \ell(x) + \epsilon)$ for $x = a$ and for $x = b$. Consider x in $[a, b]$. If $x \leq m$, then using (1) we obtain

$$\frac{f(b) - f(x)}{b - x} \geq \frac{\ell(b) - \epsilon - [\ell(x) + \epsilon]}{b - a} \geq \frac{\ell(b) - \epsilon - [\ell(m) + \epsilon]}{b - a} > n,$$

so Property R_n holds with $y = b$. If $x \geq m$, then

$$\frac{f(x) - f(a)}{x - a} \geq \frac{\ell(x) - \epsilon - [\ell(a) + \epsilon]}{b - a} \geq \frac{\ell(m) - \epsilon - [\ell(a) + \epsilon]}{b - a} > n,$$

and Property R_n holds with $y = a$.

Now suppose the slope of ℓ is less than $-2n$. Then use $-\ell$ to obtain a rhomboid R' satisfying Property R_n. Let $R = -R'$. If the

---

[7] For us, a rhomboid is a parallelogram with unequal adjacent angles.

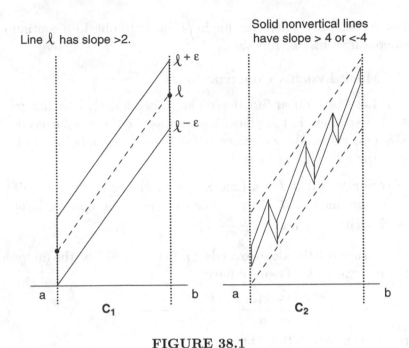

Line ℓ has slope >2.

Solid nonvertical lines
have slope > 4 or <-4

$\ell + \varepsilon$

ℓ

$\ell - \varepsilon$

a

b

C_1

a

b

C_2

FIGURE 38.1

graph of f is contained in R, then the graph of $-f$ is contained in R', and Property R_n holds for $-f$ and R'. So it holds for f and R.

Step 2. Now we construct a continuous nowhere-differentiable function on $[0,1]$. We begin with a piecewise linear function g_1 on $[0,1]$ such that the absolute value of the slope of each segment exceeds 2. Note g_1 can be selected to be linear on $[a,b]$, as illustrated in Fig. 38.1, and observe the x and y axes are not drawn to scale. Let C_1 be the union of rhomboids surrounding the line segments of g_1, where each rhomboid satisfies Property R_n for $n = 1$. Then C_1 satisfies the following property for $n = 1$.

> **Property n:** If f is a function on $[0,1]$ and if the graph of f is contained in C_n, then for each $x \in [0,1]$ there is $y \in [0,1]$ such that $0 < |y - x| \le \frac{1}{n}$ and $\left| \frac{f(y) - f(x)}{y - x} \right| > n$.

For the inductive construction, suppose we have $C_1, \ldots, C_{n-1}$ satisfying $C_{n-1} \subseteq \cdots \subseteq C_1$, where each C_k is a union of rhomboids satisfying Property k. By Lemma 38.10, there is a piecewise linear function g_n on $[0,1]$ whose graph lies in the *interior* of C_{n-1} such that the absolute value of the slope of each segment exceeds $2n$. By breaking segments of g_n into smaller segments, if necessary, we may assume their domains, i.e., their projections onto the x-axis, each have length $\leq \frac{1}{n}$. By Step 1, each such segment of g_n is included in a rhomboid satisfying Property R$_n$, with the additional conclusion $0 < |y - x| \leq \frac{1}{n}$. Therefore C_n satisfies Property n. Moreover, since the graph of g_n lies in the interior of C_{n-1}, the construction in Step 1 shows (using a smaller ϵ if necessary) we can select the rhomboids so that their union C_n is a subset of C_{n-1} and so that the vertical line segments in C_n have length $\leq \frac{1}{n}$.

Now let $C = \cap_{n=1}^{\infty} C_n$. Each set C_n is a closed and bounded set in the plane, hence compact. By Theorem 13.10 on page 89, C is compact and nonempty. Also, for each x in $[0,1]$, the sets $\{y \in [0,1] : (x,y) \in C_n\}$ are compact and nonempty and, since the vertical line segments in C_n have length $\leq \frac{1}{n}$, their intersection $\{y \in [0,1] : (x,y) \in C\}$ has exactly one point. Thus C is the graph of a function f on $[0,1]$. Since its graph is compact, Example 6 on page 171 shows f is continuous. Finally, consider x in $[0,1]$. For each n, using Property n we obtain y_n in $[0,1]$ so that $0 < |y_n - x| \leq \frac{1}{n}$ and $\left| \frac{f(y_n) - f(x)}{y_n - x} \right| > n$. Thus f is not differentiable at x. $\qquad\square$

Exercises

38.1 Why hasn't anyone constructed a nowhere-differentiable continuous function by simply finding a suitably complicated Taylor series?

38.2 Show that if f is a function on $\mathbb{R}$ and $f^{(n)}$ is a polynomial for some n, then f is itself a polynomial.

38.3 Show that there is a differentiable function on $\mathbb{R}$ whose derivative is nowhere differentiable.

38.4 If f is nowhere differentiable and g is differentiable on $\mathbb{R}$, must $f + g$ be nowhere differentiable?

38.5 Consider continuous nowhere-differentiable functions f and g on $\mathbb{R}$.

(a) Must $f + g$ be nowhere differentiable?

(b) Show fg need not be nowhere differentiable. *Hint*: Let f be a nowhere-differentiable function satisfying $d(f, 1^*) < 1$, where 1^* denotes the constant function identically equal to 1 on $\mathbb{R}$. Explain why the reciprocal g of f is nowhere-differentiable on $\mathbb{R}$.

38.6 In Exercise 31.4 it was noted, without proof, that the sum, product, etc. of infinitely differentiable functions is again infinitely differentiable. Also, f/g is infinitely differentiable if both f and g are, and if g never vanishes. Verify the last claim. *Hint*: Show the nth derivative of f/g has the form h_n/g^{2^n} where h_n is infinitely differentiable.

Appendix on Set Notation

Consider a set S. The notation $x \in S$ means x is an element of S; we might also say "x belongs to S" or "x is in S." The notation $x \notin S$ signifies x is some element but x does not belong to S. By $T \subseteq S$ we mean each element of T also belongs to S, i.e., $x \in T$ implies $x \in S$. Thus we have $1 \in \mathbb{N}$, $17 \in \mathbb{N}$, $-3 \notin \mathbb{N}$, $\frac{1}{2} \notin \mathbb{N}$, $\sqrt{2} \notin \mathbb{N}$, $\frac{1}{2} \in \mathbb{Q}$, $\frac{1}{2} \in \mathbb{R}$, $\sqrt{2} \in \mathbb{R}$, $\sqrt{2} \notin \mathbb{Q}$, and $\pi \in \mathbb{R}$. Also we have $\mathbb{N} \subseteq \mathbb{R}$, $\mathbb{Q} \subseteq \mathbb{R}$, and $\mathbb{R} \subseteq \mathbb{R}$.

Small finite sets can be listed using braces { }. For example, $\{2, 3, 5, 7\}$ is the four-element set consisting of the primes less than 10. Sets are often described by properties of their elements via the notation

$$\{ \quad : \quad \}.$$

Before the colon the variable [n or x, for instance] is indicated and after the colon the properties are given. For example,

$$\{n : n \in \mathbb{N} \text{ and } n \text{ is odd}\} \tag{1}$$

represents the set of positive odd integers. The colon is always read "such that," so the set in (1) is read "the set of all n *such that* n is

K.A. Ross, *Elementary Analysis: The Theory of Calculus*,
Undergraduate Texts in Mathematics, DOI 10.1007/978-1-4614-6271-2,
© Springer Science+Business Media New York 2013

in $\mathbb{N}$ and n is odd." Likewise

$$\{x : x \in \mathbb{R} \text{ and } 1 \le x < 3\} \tag{2}$$

represents the set of all real numbers greater than or equal to 1 and less than 3. In §4 this set is abbreviated $[1,3)$. Note that $1 \in [1,3)$ but $3 \notin [1,3)$. Just to streamline notation, the expressions (1) and (2) may be written as

$$\{n \in \mathbb{N} : n \text{ is odd}\} \quad \text{and} \quad \{x \in \mathbb{R} : 1 \le x < 3\}.$$

The first set is then read "the set of all n in $\mathbb{N}$ *such that* n is odd."

Another way to list a set is to specify a rule for obtaining its elements using some other set of elements. For example, $\{n^2 : n \in \mathbb{N}\}$ represents the set of all positive integers that are the square of other integers, i.e.,

$$\{n^2 : n \in \mathbb{N}\} = \{m \in \mathbb{N} : m = n^2 \text{ for some } n \in \mathbb{N}\} = \{1, 4, 9, 16, 25, \ldots\}.$$

Similarly $\{\sin \frac{n\pi}{4} : n \in \mathbb{N}\}$ represents the set obtained by evaluating $\sin \frac{n\pi}{4}$ for each positive integer n. Actually this set is finite:

$$\left\{\sin \frac{n\pi}{4} : n \in \mathbb{N}\right\} = \left\{\frac{\sqrt{2}}{2}, 1, 0, -\frac{\sqrt{2}}{2}, -1\right\}.$$

The set in (1) can also be written as $\{2n - 1 : n \in \mathbb{N}\}$. One more example: $\{x^3 : x > 3\}$ is the set of all cubes of all real numbers bigger than 3 and of course equals $\{y \in \mathbb{R} : y > 27\}$, i.e., $(27, \infty)$ in the notation of §5.

For sets S and T, $S \setminus T$ signifies the set $\{x \in S : x \notin T\}$. For a sequence (A_n) of sets, the union $\cup A_n$ and intersection $\cap A_n$ are defined by

$$\bigcup A_n = \{x : x \in A_n \text{ for at least one } n\},$$

$$\bigcap A_n = \{x : x \in A_n \text{ for all } n\}.$$

The *empty set* $\varnothing$ is the set with no elements at all. Thus, for example, $\{n \in \mathbb{N} : 2 < n < 3\} = \varnothing$, $\{r \in \mathbb{Q} : r^2 = 2\} = \varnothing$, $\{x \in \mathbb{R} : x^2 < 0\} = \varnothing$, and $[0,2] \cap [5, \infty) = \varnothing$.

For functions f and g, the notation $f+g$, fg, $f \circ g$, etc. is explained on page 128.

The end of a proof is indicated by a small black box. This replaces the classical QED.

Selected Hints and Answers

Notice. These hints and answers should be consulted only after serious attempts have been made to solve the problems. Students who ignore this advice will only cheat themselves.

Many problems can be solved in several ways. Your solution need not agree with that given here. Often your solution should be more elaborate.

1.1 *Hint*: The following algebra is needed to verify the induction step:

$$\frac{n(n+1)(2n+1)}{6} + (n+1)^2 = (n+1)\left[\frac{2n^2+n}{6} + n + 1\right] = \cdots$$

$$= \frac{(n+1)(n+2)(2n+3)}{6}.$$

1.3 *Hint*: Suppose the identity holds for n. Then work on the right side of the equation with $n+1$ in place of n. Since $(x+y)^2 = x^2 + 2xy + y^2$,

$$(1 + 2 + \cdots + n + (n+1)) = (1 + 2 + \cdots + n)^2$$

$$+ 2(n+1)(1 + 2 + \cdots n) + (n+1)^2.$$

Use Example 1 to show the second line has sum $(n+1)^3$; hence

$$(1+2+\cdots+(n+1))^2 = (1+2+\cdots n)^2 + (n+1)^3 = 1^3 + 2^3 + \cdots + (n+1)^3.$$

1.5 *Hint*: $2 - \frac{1}{2^n} + \frac{1}{2^{n+1}} = 2 - \frac{1}{2^{n+1}}$.

1.7 *Hint*: $7^{n+1} - 6(n+1) - 1 = 7(7^n - 6n - 1) + 36n$.

K.A. Ross, *Elementary Analysis: The Theory of Calculus*,
Undergraduate Texts in Mathematics, DOI 10.1007/978-1-4614-6271-2,
© Springer Science+Business Media New York 2013

1.9 **(a)** $n \geq 5$ and also $n = 1$.

    (b) Clearly the inequality holds for $n = 5$. Suppose $2^n > n^2$ for some $n \geq 5$. Then $2^{n+1} = 2 \cdot 2^n > 2n^2$, so $2^{n+1} > (n+1)^2$ provided $2n^2 \geq (n+1)^2$ or $n^2 \geq 2n + 1$ for $n \geq 5$. In fact, this holds for $n \geq 3$, which can be verified using calculus or directly: $n^2 \geq 3n = 2n + n > 2n + 1$.

1.11 **(a)** *Hint*: If $n^2 + 5n + 1$ is even, then so is $(n+1)^2 + 5(n+1) + 1 = n^2 + 5n + 1 + [2n + 6]$.

    (b) P_n is false for all n. *Moral*: The basis for induction (I_1) is crucial for mathematical induction.

2.1 *Hint*: Imitate Example 3. You should, of course, verify your assertions concerning nonsolutions. Note there are 16 rational candidates for solving $x^2 - 24 = 0$.

2.3 *Hint*: $\sqrt{2 + \sqrt{2}}$ is a solution of $x^4 - 4x^2 + 2 = 0$.

2.5 *Hint*: $[3 + \sqrt{2}]^{2/3}$ is a solution of $x^6 - 22x^3 + 49 = 0$.

2.7 **(a)** Show $x = \sqrt{4 + 2\sqrt{3}} - \sqrt{3}$ satisfies the quadratic equation $x^2 + 2x\sqrt{3} - 1 - 2\sqrt{3} = 0$. Alternatively, start with the observation $(\sqrt{4 + 2\sqrt{3}})^2 = 4 + 2\sqrt{3} = 1 + 2\sqrt{3} + 3 = (1 + \sqrt{3})^2$.

    (b) 2.

3.1 **(a)** A3 and A4 hold for $a \in \mathbb{N}$, but 0 and $-a$ are not in $\mathbb{N}$. Likewise M4 holds for $a \in \mathbb{N}$, but a^{-1} is not in $\mathbb{N}$ unless $a = 1$. These three properties fail for $\mathbb{N}$ since they implicitly require the numbers 0, $-a$ and a^{-1} to be in the system under scrutiny, namely $\mathbb{N}$ in this case.

    (b) M4 fails in the sense discussed in (a).

3.3 **(iv)** Apply (iii), DL, A2, A4, (ii) and A4 again to obtain

$$(-a)(-b) + (-ab) = (-a)(-b) + (-a)b = (-a)[(-b) + b]$$
$$= (-a)[b + (-b)] = (-a) \cdot 0 = 0 = ab + (-ab).$$

Now by (i) we conclude $(-a)(-b) = ab$.

    (v) Suppose $ac = bc$ and $c \neq 0$. By M4 there exists c^{-1} such that $c \cdot c^{-1} = 1$. Now (supply reasons)

$$a = a \cdot 1 = a(c \cdot c^{-1}) = (ac)c^{-1} = (bc)c^{-1} = b(c \cdot c^{-1}) = b \cdot 1 = b.$$

3.5 **(a)** If $|b| \leq a$, then $-a \leq -|b|$, so $-a \leq -|b| \leq b \leq |b| \leq a$. Now suppose $-a \leq b \leq a$. If $b \geq 0$, then $|b| = b \leq a$. If $b < 0$, then $|b| = -b \leq a$; the last inequality holds by Theorem 3.2(i) since $-a \leq b$.

(b) By (a), it suffices to prove $-|a - b| \le |a| - |b| \le |a - b|$. Each of these inequalities follows from the triangle inequality: $|b| = |(b - a) + a| \le |b - a| + |a| = |a - b| + |a|$ which implies the first inequality; $|a| = |(a - b) + b| \le |a - b| + |b|$ which implies the second inequality.

3.7 (a) Imitate the answer to Exercise 3.5(a).

(b) By (a), $|a - b| < c$ if and only if $-c < a - b < c$, and this obviously holds [see O4] if and only if $b - c < a < b + c$.

4.1 If the set is bounded above, use any three numbers $\ge$ supremum of the set; see the answers to Exercise 4.3. The sets in (h), (k) and (u) are not bounded above. Note the set in (i) is simply $[0, 1]$.

4.3 (a) 1; **(c)** 7; **(e)** 1; **(g)** 3; **(i)** 1; **(k)** No sup; **(m)** 2; **(o)** 0; **(q)** 16; **(s)** $\frac{1}{2}$; **(u)** No sup; **(w)** $\frac{\sqrt{3}}{2}$. In (s), note 1 is not prime.

4.5 Proof Since $\sup S$ is an upper bound for S, we have $\sup S \ge s$ for all $s \in S$. Also $\sup S \in S$ by assumption. Hence $\sup S$ is the maximum of S, i.e., $\sup S = \max S$.

4.7 (a) Suppose $S \subseteq T$. Since $\sup T \ge t$ for all $t \in T$ we obviously have $\sup T \ge s$ for all $s \in S$. So $\sup T$ is an upper bound for the set S. Hence $\sup T$ is $\ge$ the *least* upper bound for S, i.e., $\sup T \ge \sup S$. A similar argument shows $\inf T \le \inf S$; give it.

(b) Since $S \subseteq S \cup T$, $\sup S \le \sup(S \cup T)$ by (a). Similarly $\sup T \le \sup(S \cup T)$, so $\max\{\sup S, \sup T\} \le \sup(S \cup T)$. Since $\sup(S \cup T)$ is the least upper bound for $S \cup T$, we will have equality here provided we show: $\max\{\sup S, \sup T\}$ is an upper bound for the set $S \cup T$. This is easy. If $x \in S$, then $x \le \sup S \le \max\{\sup S, \sup T\}$ and if $x \in T$, then $x \le \sup T \le \max\{\sup S, \sup T\}$. That is, $x \le \max\{\sup S, \sup T\}$ for all $x \in S \cup T$.

4.9 (1) If $s \in S$, then $-s \in -S$, so $-s \le s_0$. Hence we have $s \ge -s_0$ by Theorem 3.2(i).

(2) Suppose $t \le s$ for all $s \in S$. Then $-t \ge -s$ for all $s \in S$, i.e., $-t \ge x$ for all $x \in -S$. So $-t$ is an upper bound for the set $-S$. So $-t \ge \sup(-S)$. That is, $-t \ge s_0$ and hence $t \le -s_0$.

4.11 Proof By 4.7 there is a rational r_1 such that $a < r_1 < b$. By 4.7 again, there is a rational r_2 such that $a < r_2 < r_1$. We continue by induction: If rationals $r_1, \ldots, r_n$ have been selected so that $a < r_n < r_{n-1} < \cdots < r_2 < r_1$, then 4.7 applies to $a < r_n$ to yield a rational r_{n+1} such that $a < r_{n+1} < r_n$. This process yields an infinite set $\{r_1, r_2, \ldots\}$ in $\mathbb{Q} \cap (a, b)$.

Alternative Proof Assume $\mathbb{Q} \cap (a, b)$ is finite. The set is nonempty by 4.7. Let $c = \min(\mathbb{Q} \cap (a, b))$. Then $a < c$, so by 4.7 there is a rational r such that $a < r < c$. Then r belongs to $\mathbb{Q} \cap (a, b)$, so $c \leq r$, a contradiction.

4.13 By Exercise 3.7(b), we have (i) and (ii) equivalent. The equivalence of (ii) and (iii) is obvious from the definition of an open interval.

4.15 Assume $a \leq b + \frac{1}{n}$ for all $n \in \mathbb{N}$ but $a > b$. Then $a - b > 0$, and by the Archimedean property 4.6 we have $n_0(a - b) > 1$ for some $n_0 \in \mathbb{N}$. Then $a > b + \frac{1}{n_0}$ contrary to our assumption.

5.1 (a) $(-\infty, 0)$; (b) $(-\infty, 2]$; (c) $[0, \infty)$; (d) $(-\sqrt{8}, \sqrt{8})$.

5.3 *Hint:* The unbounded sets are in (h), (k), (l), (o), (t) and (u).

5.5 **Proof**   Select $s_0 \in S$. Then $\inf S \leq s_0 \leq \sup S$ whether these symbols represent $\pm\infty$ or not.

5.7 Use Exercise 5.4 and the fact $-(A + B) = (-A) + (-B)$.

6.1 (a) If $s \leq t$, then clearly $s^* \subseteq t^*$. Conversely, assume $s^* \subseteq t^*$ but that $s > t$. Then $t \in s^*$ but $t \notin t^*$, a contradiction.

    (b) $s = t$ if and only if both $s \leq t$ and $t \leq s$ if and only if both $s^* \subseteq t^*$ and $t^* \subseteq s^*$ if and only if $s^* = t^*$.

6.3 (a) If $r \in \alpha$ and $s \in 0^*$, then $r + s < r$, so $r + s \in \alpha$. Hence $\alpha + 0^* \subseteq \alpha$. Conversely, suppose $r \in \alpha$. Since α has no largest element, there is a rational $t \in \alpha$ such that $t < r$. Then $r - t$ is in 0^*, so $r = t + (r - t) \in \alpha + 0^*$. This shows $\alpha \subseteq \alpha + 0^*$.

    (b) $-\alpha = \{r \in \mathbb{Q} : s \notin \alpha \text{ for some rational } s < -r\}$.

6.5 (b) No; it corresponds to $\sqrt[3]{2}$.

    (c) This is the Dedekind cut corresponding to $\sqrt{2}$.

7.1 (a) $\frac{1}{4}, \frac{1}{7}, \frac{1}{10}, \frac{1}{13}, \frac{1}{16}$

    (c) $\frac{1}{3}, \frac{2}{9}, \frac{1}{9}, \frac{4}{81}, \frac{5}{243}$

7.3 (a) converges to 1; (c) converges to 0; (e) does not converge; (g) does not converge; (i) converges to 0; (k) does not converge; (m) converges to 0 [this sequence is $(0, 0, 0, \ldots)$]; (o) converges to 0; (q) converges to 0 [see Exercise 9.15]; (s) converges to $\frac{4}{3}$.

7.5 (a) Has limit 0 since $s_n = 1/(\sqrt{n^2 + 1} + n)$.

    (c) $\sqrt{4n^2 + n} - 2n = n/(\sqrt{4n^2 + n} + 2n)$ and this is close to $\frac{n}{2n + 2n}$ for large n. So limit appears to be $\frac{1}{4}$; it is.

8.1 (a) **Formal Proof** Let $\epsilon > 0$. Let $N = \frac{1}{\epsilon}$. Then $n > N$ implies $\left|\frac{(-1)^n}{n - 0}\right| = \frac{1}{n} < \epsilon$.

    (b) *Discussion.* We want $n^{-1/3} < \epsilon$ or $\frac{1}{n} < \epsilon^3$ or $1/\epsilon^3 < n$. So for each $\epsilon > 0$, let $N = 1/\epsilon^3$. You should write out the formal proof.

(c) *Discussion.* We want $|\frac{2n-1}{3n+2} - \frac{2}{3}| < \epsilon$ or $|\frac{-7}{(3n+2)\cdot 3}| < \epsilon$ or $\frac{7}{3(3n+2)} < \epsilon$ or $\frac{7}{3\epsilon} < 3n + 2$ or $\frac{7}{9\epsilon} - \frac{2}{3} < n$. So set N equal to $\frac{7}{9\epsilon} - \frac{2}{3}$.

(d) *Discussion.* We want $(n + 6)/(n^2 - 6) < \epsilon$; we assume $n > 2$ so that absolute values can be dropped. As in Example 3 we observe $n + 6 \leq 7n$ and $n^2 - 6 \geq \frac{1}{2}n^2$ provided $n > 3$. So it suffices to get $7n/(\frac{1}{2}n^2) < \epsilon$ [for $n > 3$] or $\frac{14}{\epsilon} < n$. So try $N = \max\{3, \frac{14}{\epsilon}\}$.

8.3 *Discussion.* We want $\sqrt{s_n} < \epsilon$ or $s_n < \epsilon^2$. But $s_n \to 0$, so we can get $s_n < \epsilon^2$ for large n.

Formal Proof Let $\epsilon > 0$. Since $\epsilon^2 > 0$ and $\lim s_n = 0$, there exists N so that $|s_n - 0| < \epsilon^2$ for $n > N$. Thus $s_n < \epsilon^2$ for $n > N$, so $\sqrt{s_n} < \epsilon$ for $n > N$. That is, $|\sqrt{s_n} - 0| < \epsilon$ for $n > N$. We conclude $\lim \sqrt{s_n} = 0$.

8.5 (a) Let $\epsilon > 0$. Our goal is to show $s - \epsilon < s_n < s + \epsilon$ for large n. Since $\lim a_n = s$, there exists N_1 so that $|a_n - s| < \epsilon$ for $n > N_1$. In particular,

$$n > N_1 \quad \text{implies} \quad s - \epsilon < a_n. \tag{1}$$

Likewise there exists N_2 so that $|b_n - s| < \epsilon$ for $n > N_2$, so

$$n > N_2 \quad \text{implies} \quad b_n < s + \epsilon. \tag{2}$$

Now

$$n > \max\{N_1, N_2\} \quad \text{implies} \quad s - \epsilon < a_n \leq s_n \leq b_n < s + \epsilon;$$

hence $|s - s_n| < \epsilon$.

(b) It is easy to show $\lim(-t_n) = 0$ if $\lim t_n = 0$. Now apply (a) to the inequalities $-t_n \leq s_n \leq t_n$.

8.7 (a) Assume $\lim \cos(\frac{n\pi}{3}) = a$. Then there exists N such that $n > N$ implies $|\cos(\frac{n\pi}{3}) - a| < 1$. Consider $n > N$ and $n + 3$ where n is a multiple of 6; substituting these values in the inequality gives $|1 - a| < 1$ and $|-1 - a| < 1$. By the triangle inequality

$$2 = |(1 - a) - (-1 - a)| \leq |1 - a| + |-1 - a| < 1 + 1 = 2,$$

a contradiction.

(b) Assume $\lim(-1)^n n = a$. Then there exists N such that $n > N$ implies $|(-1)^n n - a| < 1$. For an even $n > N$ and for $n + 2$ this tells us $|n - a| < 1$ and $|n + 2 - a| < 1$. So $2 = |n + 2 - a - (n - a)| \leq |n + 2 - a| + |n - a| < 2$, a contradiction.

(c) Note the sequence takes the values $\pm\frac{\sqrt{3}}{2}$ for large n. Assume $\lim \sin(\frac{n\pi}{3}) = a$. Then there is N such that

$$n > N \quad \text{implies} \quad \left|\sin\left(\frac{n\pi}{3}\right) - a\right| < \frac{\sqrt{3}}{2}.$$

Substituting suitable $n > N$, we obtain $\left|\frac{\sqrt{3}}{2} - a\right| < \frac{\sqrt{3}}{2}$ and $\left|\frac{-\sqrt{3}}{2} - a\right| < \frac{\sqrt{3}}{2}$. By the triangle inequality

$$\sqrt{3} = \left|\frac{\sqrt{3}}{2} - \left(\frac{-\sqrt{3}}{2}\right)\right| \leq \left|\frac{\sqrt{3}}{2} - a\right| + \left|a - \left(\frac{-\sqrt{3}}{2}\right)\right|$$

$$< \frac{\sqrt{3}}{2} + \frac{\sqrt{3}}{2} = \sqrt{3},$$

a contradiction.

8.9 (a) *Hint:* There exists N_0 in $\mathbb{N}$ such that $s_n \geq a$ for $n > N_0$. Assume $s = \lim s_n$ and $s < a$. Let $\epsilon = a - s$ and select $N \geq N_0$ so that $|s_n - s| < \epsilon$ for $n > N$. Show $s_n < a$ for $n > N$; a picture might help.

9.1 (a) $\lim(\frac{n+1}{n}) = \lim(1 + \frac{1}{n}) = \lim 1 + \lim \frac{1}{n} = 1 + 0 = 1$. The second equality is justified by Theorem 9.3 and the third equality follows from Theorem 9.7(a).

(b) $\lim(3n + 7)/(6n - 5) = \lim(3 + 7/n)/(6 - 5/n) = \lim(3 + 7/n)/\lim(6-5/n) = (\lim 3 + 7 \cdot \lim(1/n))/(\lim 6 - 5 \cdot \lim(1/n)) = (3 + 7 \cdot 0)/(6 - 5 \cdot 0) = \frac{1}{2}$. The second equality is justified by Theorem 9.6, the third equality follows from Theorems 9.3 and 9.2, and the fourth equality uses Theorem 9.7(a).

9.3 First we use Theorem 9.4 twice to obtain $\lim a_n^3 = \lim a_n \cdot \lim a_n^2 = a \cdot \lim a_n^2 = a \cdot \lim a_n \cdot \lim a_n = a \cdot a \cdot a = a^3$. By Theorems 9.3 and 9.2, we have $\lim(a_n^3 + 4a_n) = \lim a_n^3 + 4 \cdot \lim a_n = a^3 + 4a$. Similarly $\lim(b_n^2 + 1) = \lim b_n \cdot \lim b_n + 1 = b^2 + 1$. Since $b^2 + 1 \neq 0$, Theorem 9.6 shows $\lim s_n = (a^3 + 4a)/(b^2 + 1)$.

9.5 *Hint:* Let $t = \lim t_n$ and show $t = (t^2 + 2)/2t$. Then show $t = \sqrt{2}$.

9.7 It has been shown that $s_n < \sqrt{2/(n-1)}$ for $n \geq 2$, and we need to prove $\lim s_n = 0$.

Discussion. Let $\epsilon > 0$. We want $s_n < \epsilon$, so it suffices to get $\sqrt{2/(n-1)} < \epsilon$ or $2/(n-1) < \epsilon^2$ or $2\epsilon^{-2} + 1 < n$.

Formal Proof. Let $\epsilon > 0$ and let $N = 2\epsilon^{-2} + 1$. Then $n > N$ implies $s_n < \sqrt{2/(n-1)} < \sqrt{2/(2\epsilon^{-2} + 1 - 1)} = \epsilon$.

9.9 **(a)** Let $M > 0$. Since $\lim s_n = +\infty$ there exists $N \geq N_0$ such that $s_n > M$ for $n > N$. Then clearly $t_n > M$ for $n > N$, since $s_n \leq t_n$ for all n. This shows $\lim t_n = +\infty$.

(c) Parts (a) and (b) take care of the infinite limits, so assume (s_n) and (t_n) converge. Since $t_n - s_n \geq 0$ for all $n > N_0$, $\lim(t_n - s_n) \geq 0$ by Exercise 8.9(a). Hence $\lim t_n - \lim s_n \geq 0$ by Theorems 9.3 and 9.2.

9.11 **(a)** *Discussion.* Let $M > 0$ and let $m = \inf\{t_n : n \in \mathbb{N}\}$. We want $s_n + t_n > M$ for large n, but it suffices to get $s_n + m > M$ or $s_n > M - m$ for large n. So select N so that $s_n > M - m$ for $n > N$.

(b) *Hint*: If $\lim t_n > -\infty$, then $\inf\{t_n : n \in \mathbb{N}\} > -\infty$. Use part (a).

9.13 If $|a| < 1$, then $\lim a^n = 0$ by Theorem 9.7(b). If $a = 1$, then obviously $\lim a^n = 1$.

Suppose $a > 1$. Then $\frac{1}{a} < 1$, so $\lim(1/a)^n = 0$ as above. Thus $\lim 1/a^n = 0$. Theorem 9.10 [with $s_n = a^n$] now shows $\lim a^n = +\infty$. [This case can also be handled by applying Exercise 9.12.]

Suppose $a \leq -1$ and assume $\lim a^n$ exists. For even n, $a^n \geq 1$ and for odd n, $a^n \leq -1$. Clearly $\lim a^n = +\infty$ and $\lim a^n = -\infty$ are impossible. Assume $\lim a^n = A$ for a real number A. There exists N such that $|a^n - A| < 1$ for $n > N$. For even n this implies $A > 0$ and for odd n this implies $A < 0$, a contradiction.

9.15 Apply Exercise 9.12 with $s_n = a^n/n!$. Then $L = \lim|s_{n+1}/s_n| = \lim \frac{a}{n+1} = 0$, so $\lim s_n = 0$.

9.17 *Discussion.* Let $M > 0$. We want $n^2 > M$ or $n > \sqrt{M}$. So let $N = \sqrt{M}$.

10.1 increasing: (c); decreasing: (a), (f); bounded: (a), (b), (d), (f).

10.3 The equality in the hint can be verified by induction; compare Exercise 1.5. Now by (1) in Discussion 10.3 we have

$$s_n = K + \frac{d_1}{10} + \cdots + \frac{d_n}{10^n} \leq K + \frac{9}{10} + \cdots + \frac{9}{10^n} < K + 1.$$

10.7 Let $t = \sup S$. For each $n \in \mathbb{N}$, $t - \frac{1}{n}$ is not an upper bound for S, so there exists s_n satisfying $t - \frac{1}{n} < s_n < t$. Now apply the squeeze lemma in Exercise 8.5.

10.9 **(a)** $s_2 = \frac{1}{2}$, $s_3 = \frac{1}{6}$, $s_4 = \frac{1}{48}$.

(b) First we prove

$$0 < s_{n+1} < s_n \leq 1 \quad \text{for all} \quad n \geq 1. \tag{1}$$

This is obvious from part (a) for $n = 1, 2, 3$. Assume (1) holds for n. Then $s_{n+1} < 1$, so

$$s_{n+2} = \frac{n+1}{n+2} s_{n+1}^2 = \left(\frac{n+1}{n+2} s_{n+1} \right) s_{n+1} < s_{n+1}$$

since $(\frac{n+1}{n+2}) s_{n+1} < 1$. Since $s_{n+1} > 0$ we also have $s_{n+2} > 0$. Hence $0 < s_{n+2} < s_{n+1} \le 1$ and (1) holds by induction. Assertion (1) shows (s_n) is a bounded monotone sequence, so (s_n) converges by Theorem 10.2.

(c) Let $s = \lim s_n$. Using limit theorems we find $s = \lim s_{n+1} = \lim \frac{n}{n+1} \cdot \lim s_n^2 = s^2$. Consequently $s = 1$ or $s = 0$. But $s = 1$ is impossible since $s_n \le \frac{1}{2}$ for $n \ge 2$. So $s = 0$.

10.11 (a) Show (t_n) is a bounded monotone sequence.

(b) The answer is not obvious! It turns out that $\lim t_n$ is a Wallis product and has value $\frac{2}{\pi}$ which is about 0.6366. Observe how much easier part (a) is than part (b).

11.1 (a) $1, 5, 1, 5, 1, 5, 1, 5$

(b) Let $\sigma(k) = n_k = 2k$. Then (a_{n_k}) is the sequence that takes the single value 5. [There are many other possible choices of σ.]

11.3 (b) For (s_n), the set S of subsequential limits is $\{-1, -\frac{1}{2}, \frac{1}{2}, 1\}$. For (t_n), $S = \{0\}$. For (u_n), $S = \{0\}$. For (v_n), $S = \{-1, 1\}$.

(c) $\limsup s_n = 1$, $\liminf s_n = -1$, $\limsup t_n = \liminf t_n = \lim t_n = 0$, $\limsup u_n = \liminf u_n = \lim u_n = 0$, $\limsup v_n = 1$, $\liminf v_n = -1$.

(d) (t_n) and (u_n) converge.

(e) (s_n), (t_n), (u_n) and (v_n) are all bounded.

11.5 (a) $[0, 1]$; (b) $\limsup q_n = 1$, $\liminf q_n = 0$.

11.7 Apply Theorem 11.2.

11.9 (a) To show $[a, b]$ is closed, we need to consider a limit s of a convergent sequence (s_n) from $[a, b]$ and show s is also in $[a, b]$. But this was done in Exercise 8.9.

(b) No! $(0, 1)$ is not closed, i.e., $(0, 1)$ does not have the property described in Theorem 11.9. For example, $t_n = \frac{1}{n}$ defines a sequence in $(0, 1)$ such that $t = \lim t_n$ does *not* belong to $(0, 1)$.

11.11 Let $t = \sup S$. There are several ways to prove the result. (1) Provide an inductive definition where $s_k \ge \max\{s_{k-1}, t - \frac{1}{k}\}$ for all k. (2) Apply Theorem 11.2 directly. (3) Use the sequence (s_n) obtained in Exercise 10.7, and show $t_n = \max\{s_1, \ldots, s_n\}$ defines an increasing sequence (t_n) in S converging to t.

12.1 Let $u_N = \inf\{s_n : n > N\}$ and $w_N = \inf\{t_n : n > N\}$. Then (u_N) and (w_N) are increasing sequences and $u_N \le w_N$ for all $N > N_0$. By Exercise 9.9(c), $\liminf s_n = \lim u_N \le \lim w_N = \liminf t_n$. The inequality $\limsup s_n \le \limsup t_n$ can be shown in a similar way or one can apply Exercise 11.8.

12.3 (a) 0; (b) 1; (c) 2; (d) 3; (e) 4; (f) 0; (g) 2.

12.5 By Exercise 12.4, $\limsup(-s_n - t_n) \le \limsup(-s_n) + \limsup(-t_n)$, so $-\limsup(-(s_n + t_n)) \ge -\limsup(-s_n) + [-\limsup(-t_n)]$. Now apply Exercise 11.8.

12.7 Let (s_{n_j}) be a subsequence of (s_n) such that $\lim_{j\to\infty} s_{n_j} = +\infty$. [We used j here instead of k to avoid confusion with the given $k > 0$.] Then $\lim_{j\to\infty} k s_{n_j} = +\infty$ by Exercise 9.10(a). Since $(k s_{n_j})$ is a subsequence of $(k s_n)$, we conclude $\limsup(k s_n) = +\infty$.

12.9 (a) Since $\liminf t_n > 0$, there exists N_1 such that $m = \inf\{t_n : n > N_1\} > 0$. Now consider $M > 0$. Since $\lim s_n = +\infty$, there exists N_2 such that $s_n > \frac{M}{m}$ for $n > N_2$. Then $n > \max\{N_1, N_2\}$ implies $s_n t_n > (\frac{M}{m})t_n \ge (\frac{M}{m})m = M$. Hence $\lim s_n t_n = +\infty$.

12.11 Partial Proof Let $M = \liminf |s_{n+1}/s_n|$ and $\beta = \liminf |s_n|^{1/n}$. To show $M \le \beta$, it suffices to prove $M_1 \le \beta$ for all $M_1 < M$. Since

$$\liminf \left| \frac{s_{n+1}}{s_n} \right| = \lim_{N\to\infty} \inf \left\{ \left| \frac{s_{n+1}}{s_n} \right| : n > N \right\} > M_1,$$

there exists N such that

$$\inf \left\{ \left| \frac{s_{n+1}}{s_n} \right| : n > N \right\} > M_1.$$

Now imitate the proof in Theorem 12.2, but note that many of the inequalities will be reversed.

12.13 Proof of $\sup A = \liminf s_n$. Consider N in $\mathbb{N}$ and observe $u_N = \inf\{s_n : n > N\}$ is a number in A, since $\{n \in \mathbb{N} : s_n < u_N\} \subseteq \{1, 2, \ldots, N\}$. So $u_N \le \sup A$ for all N and consequently $\liminf s_n = \lim u_N \le \sup A$.

Next consider $a \in A$. Let $N_0 = \max\{n \in \mathbb{N} : s_n < a\} < \infty$. Then $s_n \ge a$ for $n > N_0$. Thus for $N \ge N_0$ we have $u_N = \inf\{s_n : n > N\} \ge a$. It follows that $\liminf s_n = \lim u_N \ge a$. We have just shown that $\liminf s_n$ is an upper bound for the set A. Therefore $\liminf s_n \ge \sup A$.

13.1 (a) It is clear that d_1 and d_2 satisfy D1 and D2 of Definition 13.1. If $\boldsymbol{x}, \boldsymbol{y}, \boldsymbol{z} \in \mathbb{R}^k$, then for each $j = 1, 2, \ldots, k$,

$$|x_j - z_j| \le |x_j - y_j| + |y_j - z_j| \le d_1(\boldsymbol{x}, \boldsymbol{y}) + d_1(\boldsymbol{y}, \boldsymbol{z}),$$

so $d_1(x, z) \leq d_1(x, y) + d_1(y, z)$. So d_1 satisfies the triangle inequality and a similar argument works for d_2; give it.

(b) For the completeness of d_1 we use Theorem 13.4 and the inequalities

$$d_1(x, y) \leq d(x, y) \leq \sqrt{k}\, d_1(x, y).$$

In fact, if (x_n) is Cauchy for d_1, then the second inequality shows (x_n) is Cauchy for d. Hence by Theorem 13.4, for some $x \in \mathbb{R}^k$ we have $\lim d(x_n, x) = 0$. By the first inequality, we also have $\lim d_1(x_n, x) = 0$, i.e., (x_n) converges to x in the metric d_1. For d_2, use the completeness of d_1 and the inequalities $d_1(x, y) \leq d_2(x, y) \leq k\, d_1(x, y)$.

13.3 (b) No, because $d^*(x, y)$ need not be finite. For example, consider the elements $x = (1, 1, 1, \ldots)$ and $y = (0, 0, 0, \ldots)$.

13.7 Outline of Proof Consider an open set $U \subseteq \mathbb{R}$. Let (q_n) be an enumeration of the rationals in U. For each n, let

$$a_n = \inf\{a \in \mathbb{R} : (a, q_n] \subseteq U\}, \quad b_n = \sup\{b \in \mathbb{R} : [q_n, b) \subseteq U\}.$$

Show $(a_n, b_n) \subseteq U$ for each n and $U = \bigcup_{n=1}^{\infty}(a_n, b_n)$. Show

$$(a_n, b_n) \cap (a_m, b_m) \neq \varnothing \quad \text{implies} \quad (a_n, b_n) = (a_m, b_m).$$

Now either there will be only finitely many distinct [and disjoint] intervals or else a subsequence $\{(a_{n_k}, b_{n_k})\}_{k=1}^{\infty}$ of $\{(a_n, b_n)\}$ will consist of disjoint intervals for which $\bigcup_{k=1}^{\infty}(a_{n_k}, b_{n_k}) = U$.

13.9 (a) $\{\frac{1}{n} : n \in \mathbb{N}\} \cup \{0\}$; (b) $\mathbb{R}$; (c) $[-\sqrt{2}, \sqrt{2}]$.

13.11 Suppose E is compact, hence closed and bounded by Theorem 13.12. Consider a sequence (x_n) in E. By Theorem 13.5, a subsequence of (x_n) converges to some x in $\mathbb{R}^k$. Since E is closed, x must be *in E*; see Proposition 13.9(b).

Suppose every sequence in E has a subsequence converging to a point *in E*. By Theorem 13.12, it suffices to show E is closed and bounded. If E were unbounded, E would contain a sequence (x_n) where $\lim d(x_n, 0) = +\infty$ and then no subsequence would converge at all. Thus E is bounded. If E were nonclosed, then by Proposition 13.9 there would be a convergent sequence (x_n) in E such that $x = \lim x_n \notin E$. Since every subsequence would also converge to $x \notin E$, we would have a contradiction.

13.13 Assume, for example, that $\sup E \notin E$. The set E is bounded, so by Exercise 10.7, there exists a sequence (s_n) in E where $\lim s_n = \sup E$. Now Proposition 13.9(b) shows $\sup E \in E$, a contradiction.

13.15 (a) F is bounded because $d(\boldsymbol{x}, \boldsymbol{0}) \leq 1$ for all $\boldsymbol{x} \in F$ where $\boldsymbol{0} = (0, 0, 0, \ldots)$. To show F is closed, consider a convergent sequence $(\boldsymbol{x}^{(n)})$ in F. We need to show $\boldsymbol{x} = \lim \boldsymbol{x}^{(n)}$ is in F. For each $j = 1, 2, \ldots$, it is easy to see $\lim_{n \to \infty} x_j^{(n)} = x_j$. Since each $x_j^{(n)}$ belongs to $[-1, 1]$, x_j belongs to $[-1, 1]$ by Exercise 8.9. It follows that $\boldsymbol{x} \in F$.

    (b) For the last assertion of the hint, observe $\boldsymbol{x}^{(n)}, \boldsymbol{x}^{(m)}$ in $U(\boldsymbol{x})$ implies $d(\boldsymbol{x}^{(n)}, \boldsymbol{x}^{(m)}) \leq d(\boldsymbol{x}^{(n)}, \boldsymbol{x}) + d(\boldsymbol{x}, \boldsymbol{x}^{(m)}) < 2$ while $d(\boldsymbol{x}^{(n)}, \boldsymbol{x}^{(m)}) = 2$ for $m \neq n$. Now show no finite subfamily of $\mathcal{U}$ can cover $\{\boldsymbol{x}^{(n)} : n \in \mathbb{N}\}$.

14.1 (a), (b), (c) Converge; use Ratio Test.

    (d) Diverges; use Ratio Test or show nth terms don't converge to 0 [see Corollary 14.5].

    (e) Compare with $\sum 1/n^2$.

    (f) Compare with $\sum \frac{1}{n}$.

14.3 All but (e) converge.

14.5 (a) We assume the series begin with $n = 1$. Let $s_n = \sum_{j=1}^{n} a_j$ and $t_n = \sum_{j=1}^{n} b_j$. We are given $\lim s_n = A$ and $\lim t_n = B$. Hence $\lim(s_n + t_n) = A + B$ by Theorem 9.3. Clearly $s_n + t_n = \sum_{j=1}^{n}(a_j + b_j)$ is the nth partial sum for $\sum(a_n + b_n)$, so $\sum(a_n + b_n) = \lim(s_n + t_n) = A + B$.

    (c) The conjecture is not even reasonable for series of two terms: $a_1 b_1 + a_2 b_2 \neq (a_1 + a_2)(b_1 + b_2)$.

14.7 By Corollary 14.5, there exists N such that $a_n < 1$ for $n > N$. Since $p > 1$, $a_n^p = a_n a_n^{p-1} < a_n$ for $n > N$. Hence $\sum_{n=N+1}^{\infty} a_n^p$ converges by the Comparison Test, so $\sum a_n^p$ also converges.

14.9 *Hint:* Let $N_0 = \max\{n \in \mathbb{N} : a_n \neq b_n\} < \infty$. If $n \geq m > N_0$, then $\sum_{k=m}^{n} a_k = \sum_{k=m}^{n} b_k$.

14.11 Assume $a_{n+1}/a_n = r$ for $n \geq 1$. Then $a_2 = ra_1$, $a_3 = r^2 a_2$, etc. A simple induction argument shows $a_n = r^{n-1} a_1$ for $n \geq 1$. Thus $\sum a_n = \sum a_1 r^{n-1}$ is a geometric series.

14.13 (a) 2 and $-\frac{2}{5}$.

    (b) Note

$$s_n = \left(1 - \frac{1}{2}\right) + \left(\frac{1}{2} - \frac{1}{3}\right) + \left(\frac{1}{3} - \frac{1}{4}\right) + \cdots + \left(\frac{1}{n} - \frac{1}{n+1}\right) = 1 - \frac{1}{n+1}$$

since the intermediate fractions cancel out. Hence $\lim s_n = 1$.

(d) 2.

15.1 (a) Converges by Alternating Series Theorem.

(b) Diverges; note $\lim(n!/2^n) = +\infty$ by Exercise 9.12(b).

15.3 *Hint*: Use integral tests. Note

$$\lim_{n \to \infty} \int_3^n \frac{1}{x(\log x)^p}\, dx = \lim_{n \to \infty} \int_{\log 3}^{\log n} \frac{1}{u^p}\, du.$$

15.5 There is no smallest $p_0 > 1$, so there is no single series $\sum 1/n^{p_0}$ with which all series $\sum 1/n^p$ $[p > 1]$ can be compared.

15.7 (a) **Proof** Let $\epsilon > 0$. By the Cauchy criterion, there exists N such that $n \geq m > N$ implies $|\sum_{k=m}^n a_k| < \frac{\epsilon}{2}$. In particular,

$$n > N \quad \text{implies} \quad a_{N+1} + \cdots + a_n < \frac{\epsilon}{2}.$$

So $n > N$ implies

$$(n - N)a_n \leq a_{N+1} + \cdots + a_n < \frac{\epsilon}{2}.$$

If $n > 2N$, then $n < 2(n - N)$, so $na_n < 2(n - N)a_n < \epsilon$. This proves $\lim(na_n) = 0$.

16.1 (a) In other words, show

$$2 + 7 \cdot 10^{-1} + 4 \cdot 10^{-2} + \sum_{j=3}^{\infty} 9 \cdot 10^{-j} = 2 + 7 \cdot 10^{-1} + 5 \cdot 10^{-2} = \frac{11}{4}.$$

The series is a geometric series; see Example 1 of §14.

(b) $2.75\bar{0}$

16.3 Let A and B denote the sums of the series. By Exercise 14.5, we have $B - A = \sum(b_n - a_n)$. Since $b_n - a_n \geq 0$ for all n, and $b_n - a_n > 0$ for some n, we clearly have $B - A > 0$.

16.5 (a) $.125\bar{0}$ *and* $.124\bar{9}$; (c) $.\bar{6}$; (e) $.\overline{54}$

16.7 No.

16.9 (a) $\gamma_n - \gamma_{n+1} = \int_n^{n+1} t^{-1}\, dt - \frac{1}{n+1} > 0$ since $\frac{1}{n+1} < t^{-1}$ for all t in $[n, n+1)$.

(b) For any n, $\gamma_n \leq \gamma_1 = 1$. Also

$$\gamma_n > \sum_{k=1}^{n} \left(\frac{1}{k} - \int_k^{k+1} t^{-1}\, dt \right) > 0.$$

(c) Apply Theorem 10.2.

17.1 (a) $\text{dom}(f+g) = \text{dom}(fg) = (-\infty, 4]$, $\text{dom}(f \circ g) = [-2, 2]$, $\text{dom}(g \circ f) = (-\infty, 4]$.

(b) $f \circ g(0) = 2$, $g \circ f(0) = 4$, $f \circ g(1) = \sqrt{3}$, $g \circ f(1) = 3$, $f \circ g(2) = 0$, $g \circ f(2) = 2$.

(c) No!

(d) $f \circ g(3)$ is not, but $g \circ f(3)$ is.

17.3 (a) We are given $f(x) = \cos x$ and $g(x) = x^4$ $[p = 4]$ are continuous. So $g \circ f$ is continuous by Theorem 17.5, i.e., the function $g \circ f(x) = \cos^4 x$ is continuous. Obviously the function identically 1 is continuous [if you do not find this obvious, check it]. Hence $1 + \cos^4 x$ is continuous by Theorem 17.4(i). Finally $\log_e(1 + \cos^4 x)$ is continuous by Theorem 17.5 since this is $h \circ k(x)$ where $k(x) = 1 + \cos^4 x$ and $h(x) = \log_e x$.

(b) Since we are given $\sin x$ and x^2 are continuous, Theorem 17.5 shows $\sin^2 x$ is continuous. Similarly, $\cos^6 x$ is continuous. Hence $\sin^2 x + \cos^6 x$ is continuous by Theorem 17.4(i). Since $\sin^2 x + \cos^6 x > 0$ for all x and since x^π is given to be continuous for $x > 0$, we use Theorem 17.5 again to conclude $[\sin^2 x + \cos^6 x]^\pi$ is continuous.

(e) We are given $\sin x$ and $\cos x$ are continuous at each $x \in \mathbb{R}$. So Theorem 17.4(iii) shows $\frac{\sin x}{\cos x} = \tan x$ is continuous wherever $\cos x \neq 0$, i.e., for $x \neq$ odd multiple of $\frac{\pi}{2}$.

17.5 (a) *Remarks.* An ϵ–δ proof can be given based on the identity

$$x^m - y^m = (x - y)(x^{m-1} + x^{m-2}y + \cdots + xy^{m-2} + y^{m-1}).$$

Or the result can be proved by induction on m, as follows. It is easy to prove $g(x) = x$ is continuous on $\mathbb{R}$. If $f(x) = x^m$ is continuous on $\mathbb{R}$, then so is $(fg)(x) = x^{m+1}$ by Theorem 17.4(ii).

(b) Just use (a) and Theorems 17.4(i) and 17.3.

17.9 (a) *Discussion.* Let $\epsilon > 0$. We want $|x^2 - 4| < \epsilon$ for $|x - 2|$ small, i.e., we want $|x - 2| \cdot |x + 2| < \epsilon$ for $|x - 2|$ small. If $|x - 2| < 1$, then $|x + 2| < 5$, so it suffices to get $|x - 2| \cdot 5 < \epsilon$. Set $\delta = \min\{1, \frac{\epsilon}{5}\}$.

(c) For $\epsilon > 0$, let $\delta = \epsilon$ and observe

$$|x - 0| < \delta \quad \text{implies} \quad \left| x \sin\left(\frac{1}{x}\right) - 0 \right| < \epsilon.$$

17.11 If f is continuous at x_0 and if (x_n) is a monotonic sequence in $\mathrm{dom}(f)$ converging to x_0, then we have $\lim f(x_n) = f(x_0)$ by Definition 17.1.

Now assume

> if (x_n) is monotonic in dom(f) and $\lim x_n = x_0$, then $\lim f(x_n) = f(x_0)$, $\qquad$ (1)

but f is discontinuous at x_0. Then by Definition 17.1, there exists a sequence (x_n) in dom(f) such that $\lim x_n = x_0$ but $(f(x_n))$ does not converge to $f(x_0)$. Negating Definition 17.1, we see there exists $\epsilon > 0$ such that

> for each N there is $n > N$ satisfying $|f(x_n) - f(x_0)| \geq \epsilon$. $\qquad$ (2)

It is easy to use (2) to obtain a subsequence (x_{n_k}) of (x_n) such that

$$|f(x_{n_k}) - f(x_0)| \geq \epsilon \quad \text{for all} \quad k. \qquad (3)$$

Now Theorem 11.4 shows (x_{n_k}) has a monotonic subsequence $(x_{n_{k_j}})$. By (1) we have $\lim_{j \to \infty} f(x_{n_{k_j}}) = f(x_0)$, but by (3) we have $|f(x_{n_{k_j}}) - f(x_0)| \geq \epsilon$ for all j, a contradiction.

17.13 (a) *Hint:* Let $x \in \mathbb{R}$. Select a sequence (x_n) such that $\lim x_n = x$, x_n is rational for even n, and x_n is irrational for odd n. Then $f(x_n)$ is 1 for even n and 0 for odd n, so $(f(x_n))$ cannot converge.

17.15 We abbreviate

> (i) f is continuous at x_0,
> (ii) $\lim f(x_n) = f(x_0)$ for every sequence (x_n) in dom$(f) \setminus \{x_0\}$ converging to x_0.

From Definition 17.1 it is clear that (i) implies (ii). Assume (ii) holds but (i) fails. As in the solution to Exercise 17.11, there is a sequence (x_n) in dom(f) and an $\epsilon > 0$ such that $\lim x_n = x_0$ and $|f(x_n) - f(x_0)| \geq \epsilon$ for all n. Obviously $x_n \neq x_0$ for all n, i.e., (x_n) is in dom$(f) \setminus \{x_0\}$. The existence of this sequence contradicts (ii).

18.3 This exercise was deliberately poorly stated, as if f must have a maximum and minimum on $[0, 5)$; see the comments following Theorem 18.1. The minimum of f on $[0, 5)$ is $1 = f(0) = f(3)$, but f has *no maximum* on $[0, 5)$ though $\sup\{f(x) : x \in [0, 5)\} = 21$.

18.5 (a) Let $h = f - g$. Then h is continuous [why?] and $h(b) \leq 0 \leq h(a)$. Now apply Theorem 18.2.

(b) Use the function g defined by $g(x) = x$ for $x \in [0, 1]$.

18.7 *Hint:* Let $f(x) = xe^x$; f is continuous, $f(0) = 0$ and $f(1) = e$.

18.9 Let $f(x) = a_0 + a_1 x + \cdots + a_n x^n$ where $a_n \neq 0$ and n is odd. We may suppose $a_n = 1$; otherwise we would work with $(1/a_n)f$. Since

f is continuous, Theorem 18.2 shows it suffices to show $f(x) < 0$ for some x and $f(x) > 0$ for some other x. This is true because $\lim_{x \to \infty} f(x) = +\infty$ and $\lim_{x \to -\infty} f(x) = -\infty$ [remember $a_n = 1$], but we can avoid these limit notions as follows. Observe

$$f(x) = x^n \left[1 + \frac{a_0 + a_1 x + \cdots + a_{n-1} x^{n-1}}{x^n} \right]. \tag{1}$$

Let $c = 1 + |a_0| + |a_1| + \cdots + |a_{n-1}|$. If $|x| > c$, then

$$|a_0 + a_1 x + \cdots + a_{n-1} x^{n-1}| \le (|a_0| + |a_1| + \cdots + |a_{n-1}|)|x|^{n-1} < |x|^n,$$

so the number in brackets in (1) is positive. Now if $x > c$, then $x^n > 0$, so $f(x) > 0$. And if $x < -c$, then $x^n < 0$ [why?], so $f(x) < 0$.

19.1 *Hints*: To decide (a) and (b), use Theorem 19.2. Parts (c), (e), (f) and (g) can be settled using Theorem 19.5. Theorem 19.4 can also be used to decide (e) and (f); compare Example 6. One needs to resort to the definition to handle (d).

19.3 (a) *Discussion.* Let $\epsilon > 0$. We want

$$\left| \frac{x}{x+1} - \frac{y}{y+1} \right| < \epsilon \quad \text{or} \quad \left| \frac{x-y}{(x+1)(y+1)} \right| < \epsilon$$

for $|x - y|$ small, $x, y \in [0, 2]$. Since $x + 1 \ge 1$ and $y + 1 \ge 1$ for $x, y \in [0, 2]$, it suffices to get $|x - y| < \epsilon$. So we let $\delta = \epsilon$. **Formal Proof** Let $\epsilon > 0$ and let $\delta = \epsilon$. Then $x, y \in [0, 2]$ and $|x - y| < \delta = \epsilon$ imply

$$|f(x) - f(y)| = \left| \frac{x-y}{(x+1)(y+1)} \right| \le |x - y| < \epsilon.$$

(b) *Discussion.* Let $\epsilon > 0$. We want $|g(x) - g(y)| = |\frac{5y - 5x}{(2x-1)(2y-1)}| < \epsilon$ for $|x - y|$ small, $x \ge 1$, $y \ge 1$. For $x, y \ge 1$, we have $2x - 1 \ge 1$ and $2y - 1 \ge 1$, so it suffices to get $|5y - 5x| < \epsilon$. So let $\delta = \frac{\epsilon}{5}$. You should write out the formal proof.

19.5 (a) $\tan x$ is uniformly continuous on $[0, \frac{\pi}{4}]$ by Theorem 19.2.

(b) $\tan x$ is not uniformly continuous on $[0, \frac{\pi}{2})$ by Exercise 19.4(a), since the function is not bounded on that set.

(c) Let $\tilde{h}$ be as in Example 9. Then $(\sin x)\tilde{h}(x)$ is a continuous extension of $(\frac{1}{x}) \sin^2 x$ on $(0, \pi]$. Apply Theorem 19.5.

(e) $\frac{1}{x-3}$ is not uniformly continuous on $(3, 4)$ by Exercise 19.4(a), so it is not uniformly continuous on $(3, \infty)$ either.

(f) *Remark.* It is easy to give an ϵ–δ proof that $\frac{1}{x-3}$ is uniformly continuous on $(4, \infty)$. It is even easier to apply Theorem 19.6.

19.7 (a) We are given f is uniformly continuous on $[k, \infty)$, and f is uniformly continuous on $[0, k+1]$ by Theorem 19.2. Let $\epsilon > 0$. There exist δ_1 and δ_2 so that

$$|x - y| < \delta_1, \ x, y \in [k, \infty) \quad \text{imply} \quad |f(x) - f(y)| < \epsilon,$$

$$|x - y| < \delta_2, \ x, y \in [0, k+1] \quad \text{imply} \quad |f(x) - f(y)| < \epsilon.$$

Let $\delta = \min\{1, \delta_1, \delta_2\}$ and show

$$|x - y| < \delta, \ x, y \in [0, \infty) \quad \text{imply} \quad |f(x) - f(y)| < \epsilon.$$

19.9 (c) This is tricky, but it turns out that f is uniformly continuous on $\mathbb{R}$. A simple modification of Exercise 19.7(a) shows it suffices to show f is uniformly continuous on $[1, \infty)$ and $(-\infty, -1]$. This can be done using Theorem 19.6. Note we cannot apply Theorem 19.6 on $\mathbb{R}$ because f is not differentiable at $x = 0$; also f' is not bounded near $x = 0$.

19.11 As in the solution to Exercise 19.9(c), it suffices to show $\tilde{h}$ is uniformly continuous on $[1, \infty)$ and $(-\infty, -1]$. Apply Theorem 19.6.

20.1 $\lim_{x \to \infty} f(x) = \lim_{x \to 0+} f(x) = 1$; $\lim_{x \to 0-} f(x) = \lim_{x \to -\infty} f(x) = -1$; $\lim_{x \to 0} f(x)$ does NOT EXIST.

20.3 $\lim_{x \to \infty} f(x) = \lim_{x \to -\infty} f(x) = 0$; $\lim_{x \to 0+} f(x) = \lim_{x \to 0-} f(x) = \lim_{x \to 0} f(x) = 1$.

20.5 Let $S = (0, \infty)$. Then $f(x) = 1$ for all $x \in S$. So for any sequence (x_n) in S we have $\lim f(x_n) = 1$. It follows that $\lim_{x \to 0^S} f(x) = \lim_{x \to \infty^S} f(x) = 1$, i.e., $\lim_{x \to 0+} f(x) = \lim_{x \to \infty} f(x) = 1$. Likewise if $S = (-\infty, 0)$, then $\lim_{x \to 0^S} f(x) = \lim_{x \to -\infty^S} f(x) = -1$, so $\lim_{x \to 0-} f(x) = \lim_{x \to -\infty} f(x) = -1$. Theorem 20.10 shows $\lim_{x \to 0} f(x)$ does not exist.

20.7 If (x_n) is a sequence in $(0, \infty)$ and $\lim x_n = +\infty$, then $\lim(1/x_n) = 0$. Since $(\sin x_n)$ is a bounded sequence, we conclude $\lim(\sin x_n)/x_n = 0$ by Exercise 8.4. Hence $\lim_{x \to \infty} f(x) = 0$. Similarly $\lim_{x \to -\infty} f(x) = 0$. The remaining assertion is $\lim_{x \to 0} \frac{\sin x}{x} = 1$ which is discussed in Example 9 of §19.

20.9 $\lim_{x \to \infty} f(x) = -\infty$; $\lim_{x \to 0+} f(x) = +\infty$; $\lim_{x \to 0-} f(x) = -\infty$; $\lim_{x \to -\infty} f(x) = +\infty$; $\lim_{x \to 0} f(x)$ does NOT EXIST.

20.11 (a) $2a$; (c) $3a^2$.

20.13 First note that if $\lim_{x \to a^S} f(x)$ exists and is finite and if $k \in \mathbb{R}$, then $\lim_{x \to a^S}(kf)(x) = k \cdot \lim_{x \to a^S} f(x)$. This is Theorem 20.4(ii) where f_1 is the constant k and $f_2 = f$.

(a) The remark above and Theorem 20.4 show

$$\lim_{x \to a}[3f(x) + g(x)^2] = 3 \lim_{x \to a} f(x) + [\lim_{x \to a} g(x)]^2 = 3 \cdot 3 + 2^2 = 13.$$

(c) As in (a), $\lim_{x \to a}[3f(x) + 8g(x)] = 25$. There exists an open interval J containing a such that $f(x) > 0$ and $g(x) > 0$ for $x \in J \setminus \{a\}$. Theorem 20.5 applies with $S = J \setminus \{a\}$, $3f + 8g$ in place of f and with $g(x) = \sqrt{x}$ to give $\lim_{x \to a} \sqrt{3f(x) + 8g(x)} = \sqrt{25} = 5$.

20.15 Let (x_n) be a sequence in $(-\infty, 2)$ such that $\lim x_n = -\infty$. We contend

$$\lim(x_n - 2)^{-3} = 0. \tag{1}$$

We apply Exercises 9.10 and 9.11 and Theorems 9.9 and 9.10 to conclude $\lim(-x_n) = +\infty$, $\lim(2 - x_n) = +\infty$, $\lim(2 - x_n)^3 = +\infty$, $\lim(2 - x_n)^{-3} = 0$, and hence (1) holds.

Now consider a sequence (x_n) in $(2, \infty)$ such that $\lim x_n = 2$. We show

$$\lim(x_n - 2)^{-3} = +\infty. \tag{2}$$

Since $\lim(x_n - 2) = 0$ and each $x_n - 2 > 0$, Theorem 9.10 shows we have $\lim(x_n - 2)^{-1} = +\infty$ and (2) follows by an application of Theorem 9.9.

20.17 Suppose first that L is finite. We use (1) in Corollary 20.8. Let $\epsilon > 0$. There exist $\delta_1 > 0$ and $\delta_3 > 0$ such that

$$a < x < a + \delta_1 \quad \text{implies} \quad L - \epsilon < f_1(x) < L + \epsilon$$

and

$$a < x < a + \delta_3 \quad \text{implies} \quad L - \epsilon < f_3(x) < L + \epsilon.$$

If $\delta = \min\{\delta_1, \delta_3\}$, then

$$a < x < a + \delta \quad \text{implies} \quad L - \epsilon < f_2(x) < L + \epsilon.$$

So by Corollary 20.8 we have $\lim_{x \to a^+} f_2(x) = L$.

Suppose $L = +\infty$. Let $M > 0$. In view of Discussion 20.9, there exists $\delta > 0$ such that

$$a < x < a + \delta \quad \text{implies} \quad f_1(x) > M.$$

Then clearly

$$a < x < a + \delta \quad \text{implies} \quad f_2(x) > M,$$

and this shows $\lim_{x \to a+} f_2(x) = +\infty$. The case $L = -\infty$ is similar.

20.19 Suppose $L_2 = \lim_{x \to a^S} f(x)$ exists with $S = (a, b_2)$. Consider a sequence (x_n) in (a, b_1) with limit a. Then (x_n) is a sequence in (a, b_2) with limit a, so $\lim f(x_n) = L_2$. This shows $\lim_{x \to a^S} f(x) = L_2$ with $S = (a, b_1)$.

Suppose $L_1 = \lim_{x \to a^S} f(x)$ exists with $S = (a, b_1)$, and consider a sequence (x_n) in (a, b_2) with limit a. There exists N so that $n \geq N$ implies $x_n < b_1$. Then $(x_n)_{n=N}^{\infty}$ is a sequence in (a, b_1) with limit a. Hence $\lim f(x_n) = L_1$ whether we begin the sequence at $n = N$ or $n = 1$. This shows $\lim_{x \to a^S} f(x) = L_1$ with $S = (a, b_2)$.

21.1 Let $\epsilon > 0$. For $j = 1, 2, \dots, k$, there exist $\delta_j > 0$ such that

$$s, t \in \mathbb{R} \quad \text{and} \quad |s - t| < \delta_j \quad \text{imply} \quad |f_j(s) - f_j(t)| < \frac{\epsilon}{\sqrt{k}}.$$

Let $\delta = \min\{\delta_1, \delta_2, \dots, \delta_k\}$. Then by (1) in the proof of Proposition 21.2,

$$x, t \in \mathbb{R} \quad \text{and} \quad |s - t| < \delta \quad \text{imply} \quad d^*(\gamma(s), \gamma(t)) < \epsilon.$$

21.3 *Hint*: Show $|d(s, s_0) - d(t, s_0)| \leq d(s, t)$. Hence if $\epsilon > 0$, then

$$s, t \in S \quad \text{and} \quad d(s, t) < \epsilon \quad \text{imply} \quad |f(s) - f(t)| < \epsilon.$$

21.5 **(b)** By part (a), there is an unbounded continuous real-valued function f on E. Show $h = \frac{|f|}{1+|f|}$ is continuous, bounded and does not assume its supremum 1 on E.

21.7 **(b)** γ is continuous at t_0 if for each $t_0 \in [a, b]$ and $\epsilon > 0$ there exists $\delta > 0$ such that

$$t \in [a, b] \quad \text{and} \quad |t - t_0| < \delta \quad \text{imply} \quad d^*(\gamma(t), \gamma(t_0)) < \epsilon.$$

Note: If γ is continuous at each $t_0 \in [a, b]$, then γ is uniformly continuous on $[a, b]$ by Theorem 21.4.

21.9 **(a)** Use $f(x_1, x_2) = x_1$, say.

(b) This is definitely *not* obvious, but there do exist continuous mappings of $[0, 1]$ onto the unit square. Such functions must be "wild" and are called Peano curves [after the same Peano with the axioms]; see [16, §5.5] or [55, §6.3].

21.11 If $\mathbb{Q}$ were equal to $\cap_{n=1}^{\infty} U_n$, where each U_n is open, then $\cap_{n=1}^{\infty} \cap_{r \in \mathbb{Q}} (\mathbb{R} \setminus \{r\})$ would be an intersection of a sequence of open sets that is equal to the empty set, contrary to Theorem 21.7(a).

21.13 Suppose $\omega_f(x) = 0$. Given $\epsilon > 0$, there is $\delta > 0$ so that

$$\sup\{|f(y) - f(z)| : y, z \in (x - \delta, x + \delta)\} < \epsilon.$$

Then $|f(x) - f(y)| < \epsilon$ for all $y \in (x - \delta, x + \delta)$, i.e., $|x - y| < \delta$ implies $|f(x) - f(y)| < \epsilon$. So f is continuous at x. The proof that continuity of f at x implies $\omega_f(x) = 0$ is similar.

22.1 (a) $[0, 1]$ is connected but $[0, 1] \cup [2, 3]$ is not. See Theorem 22.2. Alternatively, apply the Intermediate Value Theorem 18.2.

22.3 Assume E is connected but E^- is not. Then there exist open sets U_1 and U_2 that separate E^- as in Definition 22.1(a). Show that U_1 and U_2 separate E, which is a contradiction. Since E satisfies (1) in Definition 22.1(a), it suffices to show $E \cap U_1 \neq \varnothing$ and $E \cap U_2 \neq \varnothing$. In fact, if $E \cap U_1 = \varnothing$, then $E^- \cap (S \setminus U_1)$ would be a closed set containing E, and it is smaller than E^- since $E^- \cap U_1 \neq \varnothing$. This contradicts the definition of E^- in Definition 13.8, so $E \cap U_1 \neq \varnothing$. Likewise $E \cap U_2 \neq \varnothing$.

22.5 (a) Assume open sets U_1 and U_2 separate $E \cup F$ as in Definition 22.1(a). Consider $x \in E \cap F$; x belongs to one of the open sets, say $x \in U_1$. Select y in $(E \cup F) \cap U_2$. Then y is in E or F, say $y \in E$. Since $x \in E \cap U_1$ and $y \in E \cap U_2$, these sets are nonempty. Thus U_1 and U_2 separate E, and E is not connected, a contradiction.

(b) No such example exists in $\mathbb{R}$ [why?], but many exist in the plane.

22.9 *Discussion.* Given $\epsilon > 0$, we need $\delta > 0$ so that

$$s, t \in \mathbb{R} \quad \text{and} \quad |s - t| < \delta \quad \text{imply} \quad d(F(s), F(t)) < \epsilon. \qquad (1)$$

Now

$$\begin{aligned}
d(F(s), F(t)) &= \sup\{|sf(x) + (1 - s)g(x) - tf(x) - (1 - t)g(x)| : x \in S\} \\
&= \sup\{|sf(x) - tf(x) - sg(x) + tg(x)| : x \in S\} \\
&\leq |s - t| \cdot \sup\{|f(x)| + |g(x)| : x \in S\}.
\end{aligned}$$

Since f and g are fixed, the last supremum is a constant M. We may assume $M > 0$, in which case $\delta = \frac{\epsilon}{M}$ will make (1) hold.

22.11 (a) Let (f_n) be a convergent sequence in $\mathcal{E}$. By Proposition 13.9(b), it suffices to show $f = \lim f_n$ is in $\mathcal{E}$. For each $x \in S$,

$$|f(x)| \le |f(x) - f_n(x)| + |f_n(x)| \le d(f, f_n) + 1.$$

Since $\lim d(f, f_n) = 0$, we have $|f(x)| \le 1$.

(b) It suffices to show $C(S)$ is path-connected. So use Exercise 22.9.

23.1 Intervals of convergence: (a) $(-1, 1)$; (c) $[-\frac{1}{2}, \frac{1}{2}]$; (e) $\mathbb{R}$; (g) $[-\frac{4}{3}, \frac{4}{3})$.

23.3 $(-\sqrt[3]{2}, \sqrt[3]{2})$.

23.5 (a) Since $|a_n| \ge 1$ for infinitely many n, we have $\sup\{|a_n|^{1/n} : n > N\} \ge 1$ for all N. Thus $\beta = \limsup |a_n|^{1/n} \ge 1$; hence $R = \frac{1}{\beta} \le 1$.

(b) Select c so that $0 < c < \limsup |a_n|$. Then $\sup\{|a_n| : n > N\} > c$ for all N. A subsequence (a_{n_k}) of (a_n) has the property that $|a_{n_k}| > c$ for all k. Since $|a_{n_k}|^{1/n_k} > (c)^{1/n_k}$ and $\lim_{k \to \infty} c^{1/n_k} = 1$ [by Theorem 9.7(d)], Theorem 12.1 shows $\limsup |a_{n_k}|^{1/n_k} \ge 1$. It follows that $\beta = \limsup |a_n|^{1/n} \ge 1$ [use Theorem 11.8]. Hence $R = \frac{1}{\beta} \le 1$.

23.9 (a) Obviously $\lim f_n(0) = 0$. Consider $0 < x < 1$ and let $s_n = nx^n$. Then $s_{n+1}/s_n = (\frac{n+1}{n})x$, so $\lim |s_{n+1}/s_n| = x < 1$. Exercise 9.12(a) shows $0 = \lim s_n = \lim nx^n = \lim f_n(x)$.

24.1 *Discussion.* Let $\epsilon > 0$. We want $|f_n(x) - 0| < \epsilon$ for all x and for large n. It suffices to arrange for $\frac{3}{\sqrt{n}} < \epsilon$ for large n. So consider $n > 9/\epsilon^2 = N$.

24.3 (a) $f(x) = 1$ for $0 \le x < 1$; $f(1) = \frac{1}{2}$; $f(x) = 0$ for $x > 1$. See Exercise 9.13.

(b) (f_n) does *not* converge uniformly on $[0, 1]$ by Theorem 24.3.

24.5 (a) $f(x) = 0$ for $x \le 1$ and $f(x) = 1$ for $x > 1$. Note $f_n(x) = 1/(1 + n/x^n)$ and $\lim_{n \to \infty} n/x^n = 0$ for $x > 1$ by Exercise 9.12 or 9.14.

(b) $f_n \to 0$ uniformly on $[0, 1]$. *Hint*: Show $|f_n(x)| \le \frac{1}{n}$ for $x \in [0, 1]$.

(c) *Hint*: Use Theorem 24.3.

24.7 (a) Yes. $f(x) = x$ for $x < 1$ and $f(1) = 0$.

(b) No, by Theorem 24.3 again.

24.9 (a) $f(x) = 0$ for $x \in [0, 1]$. For $x < 1$, $\lim_{n \to \infty} nx^n = 0$ as in Exercise 23.9(a).

(b) Use calculus to show f_n takes its maximum at $\frac{n}{n+1}$. Thus $\sup\{|f_n(x)| : x \in [0, 1]\} = f_n(\frac{n}{n+1}) = (\frac{n}{n+1})^{n+1}$. As in Example 8, it turns out $\lim f_n(\frac{n}{n+1}) = 1/e$. So Remark 24.4 shows (f_n) does not converge uniformly to 0.

(c) $\int_0^1 f_n(x)\,dx = \frac{n}{(n+1)(n+2)} \to 0 = \int_0^1 f(x)\,dx$.

24.15 (a) $f(0) = 0$ and $f(x) = 1$ for $x > 0$. (b) No. (c) Yes.

24.17 *Hint:* Use $|f_n(x_n) - f(x)| \le |f_n(x_n) - f(x_n)| + |f(x_n) - f(x)|$.

25.3 (a) Since $f_n(x) = (1 + (\cos x)/n)/(2 + (\sin^2 x)/n)$, we have $f_n \to \frac{1}{2}$ pointwise. To obtain uniform convergence, show

$$\left| f_n(x) - \frac{1}{2} \right| = \left| \frac{2\cos x - \sin^2 x}{2(2n + \sin^2 x)} \right| \le \frac{3}{2(2n)} < \epsilon$$

for all real numbers x and all $n > \frac{3}{4\epsilon}$.

(b) $\int_2^7 \frac{1}{2}\,dx = \frac{5}{2}$, by Theorem 25.2.

25.5 Since $f_n \to f$ uniformly on S, there exists $N \in \mathbb{N}$ such that $n > N$ implies $|f_n(x) - f(x)| < 1$ for all $x \in S$. In particular, we have $|f_{N+1}(x) - f(x)| < 1$ for $x \in S$. If M bounds $|f_{N+1}|$ on S [i.e., if $|f_{N+1}(x)| \le M$ for $x \in S$], then $M + 1$ bounds $|f|$ on S [why?].

25.7 Let $g_n(x) = n^{-2} \cos nx$. Then we have $|g_n(x)| \le n^{-2}$ for $x \in \mathbb{R}$ and $\sum n^{-2} < \infty$. So $\sum g_n$ converges uniformly on $\mathbb{R}$ by the Weierstrass M-Test 25.7. The limit function is continuous by Theorem 25.5.

25.9 (a) The series converges pointwise to $\frac{1}{1-x}$ on $(-1, 1)$ by (2) of Example 1 in §14. The series converges uniformly on $[-a, a]$ by the Weierstrass M-Test since $|x^n| \le a^n$ for $x \in [-a, a]$ and since $\sum a^n < \infty$.

(b) One can show directly that the sequence of partial sums $s_n(x) = \sum_{k=0}^n x^k = (1 - x^{n+1})/(1 - x)$ does not converge uniformly on $(-1, 1)$. It is easier to observe the partial sums s_n are each bounded on $(-1, 1)$, and hence if (s_n) converges uniformly, then the limit function is bounded by Exercise 25.5. But $\frac{1}{1-x}$ is not bounded on $(-1, 1)$.

25.11 (b) *Hint:* Apply the Weierstrass M-Test to $\sum h_n$, where $h_n(x) = (\frac{3}{4})^n g_n(x)$.

25.13 The series $\sum g_k$ and $\sum h_k$ are uniformly Cauchy on S and it suffices to show $\sum (g_k + h_k)$ is also; see Theorem 25.6. Let $\epsilon > 0$. There exist N_1 and N_2 such that

$$n \ge m > N_1 \quad \text{implies} \quad \left| \sum_{k=m}^n g_k(x) \right| < \frac{\epsilon}{2} \quad \text{for} \quad x \in S, \qquad (1)$$

$$n \ge m > N_2 \quad \text{implies} \quad \left| \sum_{k=m}^n h_k(x) \right| < \frac{\epsilon}{2} \quad \text{for} \quad x \in S. \qquad (2)$$

Then

$$n \geq m > \max\{N_1, N_2\} \quad \text{implies} \quad \left| \sum_{k=m}^{n} (g_k + h_k)(x) \right| < \epsilon \quad \text{for} \quad x \in S.$$

25.15 (a) Note $f_n(x) \geq 0$ for all x and n. Assume (f_n) does not converge to 0 uniformly on $[a, b]$. Then there exists $\epsilon > 0$ such that

$$\text{for each } N \text{ there exists } n > N \text{ and } x \in [a, b] \tag{1}$$
$$\text{such that } f_n(x) \geq \epsilon.$$

We claim

$$\text{for each } n \in \mathbb{N} \text{ there is } x_n \in [a, b] \text{ where } f_n(x_n) \geq \epsilon. \tag{2}$$

If not, there is $n_0 \in \mathbb{N}$ such that $f_{n_0}(x) < \epsilon$ for all $x \in [a, b]$. Since $(f_n(x))$ is decreasing for each x, we conclude $f_n(x) < \epsilon$ for all $x \in [a, b]$ and $n \geq n_0$. This clearly contradicts (1). We have now established the hint.

Now by the Bolzano-Weierstrass theorem, the sequence (x_n) given by (2) has a convergent subsequence (x_{n_k}): $x_{n_k} \to x_0$. Since $\lim f_n(x_0) = 0$, there exists m such that $f_m(x_0) < \epsilon$. Since $x_{n_k} \to x_0$ and f_m is continuous at x_0, we have $\lim_{k \to \infty} f_m(x_{n_k}) = f_m(x_0) < \epsilon$. So there exists K such that

$$k > K \quad \text{implies} \quad f_m(x_{n_k}) < \epsilon.$$

If $k > \max\{K, m\}$, then $n_k \geq k > m$, so

$$f_{n_k}(x_{n_k}) \leq f_m(x_{n_k}) < \epsilon.$$

But $f_n(x_n) \geq \epsilon$ for *all* n, so we have a contradiction.

(b) *Hint*: Show part (a) applies to the sequence g_n where $g_n = f - f_n$.

26.3 (a) Let $f(x) = \sum_{n=1}^{\infty} n x^n = x/(1-x)^2$ for $|x| < 1$. Then by Theorem 26.5

$$\sum_{n=1}^{\infty} n^2 x^{n-1} = f'(x) = \frac{d}{dx} \left[\frac{x}{(1-x)^2} \right] = (1+x)(1-x)^{-3};$$

therefore $\sum_{n=1}^{\infty} n^2 x^n = (x + x^2)(1-x)^{-3}$.

(b) 6 and $\frac{3}{2}$.

26.5 *Hint*: Apply Theorem 26.5.

26.7 No! The power series would be differentiable at each $x \in \mathbb{R}$, but $f(x) = |x|$ is not differentiable at $x = 0$.

27.1 Let ϕ be as in the hint. By Theorem 27.4, there is a sequence (q_n) of polynomials such that $q_n \to f \circ \phi$ uniformly on $[0,1]$. Note ϕ is one-to-one and $\phi^{-1}(y) = \frac{y-a}{b-a}$. Let $p_n = q_n \circ \phi^{-1}$. Then each p_n is a polynomial and $p_n \to f$ uniformly on $[a,b]$.

27.3 (a) Assume a polynomial p satisfies $|p(x) - \sin x| < 1$ for all $x \in \mathbb{R}$. Clearly p cannot be a constant function. But if p is nonconstant, then p is unbounded on $\mathbb{R}$ and the same is true for $p(x) - \sin x$, a contradiction.

  (b) Assume $|e^x - \sum_{k=0}^{n-1} a_k x^k| < 1$ for all $x \in \mathbb{R}$. For $x > 0$ we have

$$e^x - \sum_{k=0}^{n-1} a_k x^k \geq \frac{1}{n!} x^n - \sum_{k=0}^{n-1} |a_k| x^k$$

and for large x the right side will exceed 1.

27.5 (a) $B_n f(x) = x$ for all n. Use (2) in Lemma 27.2.

  (b) $B_n f(x) = x^2 + \frac{1}{n} x(1 - x)$. Use (4) in Lemma 27.2.

28.1 (a) $\{0\}$; (b) $\{0\}$; (c) $\{n\pi : n \in \mathbb{Z}\}$; (d) $\{0,1\}$; (e) $\{-1,1\}$; (f) $\{2\}$.

28.3 (b) Since $x - a = (x^{1/3} - a^{1/3})(x^{2/3} + a^{1/3}x^{1/3} + a^{2/3})$,

$$f'(a) = \lim_{x \to a} (x^{2/3} + a^{1/3}x^{1/3} + a^{2/3})^{-1} = (3a^{2/3})^{-1} = \frac{1}{3}a^{-2/3}$$

for $a \neq 0$.

  (c) f is not differentiable at $x = 0$ since the limit $\lim_{x \to 0} x^{1/3}/x$ does not exist *as a real number*. The limit does exist and equals $+\infty$, which reflects the geometric fact that the graph of f has a vertical tangent at $(0,0)$.

28.5 (c) Let

$$h(x) = \frac{g(f(x)) - g(f(0))}{f(x) - f(0)}.$$

According to Definition 20.3(a), for $\lim_{x \to 0} h(x)$ to be meaningful, h needs to be defined on $J \setminus \{0\}$ for some open interval J containing 0. But the calculation in (b) shows h is undefined at $(\pi n)^{-1}$ for $n = \pm 1, \pm 2, \ldots$.

28.7 (d) f' is continuous on $\mathbb{R}$, but f' is not differentiable at $x = 0$.

28.9 (b) $f(x) = x^4 + 13x$ and $g(y) = y^7$. Then

$$h'(x) = g'(f(x)) \cdot f'(x) = 7(x^4 + 13x)^6 \cdot (4x^3 + 13).$$

28.11 With the stated hypotheses, $h \circ g \circ f$ is differentiable at a and $(h \circ g \circ f)'(a) = h'(g \circ f(a)) \cdot g'(f(a)) \cdot f'(a)$. **Proof** By 28.4, $g \circ f$ is differentiable at a and $(g \circ f)'(a) = g'(f(a)) \cdot f'(a)$. Again by 28.4,

$$(h \circ (g \circ f))'(a) = h'((g \circ f)(a)) \cdot (g \circ f)'(a).$$

28.13 There exist positive numbers δ_1 and ϵ so that f is defined on the interval $(a - \delta_1, a + \delta_1)$ and g is defined on $(f(a) - \epsilon, f(a) + \epsilon)$. By Theorem 17.2, there exists $\delta_2 > 0$ so that

$$x \in \mathrm{dom}(f) \quad \text{and} \quad |x - a| < \delta_2 \quad \text{imply} \quad |f(x) - f(a)| < \epsilon.$$

If $|x - a| < \min\{\delta_1, \delta_2\}$, then $x \in \mathrm{dom}(f)$ and $|f(x) - f(a)| < \epsilon$, so $f(x) \in \mathrm{dom}(g)$, i.e., $x \in \mathrm{dom}(g \circ f)$.

29.1 (a) $x = \frac{1}{2}$

(c) If $f(x) = |x|$, then $f'(x) = \pm 1$ except at 0. So no x satisfies the equation $f'(x) = \frac{f(2) - f(-1)}{2 - (-1)} = \frac{1}{3}$. Missing hypothesis: f is not differentiable on $(-1, 2)$, since f is not differentiable at $x = 0$.

(e) $x = \sqrt{3}$

29.3 (a) Apply Mean Value Theorem to $[0, 2]$.

(b) By the Mean Value Theorem, $f'(y) = 0$ for some $y \in (1, 2)$. In view of this and part (a), Theorem 29.8 shows f' takes all values between 0 and $\frac{1}{2}$.

29.5 For any $a \in \mathbb{R}$ we have $|\frac{f(x) - f(a)}{x - a}| \leq |x - a|$. It follows easily that $f'(a)$ exists and equals 0 for all $a \in \mathbb{R}$. So f is constant by Corollary 29.4.

29.7 (a) Applying 29.4 to f', we find $f'(x) = a$ for some constant a. If $g(x) = f(x) - ax$, then $g'(x) = 0$ for $x \in I$, so by 29.4 there is a constant b such that $g(x) = b$ for $x \in I$.

29.9 *Hint*: Let $f(x) = e^x - ex$ for $x \in \mathbb{R}$. Use f' to show f is increasing on $[1, \infty)$ and decreasing on $(-\infty, 1]$. Hence f takes its minimum at $x = 1$.

29.13 Let $h(x) = g(x) - f(x)$ and show $h(x) \geq 0$ for $x \geq 0$.

29.15 As in Example 2, let $g(x) = x^{1/n}$. Since $\mathrm{dom}(g) = [0, \infty)$ if n is even and $\mathrm{dom}(g) = \mathbb{R}$ if n is odd, we have $\mathrm{dom}(g) = \mathrm{dom}(h) \cup \{0\}$. Also $h = g^m$. Use the Chain Rule to calculate $h'(x)$.

29.17 Suppose $f(a) = g(a)$. Then

$$\lim_{x \to a^+} \frac{h(x) - h(a)}{x - a} = g'(a) \quad \text{and} \quad \lim_{x \to a^-} \frac{h(x) - h(a)}{x - a} = f'(a). \quad (1)$$

If also $f'(a) = g'(a)$, then Theorem 20.10 shows $h'(a)$ exists and, in fact, $h'(a) = f'(a) = g'(a)$.

Now suppose h is differentiable at a. Then h is continuous at a and so $f(a) = \lim_{x \to a^-} f(x) = \lim_{x \to a^-} h(x) = h(a) = g(a)$. Hence (1) holds. But the limits in (1) both equal $h'(a)$, so $f'(a) = g'(a)$.

30.1 (a) 2; (b) $\frac{1}{2}$; (c) 0; (d) 1. Sometimes L'Hospital's rule can be avoided. For example, for (d) note that

$$\frac{\sqrt{1+x} - \sqrt{1-x}}{x} = \frac{2}{\sqrt{1+x} + \sqrt{1-x}}.$$

30.3 (a) 0; (b) 1; (c) $+\infty$ (d) $-\frac{2}{3}$.

30.5 (a) e^2; (b) e^2; (c) e.

31.1 Differentiate the power series for $\sin x$ term-by-term and cite Theorem 26.5.

31.3 The derivatives do not have a *common* bound on any interval containing 1.

31.5 (a) $g(x) = f(x^2)$ for $x \in \mathbb{R}$ where f is as in Example 3. Use induction to prove there exist polynomials p_{kn}, $1 \le k \le n$, so that

$$g^{(n)}(x) = \sum_{k=1}^{n} f^{(k)}(x^2) p_{kn}(x) \quad \text{for} \quad x \in \mathbb{R}, n \ge 1.$$

31.9 Use $x_n = x_{n-1} - \frac{x_{n-1} - \cos(x_{n-1})}{1 + \sin(x_{n-1})}$. To six places, the answer is 0.739085. As always, use radians.

31.11 For some $C > 0$, we have $|f'(x)| \le C$ for x in (a, b). Then $|f(x_{n-1})| \le |x_n - x_{n-1}| \cdot C$ for all n. Let $n \to \infty$.

32.1 Use the partition P in Example 1 to calculate $U(f, P) = b^4 n^2 (n + 1)^2 / (4n^4)$ and $L(f, P) = b^4 (n - 1)^2 n^2 / (4n^4)$. Conclude $U(f) = b^4 / 4$ and $L(f) = b^4 / 4$.

32.3 (a) The upper sums are the same as in Example 1, so $U(g) = b^3 / 3$. Show $L(g) = 0$.

(b) No.

32.5 S is all the numbers $L(f, P)$, and T is all $U(f, P)$.

32.7 Assume $f(x) = 0$ for all $x \in [a, b]$. A simple induction shows we may assume $g(x) = 0$ except at one point $u \in [a, b]$. Clearly all lower sums $L(g, P) = 0$, so $L(g) = 0$. Since u belongs to at most two intervals of any partition P, we have

$$U(g, P) \le 2|g(u)| \max_k [t_k - t_{k-1}].$$

Infer $U(g) = 0$; hence $\int_a^b g = 0 = \int_a^b f$.

33.1 If f is decreasing on $[a, b]$, then $-f$ is increasing on $[a, b]$, so $-f$ is integrable as proved in Theorem 33.1. Now apply Theorem 33.3 with $c = -1$.

33.3 (b) $4A + 6B$

33.7 (a) For any set $S \subseteq [a,b]$ and $x_0, y_0 \in S$, we have

$$f(x_0)^2 - f(y_0)^2 \leq |f(x_0) + f(y_0)| \cdot |f(x_0) - f(y_0)|$$
$$\leq 2B|f(x_0) - f(y_0)| \leq 2B[M(f,S) - m(f,S)].$$

It follows that $M(f^2, S) - m(f^2, S) \leq 2B[M(f,S) - m(f,S)]$. Use this to show $U(f^2, P) - L(f^2, P) \leq 2B[U(f,P) - L(f,P)]$.

(b) Use Theorem 32.5 and part (a).

33.9 Select $m \in \mathbb{N}$ so that $|f(x) - f_m(x)| < \frac{\epsilon}{2(b-a)}$ for all $x \in [a,b]$. Then for any partition P

$$-\frac{\epsilon}{2} \leq L(f - f_m, P) \leq U(f - f_m, P) \leq \frac{\epsilon}{2}.$$

Select a partition P_0 so that $U(f_m, P_0) - L(f_m, P_0) < \frac{\epsilon}{2}$. Since $f = (f - f_m) + f_m$, we can use inequalities from the proof of Theorem 33.3 to conclude $U(f, P_0) - L(f, P_0) < \epsilon$. Now Theorem 32.5 shows f is integrable. To complete the exercise, proceed as in the proof of Theorem 25.2.

33.11 (a) and (b): Show f is neither continuous nor monotonic on any interval containing 0.

(c) Let $\epsilon > 0$. Since f is piecewise continuous on $[\frac{\epsilon}{8}, 1]$, there is a partition P_1 of $[\frac{\epsilon}{8}, 1]$ such that $U(f, P_1) - L(f, P_1) < \frac{\epsilon}{4}$. Likewise there is a partition P_2 of $[-1, -\frac{\epsilon}{8}]$ such that $U(f, P_2) - L(f, P_2) < \frac{\epsilon}{4}$. Let $P = P_1 \cup P_2$, a partition of $[-1, 1]$. Since

$$\left\{ M\left(f, \left[-\frac{\epsilon}{8}, \frac{\epsilon}{8}\right]\right) - m\left(f, \left[-\frac{\epsilon}{8}, \frac{\epsilon}{8}\right]\right) \right\} \cdot \left\{ \frac{\epsilon}{8} - \left(-\frac{\epsilon}{8}\right) \right\} \leq \frac{\epsilon}{2},$$

we conclude $U(f, P) - L(f, P) < \epsilon$. Now Theorem 32.5 shows f is integrable.

33.13 Apply Theorem 33.9 to $f - g$.

34.3 (a) $F(x) = 0$ for $x < 0$; $F(x) = x^2/2$ for $0 \leq x \leq 1$; $F(x) = 4x - \frac{7}{2}$ for $x > 1$.

(c) F is differentiable except possibly at $x = 1$ by Theorem 34.3. To show F is not differentiable at $x = 1$, use Exercise 29.17.

34.5 $F'(x) = f(x+1) - f(x-1)$.

34.9 Use $a = 0$, $b = \frac{\pi}{6}$ and $g(x) = \sin x$.

35.3 (a) 21; (b) 14; (c) 0.

35.5 (a) Every upper sum is $F(b) - F(a)$ and every lower sum is 0. Hence $U_F(f) = F(b) - F(a) \neq 0 = L_F(f)$.

35.7 (a) Imitate solution to Exercise 33.7.

(b) and (c): Use hints in Exercise 33.8.

35.9 (a) Let m and M be the [assumed] minimum and maximum of f on $[a, b]$. Then $\int_a^b m\, dF \le \int_a^b f\, dF \le \int_a^b M\, dF$ or $m \le [F(b) - F(a)]^{-1} \int_a^b f\, dF \le M$. Apply Theorem 18.2.

(b) Consider f and g as in Exercise 33.14, and let F be as in Exercise 35.8. By part (a), for some $x \in [a, b]$ we have

$$\int_a^b f(t)g(t)\, dt = \int_a^b f\, dF = f(x)[F(b) - F(a)] = f(x)\int_a^b g(t)\, dt.$$

35.11 Let $\epsilon > 0$ and select a partition

$$P = \{a = t_0 < t_1 < \cdots < t_n = b\}$$

satisfying $U_F(f, P) - L_F(f, P) < \epsilon$. Let $u_k = \phi^{-1}(t_k)$ and

$$Q = \{c = u_0 < u_1 < \cdots < u_n = d\}.$$

Show $U_G(g, Q) = U_F(f, P)$ and $L_G(g, Q) = L_F(f, P)$. Then $U_G(g, Q) - L_G(g, Q) < \epsilon$, so g is G-integrable. The equality of the integrals follows easily.

36.1 *Hint:* If B bounds $|f|$, then

$$\left| \int_a^d f(x)\, dx - \int_a^b f(x)\, dx \right| \le B(b - d).$$

36.3 (b) Use part (a) and Examples 1 and 2.

36.7 (a) It suffices to show $\int_1^\infty e^{-x^2}\, dx < \infty$. But $e^{-x^2} \le e^{-x}$ for $x \ge 1$ and $\int_1^\infty e^{-x}\, dx = \frac{1}{e}$.

(b) The double integral equals $[\int_{-\infty}^\infty e^{-x^2}\, dx]^2$, and it also equals

$$\int_0^\infty \int_0^{2\pi} e^{-r^2} r\, d\theta\, dr = 2\pi \int_0^\infty e^{-r^2} r\, dr = \pi.$$

36.9 (a) *Hint:* Use Theorem 35.13.

(b) 1; (c) $+\infty$; (d) $\sqrt{2/\pi}$; (e) 0.

36.13 *Claim:* If f is continuous on $\mathbb{R}$ and $\int_{-\infty}^\infty |f|\, dF < \infty$, then f is F-integrable. **Proof** Since $0 \le f + |f|$, the integral $\int_{-\infty}^\infty [f + |f|]\, dF$ exists, and since $f + |f| \le 2|f|$, this integral is finite, i.e., $f + |f|$ is F-integrable. Since $-|f|$ is F-integrable, Exercise 36.10 shows the sum of $f + |f|$ and $-|f|$ is F-integrable.

36.15 (a) For example, let $f_n(x) = \frac{1}{n}$ for $x \in [0, n]$ and $f_n(x) = 0$ elsewhere.

(b) **Outline of Proof** First, f is F-integrable on each $[a, b]$ by Exercise 35.6. An elaboration of Exercise 25.5 shows there is a *common* bound B for $|f|$ *and* all $|f_n|$. Consider any $b > 0$

such that $1 - F(b) < \frac{\epsilon}{2B}$. There exists a number N so that $|\int_0^b f\, dF - \int_0^b f_n\, dF| < \frac{\epsilon}{2}$ for $n > N$. Then

$$n > N \quad \text{implies} \quad \left| \int_0^b f\, dF - \int_0^\infty f_n\, dF \right| < \epsilon. \qquad (1)$$

In particular, $m, n > N$ implies $|\int_0^\infty f_n\, dF - \int_0^\infty f_m\, dF| < 2\epsilon$, so $(\int_0^\infty f_n\, dF)_{n \in \mathbb{N}}$ is a Cauchy sequence with a finite limit L. From (1) it follows that

$$1 - F(b) < \frac{\epsilon}{2B} \quad \text{implies} \quad \left| \int_0^b f\, dF - L \right| \leq \epsilon,$$

so $\lim_{b \to \infty} \int_0^b f\, dF = L$. Hence $\int_0^\infty f\, dF$ exists, is finite, and equals $\lim_{n \to \infty} \int_0^\infty f_n\, dF$. A similar argument handles $\int_{-\infty}^0 f\, dF$.

37.1 *Hint:*

$$\int_1^{yz} \frac{1}{t}\, dt - \int_1^y \frac{1}{t}\, dt = \int_y^{yz} \frac{1}{t}\, dt.$$

37.7 (a) $B(x) = E(xL(b))$, so by the Chain Rule, we have $B'(x) = E(xL(b)) \cdot L(b) = L(b)b^x = (\log_e b)b^x$.

37.9 (a) $\log_e y = L(y) = \int_1^y \frac{1}{t}\, dt \leq y - 1 < y$.

38.3 Apply the Fundamental Theorem of Calculus 34.3 to a continuous nowhere-differentiable function f.

A Guide to the References

There are many books with goals similar to ours, including: Brannan [12], Clark [16], and Pedrick [51]. Zorn [73] is a gentle, well written introduction with a strong first chapter on "numbers, sets and proofs." Lay [41] also starts with sections on logic and proofs, and provides good answers. Abbott [1] appears to be a fine book, though lots of the text is left to the exercises. Bauldry [3] is a nice book designed for a last course in real analysis for those intending to go into teaching. Morgan [47] gives a concise introduction to the key concepts.

Beardon [6] gives a coherent treatment of limits by defining the notion just once, in terms of directed sets. Bressoud [13] is a wonderful book that introduces analysis through its history. Gardiner [23] is a more challenging book, though the subject is at the same level; there is a lot about numbers and there is a chapter on geometry. Hijab [32] is another interesting, idiosyncratic and challenging book. Note many of these books do not provide answers to some of the exercises, though [32] provides answers to **all** of the exercises.

Rotman [61] is a very nice book that serves as an intermediate course between the standard calculus sequence and the first course in **both** abstract algebra and real analysis. I always shared Rotman's doubts about inflicting logic on undergraduates, until I saw Wolf's wonderful book [70] that introduces logic as a mathematical tool in an **interesting** way.

More encyclopedic real analysis books that are presented at the same level with great detail and numerous examples are: Lewin and Lewin [43], Mattuck [46] and Reed [56]. I find Mattuck's book [46] very thoughtfully written; he shares many of his insights (acquired over many years) with the reader.

There are several superb texts at a more sophisticated level: Beals [4], Bear [5], Hoffman [33], Johnsonbaugh and Pfaffenberger [34], Protter and Morrey [53], Rudin [62] and Stromberg [65]. Any of these books can be used to obtain a really thorough understanding of analysis and to prepare for various advanced graduate-level topics in analysis. The possible directions for study after this are too numerous to enumerate here. However, a reader who has no specific needs or goals but who would like an introduction to several important ideas in several branches of mathematics would enjoy and profit from Garding [24].

References

[1] Abbott, S.D.: Understanding Analysis. Springer, New York (2010)

[2] Bagby, R.J.: Introductory Analysis – A Deeper View of Calculus. Academic, San Diego (2001)

[3] Bauldry, W.C.: Introduction to Real Analysis – An Educational Approach. Wiley (2010)

[4] Beals, R.: Advanced Mathematical Analysis. Graduate Texts in Mathematics, vol. 12. Springer, New York/Heidelberg/Berlin (1973). Also, Analysis–an Introduction, Cambridge University Press 2004

[5] Bear, H.S.: An Introduction to Mathematical Analysis. Academic, San Diego (1997)

[6] Beardon, A.F.: Limits – A New Approach to Real Analysis. Undergraduate Texts in Mathematics. Springer, New York/Heidelberg/Berlin (1997)

[7] Berberian, S.K.: A First Course in Real Analysis. Springer, New York (1994)

[8] Birkhoff, G., Mac Lane, S.: A Survey of Modern Algebra. Macmillan, New York (1953). A. K. Peters/CRC 1998

[9] Boas, R.P. Jr.: A Primer of Real Functions, 4th edn. Revised and updated by Harold P. Boas. Carus Monograph, vol. 13. Mathematical Association of America, Washington, DC (1996)

K.A. Ross, *Elementary Analysis: The Theory of Calculus*,
Undergraduate Texts in Mathematics, DOI 10.1007/978-1-4614-6271-2,
© Springer Science+Business Media New York 2013

[10] Borman, J.L.: A remark on integration by parts. Amer. Math. Monthly **51**, 32–33 (1944)

[11] Botsko, M.W.: Quicky problem. Math. Mag. **85**, 229 (2012)

[12] Brannan, D.: A First Course in Mathematical Analysis. Cambridge University Press, Cambridge/New York (2006)

[13] Bressoud, D.: A Radical Approach to Real Analysis, 2nd edn. The Mathematical Association of America, Washington, DC (2007)

[14] Burckel, R.B.: An Introduction to Classical Complex Analysis, vol. 1. Birkhäuser, Basel (1979)

[15] Burgess, C.E.: Continuous functions and connected graphs. Amer. Math. Monthly **97**, 337–339 (1990)

[16] Clark, C.: The Theoretical Side of Calculus. Wadsworth, Belmont (1972). Reprinted by Krieger, New York 1978

[17] Corominas, E., Sunyer Balaguer, F.: Conditions for an infinitely differentiable function to be a polynomial, Rev. Mat. Hisp.-Amer. (4) **14**, 26–43 (1954). (Spanish)

[18] Cunningham, F. Jr.: The two fundamental theorems of calculus. Amer. Math. Monthly **72**, 406–407 (1975)

[19] Dangello, F., Seyfried, M.: Introductory Real Analysis. Houghton Mifflin, Boston, (2000)

[20] Donoghue, W.F. Jr.: Distributions and Fourier Transforms. Academic, New York (1969)

[21] Dunham, W.: The Calculus Gallery: Masterpieces from Newton to Lebesgue. Princeton University Press, Princeton/Woodstock (2008)

[22] Fitzpatrick, P.M.: Real Analysis. PWS, Boston (1995)

[23] Gardiner, A.: Infinite Processes, Background to Analysis. Springer, New York/Heidelberg/Berlin (1982). Republished as Understanding Infinity – The Mathematics of Infinite Processes. Dover 2002

[24] Garding, L.: Encounter with Mathematics. Springer, New York/Heidelberg/Berlin (1977)

[25] Gaskill, H.S., Narayanaswami, P.P.: Elements of Real Analysis. Prentice-Hall, Upper Saddle River (1998)

[26] Gaughan, E.D.: Introduction to Analysis, 5th edn. American Mathematical Society, Providence (2009)

[27] Gordon, R.A.: Real Analyis – A First Course, 2nd edn. Addison-Wesley, Boston (2002)

[28] Greenstein, D.S.: A property of the logarithm. Amer. Math. Monthly **72**, 767 (1965)

[29] Hewitt, E.: Integration by parts for Stieltjes integrals. Amer. Math. Monthly **67**, 419–423 (1960)

[30] Hewitt, E.: The role of compactness in analysis. Amer. Math. Monthly **67**, 499–516 (1960)

[31] Hewitt, E., Stromberg, K.: Real and Abstract Analysis. Graduate Texts in Mathematics, vol. 25. Springer, New York/Heidelberg/Berlin (1975)

[32] Hijab, O.: Introduction to Calculus and Classical Analysis. Undergraduate Texts in Mathematics, 2nd edn. Springer, New York/Heidelberg/Berlin (2007)

[33] Hoffman, K.: Analysis in Euclidean Space. Prentice-Hall, Englewood Cliffs (1975). Republished by Dover 2007

[34] Johnsonbaugh, R., Pfaffenberger, W.E.: Foundations of Mathematical Analysis. Marcel Dekker, New York (1980). Republished by Dover 2010

[35] Kantrowitz, R.: Series that converge absolutely but don't converge. Coll. Math. J. **43**, 331–333 (2012)

[36] Kenton, S.: A natural proof of the chain rule. Coll. Math. J. **30**, 216–218 (1999)

[37] Kosmala, W.: Advanced Calculus – A Friendly Approach. Prentice-Hall, Upper Saddle River (1999)

[38] Krüppel, M.: On the zeros of an infinitely often differentiable function and their derivatives. Rostock. Math. Kolloq. **59**, 63–70 (2005)

[39] Landau, E.: Foundations of Analysis. Chelsea, New York (1951). Republished by American Mathematical Society 2001

[40] Lang, S.: Undergraduate Analysis. Undergraduate Texts in Mathematics, 2nd edn. Springer, New York/Heidelberg/Berlin (2010)

[41] Lay, S.R.: Analysis – An Introduction to Proof, 4th edn. Prentice-Hall (2004)

[42] Lewin, J.: A truly elementary approach to the bounded convergence theorem. Amer. Math. Monthly **93**, 395–397 (1986)

[43] Lewin, J., Lewin, M.: An Introduction to Mathematical Analysis, 2nd edn. McGraw-Hill, New York (1993)

[44] Lynch, M.: A continuous nowhere differentiable function. Amer. Math. Monthly **99**, 8–9 (1992)

[45] Lynch, M.: A continuous function which is differentiable only at the rationals. Math. Mag. **86**, April issue (2013)

[46] Mattuck, A.: Introduction to Analysis. Prentice-Hall, Upper Saddle River (1999)

[47] Morgan, F.: Real Analysis. American Mathematical Society, Providence (2005)

[48] Newman, D.J.: A Problem Seminar. Springer, New York/Berlin/Heidelberg (1982)

[49] Niven, I.: Irrational Numbers. Carus Monograph, vol. 11. Mathematical Association of America, Washington, DC (1956)

[50] Niven, I., Zuckerman, H.S., Montgomery, H.I.: An Introduction to the Theory of Numbers, 5th edn. Wiley, New York (1991)

[51] Pedrick, G.: A First Course in Analysis. Undergraduate Texts in Mathematics. Springer, New York/Heidelberg/Berlin (1994)

[52] Phillips, E.: An Introduction to Analysis and Integration Theory. Intext Educational Publishers, Scranton/Toronto/London (1971)

[53] Protter, M.H., Morrey, C.B.: A First Course in Real Analysis. Undergraduate Texts in Mathematics, 2nd edn. Springer, New York/Heidelberg/Berlin (1997)

[54] Pugh, C.: Real Mathematical Analysis. Springer, New York/Heidelberg/Berlin (2002)

[55] Randolph, J.F.: Basic Real and Abstract Analysis. Academic, New York (1968)

[56] Reed, M.: Fundamental Ideas of Analysis. Wiley, New York (1998)

[57] Robdera, M.A.: A Concise Approach to Mathematical Analysis. Springer, London/New York (2003)

[58] Rosenlicht, M.: Introduction to Analysis. Dover, New York (1985)

[59] Ross, K.A.: First digits of squares and cubes. Math. Mag. **85**, 36–42 (2012)

[60] Ross, K.A., Wright, C.R.B.: Discrete Mathematics, 5th edn. Prentice-Hall, Upper Saddle River (2003)

[61] Rotman, J.: Journey into Mathematics – An Introduction to Proofs. Prentice-Hall, Upper Saddle River (1998)

[62] Rudin, W.: Principles of Mathematical Analysis, 3rd edn. McGraw-Hill, New York (1976)

[63] Schramm, M.J.: Introduction to Real Analysis. Prentice-Hall, Upper Saddle River (1996). Dover 2008

[64] Stolz, O.: Über die Grenzwerthe der Quotienten. Math. Ann. **15**, 556–559 (1879)

[65] Stromberg, K.: An Introduction to Classical Real Analysis. Prindle, Weber & Schmidt, Boston (1980)

[66] Thim, J.: Continuous Nowhere Differentiable Functions, Master's thesis (2003), Luleå University of Technology (Sweden). http://epubl.luth.se/1402-1617/2003/320/LTU-EX-03320-SE.pdf

[67] Thomson, B.S.: Monotone convergence theorem for the Riemann integral. Amer. Math. Monthly **117**, 547–550 (2010)

[68] van der Waerden, B.L.: Ein einfaches Beispiel einer nicht-differenzierbare Stetige Funktion. Math. Z. **32**, 474–475 (1930)

[69] Weierstrass, K.: Über continuirliche Funktionen eines reellen Arguments, die für keinen Werth des letzteren einen bestimmten Differentialquotienten besitzen, Gelesen Akad. Wiss. 18 July 1872, and J. für Mathematik **79**, 21–37 (1875)

[70] Wolf, R.S.: Proof, Logic, and Conjecture: The Mathematician's Toolbox. W. H. Freeman, New York (1998)

[71] Wolfe, J.: A proof of Taylor's formula. Amer. Math. Monthly **60**, 415 (1953)

[72] Zhou, L., Markov, L.: Recurrent proofs of the irrationality of certain trigonometric values. Amer. Math. Monthly **117**, 360–362 (2010)

[73] Zorn, P.: Understanding Real Analysis. A. K. Peters, Natick (2010)

Symbols Index

K.A. Ross, *Elementary Analysis: The Theory of Calculus,*
Undergraduate Texts in Mathematics, DOI 10.1007/978-1-4614-6271-2,
© Springer Science+Business Media New York 2013

Index

Abel's theorem, 212
absolute value, 17
absolutely convergent series, 96
algebraic number, 8
alternating series theorem, 108
Archimedean property, 25
associative laws, 14

Baire Category Theorem, 172, 173
basic examples, limits, 48
basis for induction, 3
Bernstein polynomials, 218
binomial series theorem, 255
binomial theorem, 6
Bolzano-Weierstrass theorem, 72
  for $\mathbb{R}^k$, 86
boundary of a set, 88
bounded function, 133
bounded sequence, 45
bounded set, 21
  in $\mathbb{R}^k$, 86
  in a metric space, 94

Cantor set, 89
Cauchy criterion
  for integrals, 274, 275
  for series, 97
  for series of functions, 205
Cauchy principal value, 333
Cauchy sequence, 62
  in a metric space, 85
  uniformly, 202
Cauchy's form of the remainder of
    a Taylor series, 254
cell in $\mathbb{R}^k$, 91
chain rule, 227
change of variable, 295, 330
closed interval, 20, 28
closed set, 75
  in a metric space, 88, 171
closure of a set, 88, 171
coefficients of a power series, 187
commutative laws, 14
compact set, 90
comparison test
  for integrals, 336
  for series, 98

K.A. Ross, *Elementary Analysis: The Theory of Calculus*,
Undergraduate Texts in Mathematics, DOI 10.1007/978-1-4614-6271-2,
© Springer Science+Business Media New York 2013

Index